WINDOWS® PHONE 7 PROGRAMMING FOR ANDROID™ AND iOS DEVELOPERS

FOREWORD xxv

INTRODUCTION xxvii

CHAPTER 1 What's New in Windows Phone 7 1

CHAPTER 2 The Development Environment 15

CHAPTER 3 Fundamentals 39

CHAPTER 4 User Interfaces 63

CHAPTER 5 Application Data Storage 95

CHAPTER 6 Web Services and Push Notifications 113

CHAPTER 7 Leveraging Location and Maps 155

CHAPTER 8 Graphics 183

CHAPTER 9 Multimedia 205

CHAPTER 10 Utilizing System Hardware 235

CHAPTER 11 What You Need to Do about Security 257

APPENDIX A An Introduction to Smartphone Chipset 277

APPENDIX B An Introduction to Microsoft Expression Blend for Windows Phone 281

APPENDIX C Sample Applications Guide 293

INDEX 307

Praise for

Windows Phone 7 Programming for Android and iOS Developers

"Among quite a few Windows Phone 7 programming books on the market, this book is one of the best for iOS and Android developers who look into creating applications on the new Windows Phone 7 platform. It is also an excellent desktop reference for mobile computing courses in schools, as it provides architectural overview of many popular mobile technologies on the three major mobile platforms, as well as sample code that will help students understand the technologies and develop programming skills for Windows Phone quickly.

—Dr. Lionel M. Ni
Chair Professor and Dean of HKUST
Fok Ying Tung Graduate School
Hong Kong University of Science and Technology

"Mobile is a rapidly growing space that is yet to reach its full potential and reach worldwide. The combination of mobility, form factor, price, social media, and a global market is enabling scenarios that previously were either hard to imagine or to develop. Developers, both professional and otherwise, are flocking to mobile platforms to create applications and games. An approachable, effective, integrated application platform is a goal that every software vendor should be aiming for. The Windows Phone 7 application platform enables developers to create engaging consumer experiences running on a Windows Phone. It is built upon existing Microsoft tools and technologies such as Visual Studio, Expression Blend, Silverlight, and the XNA framework. Developers already familiar with those technologies will be able to create new applications for the Windows Phone quickly and easily.

"This book helps you to understand the Windows Phone application framework. It focuses on Windows Phone design guidelines, provides guidance architecture, and solidifies it with hands-on, real-world development scenarios. It is comprehensive, highly readable, and replete with useful real-world examples. I highly recommend this book for smartphone developers, IT professionals, and engineering managers. Whether you are new, or one who has experience on the iOS and Android platforms, this book serves as the ideal reference for development on the Windows Phone."

—Saty Bahadur
Principal Development Manager
Windows Phone Engineering
Microsoft Corporation

"With Microsoft back in the game with Windows Phone, and with a growing number of Windows Phone applications in end-user hands, more and more smartphone developers are looking for books to learn how development on Windows Phone works.

"This book provides much-needed architecture guidance, theory, and hands-on practice for real-world cases aimed at developers. It is comprehensive, highly readable, and replete with useful examples. This book is exceedingly useful for mobile developers, mobile users, IT engineers, and managers."

—Quincy Milton
Principal Test Manager
Windows Phone Customer Experience Engineering
Microsoft Corporation

"This timely book will be invaluable to the many individuals and organizations that wish to extend existing development skills in iOS and/or Android onto the Windows Phone 7 platform. The book's brilliant approach of focusing on the differences between Windows Phone 7 and the mobile platforms the reader already knows makes for remarkably quick and efficient learning."

—Ib Green
Head of Capacity
Teleca USA, Inc.

Windows® Phone 7 Programming for Android™ and iOS Developers

Windows® Phone 7 Programming for Android™ and iOS Developers

Zhinan Zhou
Robert Zhu
Pei Zheng
Baijian Yang

WILEY

John Wiley & Sons, Inc.

Windows® Phone 7 Programming for Android™ and iOS Developers

Published by
John Wiley & Sons, Inc.
10475 Crosspoint Boulevard
Indianapolis, IN 46256
www.wiley.com

Published simultaneously in Canada

ISBN: 978-1-118-02197-2
ISBN: 978-1-118-15420-5 (ebk)
ISBN: 978-1-118-15422-9 (ebk)
ISBN: 978-1-118-15421-2 (ebk)

Manufactured in the United States of America

10 9 8 7 6 5 4 3 2 1

For general information on our other products and services please contact our Customer Care Department within the United States at (877) 762-2974, outside the United States at (317) 572-3993 or fax (317) 572-4002.

Wiley also publishes its books in a variety of electronic formats. Some content that appears in print may not be available in electronic books.

Library of Congress Control Number: 2011931747

To my son, Vincent, who is the first reader of this book and loves smartphones much more than me. To my wife, Xu, for her support and tolerance.

—Zhinan Zhou

To my wife, Jane, and my daughters, Jacqueline and Angie, for their great understanding and endless support.

—Robert Zhu

To my wife, Ning Liu, for her encouragement and support.

—Pei Zheng

In memory of my parents.

—Baijian Yang

ABOUT THE AUTHORS

ZHINAN ZHOU is a senior software engineer with 10 years of professional experience in the R&D of mobile applications and wireless communications. He is currently with Samsung Telecommunications America, responsible for creating cutting-edge technologies for the mobile-device space and for supporting other Samsung development groups bringing new functionality to market. Zhou has a rich hands-on experience in mobile-platform development on Android, Windows Phone, and J2ME. Zhou is also an author of numerous cited publications on IEEE and ACM journals and conferences. He received a Ph.D. in computer science from Michigan State University in 2006.

ROBERT ZHU is a principal development lead with Microsoft, developing Windows Phone software products, providing hands-on design in computer engineering such as kernel, device driver, and board support packages, and driving the technical partnership with mobile carriers and OEM partners. Zhu also gave training classes to OEMs on driver development, and Windows Mobile OS development. Before working for Microsoft, he was with Digital Equipment Corporation (DEC), U.S.A., as senior software engineer on the 64-bit DEC Alpha platform for workstation server optimization and performance tuning for Windows, and was also a software lead with Motorola Wireless Division, Canada. He obtained a master of computer science degree at the University of Washington; a master of computing and electrical engineering degree from Simon Fraser University, Canada; and a bachelor of engineering degree from Tsinghua University. He was in a Ph. D. program with the SFU School of Engineering Science, Canada.

PEI ZHENG is a senior software architect with 10 years of experience in the mobile wireless industry. He is currently with Sony Ericsson, responsible for overall device platform software architecture and key software differentiations on Android and Windows Phone. Before that, he was with Microsoft and Lucent Technologies. Zheng is the author of two books in the mobile computing area, *Smart Phone and Next Generation Mobile Computing* from Morgan Kaufmann and *Professional Smartphone Programming* from Wiley/Wrox, as well as numerous cited publications in IEEE journals and conferences. Zheng received a Ph.D. in computer science from Michigan State University in 2003.

BAIJIAN YANG is currently an associate professor in the Department of Technology, Ball State University. He has extensive industry and academic experience in mobile computing, distributed computing, and information security. His current industry certifications include MCSE, CISSP, and Six Sigma Black Belt. Yang is also a contributing author of *Professional Smartphone Programming* from Wiley/Wrox and numerous refereed publications. Yang received his Ph.D. in computer science from Michigan State University in 2002.

ABOUT THE TECHNICAL EDITOR

JOHN MUELLER is a freelance author and technical editor. He has writing in his blood, having produced 87 books and over 300 articles to date. His technical editing skills have helped more than 60 authors refine the content of their manuscripts. Mueller has provided technical editing services to both *Data Based Advisor* and *Coast Compute* magazines. He's also contributed articles to such magazines as *Software Quality Connection*, *DevSource*, *InformIT*, *SQL Server Professional*, *Visual C++ Developer*, *Hard Core Visual Basic*, *asp.netPro*, *Software Test and Performance*, and *Visual Basic Developer*.

CREDITS

ACQUISITIONS EDITOR
Paul Reese

PROJECT EDITOR
William Bridges

TECHNICAL EDITOR
John Mueller

PRODUCTION EDITOR
Daniel Scribner

EDITORIAL MANAGER
Mary Beth Wakefield

FREELANCER EDITORIAL MANAGER
Rosemarie Graham

ASSOCIATE DIRECTOR OF MARKETING
David Mayhew

BUSINESS MANAGER
Amy Knies

PRODUCTION MANAGER
Tim Tate

VICE PRESIDENT AND EXECUTIVE GROUP PUBLISHER
Richard Swadley

VICE PRESIDENT AND EXECUTIVE PUBLISHER
Neil Edde

ASSOCIATE PUBLISHER
Jim Minatel

PROJECT COORDINATOR, COVER
Katie Crocker

PROOFREADER
James Saturnio, Word One

INDEXER
Robert Swanson

COVER DESIGNER
LeAndra Young

COVER IMAGE
© iStock / Andrew Rich

ACKNOWLEDGMENTS

THE AUTHORS OF THE BOOK WOULD LIKE TO THANK Eric Hautala, Saty Bahadur, and Quincy Milton from the Windows Phone leadership team for their support of this book project. Our special thanks go to Dr. Lionel M. Ni, chair professor of the Hong Kong University of Science and Technology, for his inspiration and long-term support of our work in the mobile computing space. We also thank Ib Green, head of capacity for Teleca USA, for supporting this book project and reviewing the content.

We also thank Paul Reese from Wrox/Wiley for guiding us through the book publication process. Thanks also go to John Mueller and Bill Bridges, who have provided tons of valuable comments and suggestions, including insightful technical feedback as well as writing guidance. Our appreciation also extends to other members of the Wrox/Wiley team for their hard-working and consistent contributions to the book project.

We also want to thank our coworkers and friends in the mobile space, who have reviewed the book and provided valuable feedback. Ying Chin has given a lot of inspiring technical input, offered mobile programming expertise, and reviewed multiple chapters of the book, and many people reviewed some chapters of the book or provided insightful comment on mobile software development to the authors. These include Ciaran Rochford, Kevin Lee, Madelyn Bryant McIntire, Paul Wu, Mark Leiter, Meishan Li, Anil Hashia, Brad Belanger, Alex Su, Hailin Jiang, Ming Wang, John Zhu, Moataz Mohamed, John Basacchi, Neil McDonald, Aimin Pan, Richard Zhang, Yutao Xi, Haifeng Zheng, Joy Fan, Junqun Dai, Shigenori Tanaka, Yana Wang, Lingzhi Cao, Bolian Yin, Landong Wei, Yi Yang, Yong Chen, Houston Wong, Kevin Wu, Andrew Lin, Richen Wei, Peng Huang, Chen Wang, Emeel Noohi, Scott Leatham, Brian Stern, Youngji Kim, Jason Whitehorn, and Kevin Son.

We would like to thank our family members for their encouragement and support of the book project. In particular, Baijian Yang would like to thank his wife, Chen, and daughter, Ashley, for their love, support, and tolerance. He would also like to thank his co-authors for their inspiration and encouragement.

CONTENTS

FOREWORD *xxv*

INTRODUCTION *xxvii*

CHAPTER 1: WHAT'S NEW IN WINDOWS PHONE 7 **1**

An Overview of Windows Phone 7 **2**
A Brief History 2
The Big Ideas 3
System Architecture 4
Application Framework 6
Windows Phone Marketplace 7
Limitations and Road Map 7
Side-By-Side Comparisons with Android and iPhone **8**
Operating System 8
Application Framework 10
Application Store Process 12
Summary **13**

CHAPTER 2: THE DEVELOPMENT ENVIRONMENT **15**

Overview **15**
Using Windows Phone 7 Developer Tools **17**
Creating WP7 Applications with Visual Studio 17
Testing WP7 Applications in the Windows Phone Emulator 21
Testing WP7 Applications on the Actual Windows Phone Device 26
Publishing Windows Phone 7 Applications to the Marketplace **28**
Preparation 28
Submission 30
Post-Submission Actions 35
Comparisons with Android and iPhone 35
Summary **36**

CHAPTER 3: FUNDAMENTALS **39**

Basic Application Project Structure **40**
Application Project Structure for Windows Phone 7 40
Comparing Application Project Structure for Android and iOS 42

Application Execution Model and Life Cycles **43**
Application Execution Model and Life Cycles in Windows Phone 7 44
Comparing Application Model and Life Cycles in Android and iOS 47
Preserve/Restore Application and Page Transient States for Windows Phone 7 50
Using Launchers and Choosers **59**
Work-Around Solutions in Windows Phone 7 **60**
Multitasking 60
Background Service 61
Data Sharing 61
System Event Hooks 61
Summary **62**

CHAPTER 4: USER INTERFACES **63**

UI Design Principles **64**
UI Design Resources 64
Platform Characteristics 64
Building the WP7 UI **66**
Defining WP7 UI with XAML 66
Defining WP7 UI Programmatically 67
Pages and Navigation Among Pages 68
Using Controls 74
Data Binding and MVVM 81
Handling UI Events 90
Other UI Considerations 91
UI Customization **92**
Summary **94**

CHAPTER 5: APPLICATION DATA STORAGE **95**

Application Storage on Mobile Devices **95**
Local and Cloud 96
Local Files and Databases 97
Using Isolated Storage **98**
Where Is Isolated Storage? 98
Saving Data to the Cloud **105**
Building a Cloud Data Service 105
Using Cloud Data Services 108
Data Storage Design Considerations **110**
Summary **112**

CHAPTER 6: WEB SERVICES AND PUSH NOTIFICATIONS 113

Using Web Services 113
A Primer of Web Services 114
Consuming Web Services on WP7 117
WP7 Push Notifications 131
Why Push Notifications? 131
Push Notification Architecture 131
Push Notifications on WP7 136
Mobile Advertising 146
Mobile Advertising Basics 147
Mobile Advertising Providers 148
Adding Ads to WP7 Apps 150
Summary 152

CHAPTER 7: LEVERAGING LOCATION AND MAPS 155

Location Frameworks Roundup 156
Getting Current Location 159
Geographical Data 159
Civilian Data 166
Using Maps 173
Using the Bing Map Control 173
Combining the Location Service and Bing Map 176
Summary 180

CHAPTER 8: GRAPHICS 183

Graphics Fundamentals 184
Basic 2D and 3D Graphics Glossary 184
Drawing Graphics on Smartphones 187
Drawing Graphics with XNA 189
When Do You Use XNA? 189
Game Loop 190
Textures and Sprites 194
Animation 198
3D Graphics 200
Summary 204

CHAPTER 9: MULTIMEDIA 205

Multimedia Overview 205
Multimedia Architectural Overview 206
WP7 Multimedia 206

iOS Multimedia 208
Android Multimedia 209
Supported Media Codecs 212
Playing Audio on WP7 215
Playing Sounds Using MediaElement 215
Playing Sounds Using SoundEffect 222
Sound, Picture, and Graphics Integration 223
Playing Video on WP7 229
Playing Video Using MediaPlayerLauncher 229
Playing Video Using MediaElement 230
Reusable Media Player Controls 232
Summary 232

CHAPTER 10: UTILIZING SYSTEM HARDWARE 235

An Overview of Smartphone Hardware 235
ARM Processor Architecture 236
Smartphone Hardware Components 237
Accessing the Microphone 238
Accessing the Microphone on iOS 239
Accessing the Microphone on Android 239
Accessing the Microphone on WP7 240
Accessing the Camera 242
Accessing the Camera on iOS 242
Accessing the Camera on Android 243
Accessing the Camera on WP7 244
Using Sensors 245
Accessing the Sensors on iOS 246
Accessing the Sensors on Android 246
Accessing Sensors on WP7 248
Utilizing FM Radio and Bluetooth 254
Utilizing the FM Radio On WP7 254
Utilizing Bluetooth 255
Summary 255

CHAPTER 11: WHAT YOU NEED TO DO ABOUT SECURITY 257

Understanding Mobile Application Security Models 258
Windows Phone 7 Security Overview 258
iOS Security Overview 260
Android Security Overview 261
Security Model Comparisons 262
Using Windows Phone 7 Security APIs 262

.NET Security Namespace 263
Protecting Data Confidentiality with Encryption 263
Ensuring Data Integrity with Hashing 270
Security Best Practices 274
Summary 275

APPENDIX A: AN INTRODUCTION TO SMARTPHONE CHIPSET 277

APPENDIX B: AN INTRODUCTION TO MICROSOFT EXPRESSION BLEND FOR WINDOWS PHONE 281

Microsoft Expression Blend IDE 282
Creating Animation with Expression Blend 284
Customizing UI with Expression Blend 288
Summary 291

APPENDIX C: SAMPLE APPLICATIONS GUIDE 293

Chapter 2 295
HelloWP7 295
Chapter 3 295
WP7Lifecycles 295
Chapter 4 295
WP7Controls 295
WP7Navigation 296
Chapter 5 296
AppDataSample 296
SampleCloudService 297
Chapter 6 297
SocialDemo 297
Push Notification 297
AdControlDemo 298
Chapter 7 298
CurrentLocationTxt 298
SimpleMap 299
Street2Geo 299
TrackMe 299
Chapter 8 300
Cube3D 300
Rolling2D 301
ToyAnimation 301

Chapter 9 **301**
WP7AudioPlayerDemo 301
MediaPicker 301
WP7VideoPlayerDemo 302
WP7EnrichedMoviePlayerDemo 302
Chapter 10 **303**
WindowsPhoneMic 303
WPCamera 303
WP7FMRadio 303
MediaPicker 304
Chapter 11 **304**
AES 304
Hash 305
BlendWP7App1 305

INDEX ***307***

FOREWORD

In 2010, Microsoft launched Windows Phone 7 globally. Windows Phone 7 came to market with a new application programming model and a suite of services to help developers write applications and then control the pricing, sales, and improvement of their applications. I was the Director of Test for Windows Phone 7 Services leading up to and including the launch of Windows Phone 7. Our focus on the developer as a critical part of the Windows Phone 7 ecosystem has produced a growing commercial opportunity for software developers. It's also created a growing and innovative variety of applications for users that show off the capabilities of Windows Phone 7. Knowing how to take advantage of Windows Phone 7's capabilities, regardless of your past experience, is the first step in your Windows Phone 7 development journey.

This book is written to help you initially understand Windows Phone 7's application framework. If you are familiar with Windows Mobile's programming framework, you'll notice an entirely new managed application approach and the emergence of the Silverlight and XNA. If you are entirely new to a Windows Phone, you'll find this book is written to provide you a conceptual map and bridge you over to Windows Phone 7. It's full of comparisons and mappings (e.g., UI controls, compliance rules, etc.) from Android and iOS to Windows Phone 7. It also highlights the UI, platform, and service innovations. This book will be an important desk reference for those developers adopting Windows Phone 7 after doing projects for iOS or Android.

It feels like the "early days" of Windows Phone. Except these early days are being built on decades of experience with developer-friendly platforms and tools designed to delight users and make you successful. If you are starting your journey with us by reading this book, let me welcome you to Windows Phone.

—Eric Hautala
General Manager, Customer Experience Engineering
Windows Phone 7
Microsoft Corp., Inc.
June 2011

INTRODUCTION

ONE OF THE MAJOR DRIVING FORCES BEHIND the boom in smartphones and tablet devices is mobile applications. Since the Apple iPhone was launched in 2007, the mobile developer community has created a vast number of ubiquitous applications for iOS devices and Google Android devices. Mobile applications for Windows Phone 7 (WP7) are poised to grow in the next several years, driven by the software giant's mobile strategy and collaborations with handset partners such as Nokia, HTC, and Samsung.

In order for many of the iOS and Android developers to port their applications to WP7 or to create new applications, it's important to understand the architecture of the new WP7 operating system, and to become familiar with various application development patterns from an iOS-Android-WP7 comparison standpoint. This book aims at addressing this need by providing essential information, technical analysis, and working samples to help iOS and Android developers create applications on WP7.

WHOM THIS BOOK IS FOR

The book targets mainly experienced mobile application developers with Android and iOS programming background. The audience may include:

- Industry professionals such as software architects and engineers with Independent Software Vendors (ISVs), device handset makers, and mobile operators
- College students who have built iOS and Android applications
- Freelance software developers who want to make a fortune with their mobile applications

To use this book, you should have some programming experience using either Java, Objective-C, or both, to develop on iOS or Android. In addition, you should be familiar with C#, the most popular programming language for WP7.

The book not only provides key programming coverage on WP7, but also presents coverage of similar topics on iOS and Android. This makes the book a good reference for those developers who have no mobile application development experience but want to start developing mobile applications on one or more platforms.

WHAT THIS BOOK COVERS

The book covers core mobile application development concepts and a list of essential topics of WP7 from the Android/iOS developer's point of view, including WP7 system architecture, application frameworks, development environment, application model, UI design, application data storage, web services and push notifications, location and maps, multimedia, 2D and 3D graphics, system services and sensors, and application security.

The book is mainly focused on WP7 Silverlight-based application development instead of XNA game development, with Chapters 8 and 9 discussing related XNA framework usages.

There is no doubt that WP7 will continue to evolve with new features and new API in the next several years. To cover the latest releases of WP7 (such as the WP7.5 release codenamed "Mango"), online update articles will be provided at the book's website.

HOW THIS BOOK IS STRUCTURED

The book is organized as 11 chapters. In order to build a solid foundation of WP7 application development, it starts with such essential topics as overview of system architecture and basic development environment setup. Then application fundamentals such as application model and application life cycle are discussed. UI design and application data follow, after which comes a set of key topics ranging from web services, to location and maps, to multimedia. Accompanying sample projects, which are available for download at the `Wrox.com` website, are referenced in those chapters to illustrate certain programming patterns and Application Programming Interface (API) usage.

It is recommended that new WP7 developers start from Chapter 1 and go through the first three chapters to gain a basic understanding of the big picture. After that, developers can read any chapter of interest and don't need to read the chapters one by one.

The following is a brief description of each chapter:

Chapter 1: "What's New in Windows Phone 7" provides an overview of the new WP7 operating system, the application framework, the MarketPlace application store, and WP7 capabilities and limitations. It also features a comparison of the three operating systems.

Chapter 2: "The Development Environment" describes basic steps to set up a WP7 development environment, including preparing system prerequisites, downloading and installing required tools, and accessing online documentation. This chapter also covers publishing an application on MarketPlace.

Chapter 3: "Fundamentals" is concerned with key concepts surrounding application execution model and life cycle, basic application structure, and common system tasks. Those topics are discussed in the context of comparisons with iOS and Android.

Chapter 4: "User Interfaces" covers the Metro UI style, application UI design guidelines, basic page structure, the eXtensible Application Markup Language (XAML), and unique controls such as the pivot control and the panorama control, with examples.

Chapter 5: "Application Data Storage" discusses using data storage in a WP7 application, as well as leveraging cloud storage. The chapter starts with a general application data introduction on iOS and Android, and then provides WP7 details such as isolated storage classes, data serialization, and using Windows Azure cloud storage.

Chapter 6: "Web Services and Push Notifications" covers the consumption of public web services using HTTP methods, as well as using push notifications in a WP7 application. The chapter also discusses JavaScript Object Notation (JSON) and eXtensible Markup Language (XML) parsing and serialization, Language-INtegrated Query (LINQ), and the mobile advertising control for WP7.

Chapter 7: "Leveraging Location and Maps" presents the WP7 location data framework, including the location API and the Bing map control. Instructions on how to integrate maps and navigation into a WP7 application will be provided.

Chapter 8: "Graphics" covers application graphics basics, 2D and 3D graphics framework in WP7, and using the XNA (Xbox New Architecture, or XNA is Not an Acronym) framework to build animations. It also highlights the WP7 graphics engine, as compared with iOS and Android.

Chapter 9: "Multimedia" discusses typical image-, audio-, and video-related scenarios in a WP7 application. It starts with an overview of the system-level support for multimedia in WP7, followed by descriptions of common multimedia playback and editing tasks enabled by Silverlight and XNA.

Chapter 10: "Utilizing System Hardware" covers an application's interaction with the underlying phone system. It covers the access to device microphone, camera, and accelerometer sensor for a variety of usage scenarios. The chapter also highlights WP7 limitations in terms of providing access to such peripherals as Bluetooth and digital compass.

Chapter 11: "What You Need to Do about Security" discusses security application frameworks in iOS, Android, and WP7, and key concepts in the domain of mobile application security such as sandbox and security chamber. Then the chapter discusses the WP7 application security API and common scenarios such as data encryption and hashing.

Appendix A: "An Introduction to Smartphone Chipset"

Appendix B: "An Introduction to Microsoft Expression Blend for Windows Phone"

Appendix C: "Sample Applications Guide"

WHAT YOU NEED TO USE THIS BOOK

You will need a computer running Windows 7 or Windows Vista to develop for WP7. In addition, you will need to download and install the latest Windows Phone 7 developer tools from a Microsoft website (`http://create.msdn.com/en-us/home/getting_started`). You can use the device emulator that comes with the tools or test your application on a real WP7 device.

CONVENTIONS

To help you get the most from the text and keep track of what's happening, we've used a number of conventions throughout the book.

> *Boxes with a warning icon like this one hold important, not-to-be-forgotten information that is directly relevant to the surrounding text.*

The pencil icon indicates notes, tips, hints, tricks, or and asides to the current discussion.

As for styles in the text:

- We *italicize* new terms and important words when we introduce them.
- We show file names, URLs, and code within the text like so: `persistence.properties`.
- We present code in two different ways. The first is as "listings" with a number and other identification that will help you download the code from `Wrox.com`. The second way is code snippets, which — if they are downloadable — have an identifying CodeNote at the end. Code is presented in a monofont style, like the following:

```
We use a monofont type with no highlighting for most code examples.
```

SOURCE CODE

As you work through the examples in this book, you may choose either to type in all the code manually, or to use the source code files that accompany the book. All the source code used in this book is available for download at `www.wrox.com`. When at the site, simply locate the book's title (use the Search box or one of the title lists) and click the Download Code link on the book's detail page to obtain all the source code for the book. Code that is included on the Web site is highlighted by the following icon:

Available for download on Wrox.com

Listings include a number and usually the filename in the title. If it is just a downloadable code snippet, you'll find the filename in a code note such as this:

Code snippet filename

Because many books have similar titles, you may find it easiest to search by ISBN; this book's ISBN is 978-1-118-02197-2.

Once you download the code, just decompress it with your favorite compression tool. Alternately, you can go to the main Wrox code download page at `www.wrox.com/dynamic/books/download.aspx` to see the code available for this book and all other Wrox books.

ERRATA

We make every effort to ensure that there are no errors in the text or in the code. However, no one is perfect, and mistakes do occur. If you find an error in one of our books, like a spelling mistake or faulty piece of code, we would be very grateful for your feedback. By sending in errata, you may save another reader hours of frustration, and at the same time, you will be helping us provide even higher quality information.

To find the errata page for this book, go to `www.wrox.com` and locate the title using the Search box or one of the title lists. Then, on the book details page, click the Book Errata link. On this page, you can view all errata that has been submitted for this book and posted by Wrox editors. A complete book list, including links to each book's errata, is also available at `www.wrox.com/misc-pages/booklist.shtml`.

If you don't spot "your" error on the Book Errata page, go to `www.wrox.com/contact/techsupport.shtml` and complete the form there to send us the error you have found. We'll check the information and, if appropriate, post a message to the book's errata page and fix the problem in subsequent editions of the book.

P2P.WROX.COM

For author and peer discussion, join the P2P forums at `p2p.wrox.com`. The forums are a Web-based system for you to post messages relating to Wrox books and related technologies and interact with other readers and technology users. The forums offer a subscription feature to e-mail you topics of interest of your choosing when new posts are made to the forums. Wrox authors, editors, other industry experts, and your fellow readers are present on these forums.

At `http://p2p.wrox.com`, you will find a number of different forums that will help you, not only as you read this book, but also as you develop your own applications. To join the forums, just follow these steps:

1. Go to `p2p.wrox.com` and click the Register link.
2. Read the terms of use and click Agree.
3. Complete the required information to join, as well as any optional information you wish to provide, and click Submit.
4. You will receive an e-mail with information describing how to verify your account and complete the joining process.

> *You can read messages in the forums without joining P2P, but in order to post your own messages, you must join.*

Once you join, you can post new messages and respond to messages other users post. You can read messages at any time on the Web. If you would like to have new messages from a particular forum e-mailed to you, click the Subscribe to this Forum icon by the forum name in the forum listing.

For more information about how to use the Wrox P2P, be sure to read the P2P FAQs for answers to questions about how the forum software works, as well as many common questions specific to P2P and Wrox books. To read the FAQs, click the FAQ link on any P2P page.

1

What's New in Windows Phone 7

WHAT'S IN THIS CHAPTER

- An architectural overview of WP7
- Comparisons with iOS and Android

Mobile application developers will find it useful to have an architectural understanding of the underlying Windows Phone 7 (WP7) software platform. In particular, developers need to understand the application framework, its capabilities and limitations, and platform extensibility. It's also important to know potential technical approaches to common development tasks. For Android or iPhone app developers, it's vital to obtain a big picture of the new WP7 operating system.

> *Some terms can prove confusing because different people use them in different ways. For the purposes of this book, you'll see iPhone Operating System (iOS) when the text discusses applications or code. The book uses iPhone when it asks you to perform an action with the physical device (or an emulator), such as clicking a button. In addition, we'll use WP7 and Windows Phone 7 interchangeably throughout the book.*

The first chapter of this book provides an overview of the new WP7 operating system, the history of Windows Mobile, the WP7 hardware platform (also known as the chassis specification), the new Windows Compact Edition (CE) kernel, the application framework, the application store, and WP7 capabilities and limitations. Readers who aren't familiar with Windows phone technologies will see examples that use Android or iPhone technologies. It's important to understand the overall design philosophy of WP7 and its potential impact on the competition.

The chapter will outline a roadmap of Microsoft's Windows Phone offerings in the next 12-18 months. The chapter also compares the three major mobile platforms from different perspectives:

- **Underlying operating system origins:** MacOS, Linux, or Windows CE
- **Application frameworks:** Xcode on iPhone, Java on Android framework, or Silverlight and Xbox/DirectX New generation Architecture (XNA) on the WP7 app framework
- **App store process:** iPhone App Store, Android Market, or Windows Marketplace

AN OVERVIEW OF WINDOWS PHONE 7

This section will present an overview of WP7, including a brief introduction to Windows Mobile history, the design rationale of WP7 and its system architecture, as well as the new application framework and application store.

A Brief History

Microsoft's mobile operating system originated with the Pocket PC 2000 release in 2000, which was targeting Personal Digital Assistant (PDA) devices without any cellular capability. It was built on top of the Windows CE 3.0 kernel and supported multiple processor architectures, including Acorn RISC Machine (ARM), where RISC stands for Reduced Instruction Set Computer, Microprocessor without Interlocked Pipeline Stages (MIPS), and x86. The rationale was to provide scaled-down desktop experience on a mobile form factor, with a stylus mimicking the computer mouse interface, and a resistive touch screen that reacts to stylus tapping.

In 2003, Microsoft released Windows Mobile 2003 and Windows Mobile 2003 Second Edition that started to offer phone capability. This release also offered strong enterprise-oriented features such as Pocket Outlook, Virtual Private Network (VPN) support, and ActiveSync.

Then a major upgrade, Windows Mobile 5, was released in 2005. Windows Mobile 5 allowed developers to write managed applications that ran on top of the .NET Compact Framework. It also provided a Direct Push technology where Microsoft Exchange e-mails can be pushed to the Pocket Outlook client on the device as they arrive. The GUI was essentially similar to the previous releases.

Windows Mobile 6 and Windows Mobile 6.1 were released in 2007 and 2008. Both are built on top of Windows CE 5.2. The focus was still on providing a rich set of features rather than a compelling user interface (UI).

With all the Windows Mobile releases until Windows Mobile 6.1, Microsoft managed to build a strong mobile product line, targeting enterprise professionals. Its major competitor was Research in Motion (RIM). Microsoft's development efforts turned out to be quite a success from 2006 to early 2007. During this time, Windows Mobile took over 20 percent of the smartphone market and shipped 12 million devices.

The iPhone arrived in June 2007. iPhone's "Touching is believing" user experience was undoubtedly a tremendous innovation compared to any other smartphone on the market at that time. The unique multi-touch, finger-friendly user interface changed the public's opinion that smartphones were designed for professionals; as a result the smartphone market grew rapidly in the following years.

Initially Microsoft didn't realize the mobile market was undergoing a major overhaul. It failed to react quickly to accommodate the dramatic growth of the smartphone market driven by explosive adoption of the iPhone among average consumers. In 2007 and 2008, Microsoft worked on Windows Mobile 7, which for the most part resembled Windows Mobile 6 from a user interface perspective but with multi-touch support. In the interim, Microsoft released Windows Mobile 6.5, which provided a minor update with finger-friendly tiles and menus. Unsurprisingly, it failed to impress the market.

Google entered mobile space with Android in 2008, and has enjoyed rapid growth since then, partly because Microsoft has failed to release a major update for about three years (since Windows Mobile 6.1). Google has formed the Open Handset Alliance (OHA) with major handset makers, silicon vendors, and mobile operators to create the Android open platform. As Microsoft struggled to build Windows Mobile 7, handset makers turned to Google Android.

Feeling the pressure from Apple and Google, Microsoft has shuffled its mobile business division, reset the Windows Mobile 7 effort, and started WP7 from scratch. WP7 sports a new tile interface, Marketplace application store, Silverlight- and XNA-based application framework, and Xbox LIVE and Zune integration. The effort has finally paid off. WP7 was launched in Europe, Singapore, and Australia in October 2010, and in the U.S. and Canada in November 2010. Microsoft shipped 1.5 million WP7 devices in the first six weeks. It's still too early to project WP7's future in terms of market share. Nonetheless, WP7 is unique in many ways compared to iOS and Android, and thus offers another choice for smartphone users. Microsoft continues to invest in mobile technology and keeps improving Windows Phone. It'll be quite interesting to see the competition among the three major mobile operating systems for the next few years.

The Big Ideas

WP7 is the outcome of Microsoft's new mobile strategy, which is to shift from enterprise-oriented mobile product design to consumer-focused design. As Andy Lees, Microsoft's president of the mobile and embedded division, put it in an interview:

> *We made a very big decision to re-examine everything, because the industries surrounding mobile are at an inflection point. . . . The technological advances over the past few years enable us to do bold new things we've never done before. But the most important thing is that we are bringing it all together with an almost maniacal focus on the consumer.*
>
> `www.microsoft.com/presspass/features/2010/feb10/02-15windowsphone7.mspx`

The following list describes the overall goals that Microsoft tried to achieve when developing WP7:

- **Consumer Focused:** Microsoft reviewed its competitors' offerings in order to understand what the consumer wants in terms of mobile user experience. For example, consumers want to touch the screen using their fingers, rather than using a stylus. Therefore, the developer must create a graphical user interface (GUI) that's finger-friendly, with enlarged actionable components that support tapping (briefly using a finger to touch the surface), dragging (pressing and holding an item, and moving it on the surface), flicking (briefly brushing the surface), pinching (pressing and holding, using two fingers and moving them closer), spreading (pressing and holding, using two fingers and moving them apart), and so on. In addition, WP7 applications can enable unique user experiences such as Panorama and Pivots, which are discussed in Chapter 4. Another example of consumer-focused design is the seamless integration with Microsoft's other computing assets, such as Zune media service, Xbox LIVE , Office Live, and Bing search service. This integration makes it possible for consumers to enjoy these services across different screens on different devices.

- **Life in Motion:** The rationale behind the WP7 user experience is "life in motion," where the phone keeps pace with events happening in people's life in a well-integrated, effortless way. For example, live tiles on the Home screen show real-time updates of the user's contacts, calendars, games, messages, and phone calls. A quick glance gives the user all the needed information without the user's touching anything. And if the user touches any of those tiles, WP7 displays a hub screen where events of the selected type are aggregated into a single view from various applications, web services, and other sources.
- **Consistent experience:** The Windows Phone user experience is consistent across applications and services on assorted devices. Any third-party hardware or software innovations must be in line with the unified model to avoid fragmentation.
    - **Hardware:** Microsoft and its partners defined a set of specifications where all WP7 devices rely on a few chipsets. The reason these chipsets are so important is that Qualcomm and Microsoft have performed all the major work on the Board Support Packages (BSPs), which are driver and hardware configurations. During the Windows Mobile era, original equipment manufacturers (OEMs) had the opportunity to choose any chipset. Supporting all the chipset variants with Windows Mobile was a big challenge for Microsoft. Now, with the unique chassis specifications, a large part of the device BSPs are provided for OEMs — they only need to select some peripherals and create drivers for them. The unified hardware design is actually good for developers; there is no need to consider different CPU speeds, memory capacity, and screen sizes. They are the same on all WP7 devices.
    - **Software:** All applications are either Silverlight- or XNA-based, leveraging the same set of .NET Framework APIs. All third-party applications must pass the Marketplace certification before the user can install them on a device.

System Architecture

The WP7 OS is based on a variant of Microsoft Embedded OS, Windows CE 6 (also known as Windows Embedded CE 6), while the Windows Mobile 6.x variants are all based on Windows CE 5. Generally, Windows CE provides a 32-bit kernel that is designed for embedded devices, and a set of system services such as memory management, networking and connection management, I/O, and graphics. On the other hand, the Windows Phone OS is built on top of the CE kernel with added specific system services and an application framework for mobile phones.

The major differences between CE 6 and CE 5 are listed below:

- Process address space is increased from 32MB to 2GB. On Windows CE 5, every process can occupy 32MB of address space. Windows CE 6 increases process storage to 2GB.

- The number of processes has been increased from 32 to 32K (32,768). This is important to application developers. On Windows Mobile 6.x, only 32 processes can be active at the same time. If a user wants to launch another process, the system will reject it. WP7 eliminates the 32-process limitation.
- User mode and kernel mode device drivers are possible.
- `device.exe`, `filesys.exe`, and `GWES.exe` have been moved to Kernel mode, to improve device performance.

Overall, the operating system in WP7 devices is more secure and stable, and offers better performance.

Figure 1-1 illustrates the WP7 OS architecture. As shown in the figure, the operating system contains three layers (from bottom to top): hardware, kernel space, and user space. All .NET Framework applications run in the user space. The OS kernel, drivers, and system services execute in kernel space. Compared to the architecture of Windows Mobile 6.5, on which you can execute both native and managed applications, WP7 OS enforces managed application development only. Furthermore, managed applications can use only the features provided by Silverlight, XNA, and Phone APIs; nothing else is accessible from within applications.

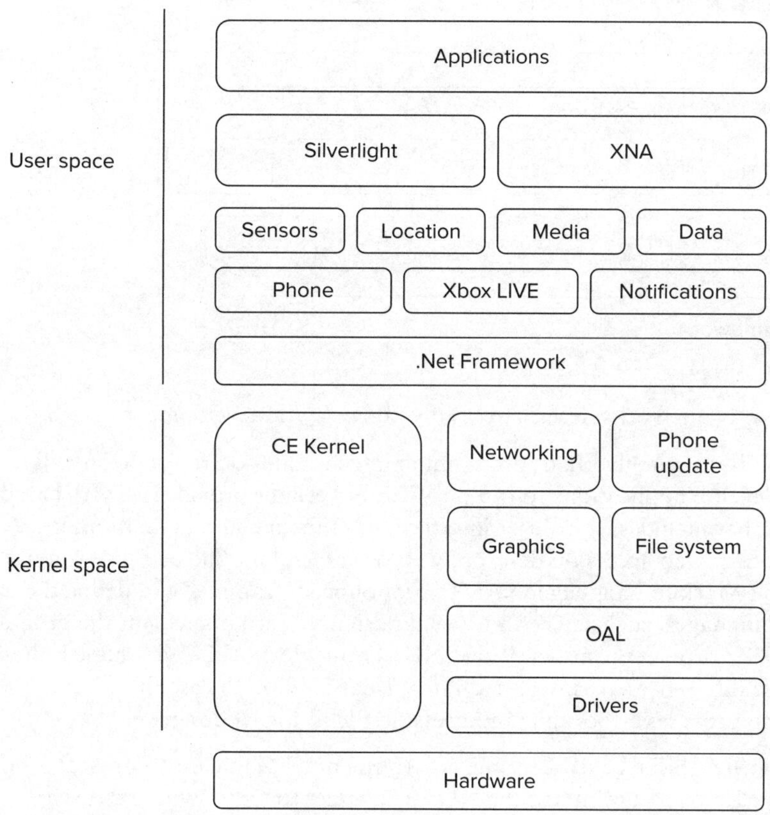

FIGURE 1-1: WP7 architecture

Application Framework

Mobile application developers are mainly concerned with changes to the application framework. When targeting Windows Mobile 6.x, developers can use either native Win32 APIs to write C/C++ code or C# and Visual Basic .NET to write managed code. The managed code runs on top of the .NET Compact Framework. On WP7, however, all applications are managed applications, and Microsoft provides two application frameworks: Silverlight and XNA, as shown in Figure 1-2 (source: `http://msdn.microsoft.com/en-us/library/ff402531(v=vs.92).aspx`). Microsoft suggests using Silverlight for developing event-based applications and XNA for game development.

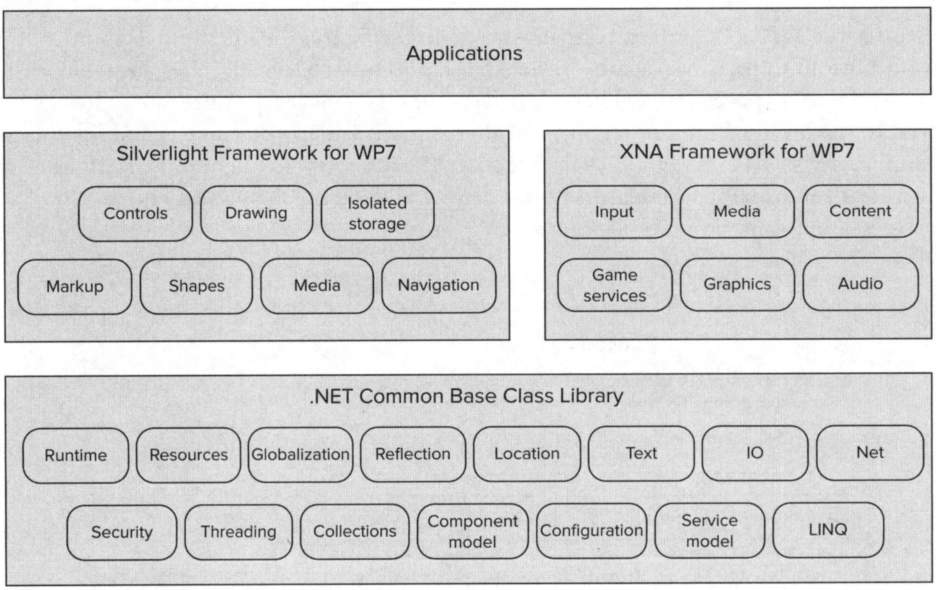

FIGURE 1-2: WP7 application framework

Some details follow on the two frameworks described briefly above (with development tools):

- **Silverlight:** People familiar with Silverlight programming techniques on the desktop will find it fairly easy to develop applications to run on WP7. Silverlight provides a .NET-based runtime environment that includes a rich user interface, multimedia, and animation. In addition, Silverlight offers web access on desktop, web server, and mobile devices. You use the Extensible Application Markup Language (XAML, pronounced "zammel") to define the user interface, and .NET languages, such as C# and Visual Basic .NET, to implement the program logic. Silverlight on WP7 provides a subset of the .NET Framework APIs, with added phone-specific APIs. All these differences are integrated into Visual Studio. To ease the design of a rich UI, Microsoft also provides a tool called Expression Blend for UI designers.
- **XNA:** XNA is yet another .NET-based runtime environment available on Microsoft Xbox, Windows, and WP7. Microsoft optimized the XNA run time, together with extensive classes and libraries for game development. XNA provides a foundation for game developers

to create cross-platform games as long as the XNA run time runs on those platforms. You can perform XNA-based development for WP7 using C# and XNA Game Studio Express, a free tool provided by Microsoft.

- **Development Tools:** Microsoft provides Visual Studio 2010 Express for Windows Phone and Expression Blend for WP7 application development free of charge. Developers can use the built-in emulator in Visual Studio 2010 Express to debug and test an application. A Zune client is required on the desktop when you want to develop applications using a device.

Windows Phone Marketplace

Microsoft provides a Windows Phone Marketplace for obtaining WP7 applications, which is similar to the iPhone and Android offerings. Any third-party applications must pass the Marketplace certification before Microsoft will publish them in the Marketplace store. The user is able to browse store applications with the Marketplace application on the device, download free applications, or purchase paid applications. Unlike Windows Mobile 6.x, application side-loading on a retail device isn't allowed officially.

Mobile developers who want to create and publish WP7 applications and Xbox LIVE applications must register at `http://create.msdn.com/` to obtain a Marketplace developer account. The registration fee is US$99. When a developer is ready to submit an application for certification, Microsoft validates the application's Silverlight Application Package (XAP, pronounced "zap") against published application content guidelines. If the XAP validation succeeds, Microsoft tests the application on a real device both automatically and manually to ensure it meets stability and performance policies. When all the tests are passed, the application is signed with the developer's certificate (assigned by Microsoft to the developer during Marketplace registration) and is ready for public release in the Marketplace store.

Limitations and Road Map

WP7 is a new mobile operating system that is different in many aspects from previous Microsoft Windows Mobile releases. Essentially this is a version-one product representing Microsoft's new thrust into the mobile space. Needless to say, being a version-one product implies limitations and work-in-progress to the computing world. WP7 is no exception. Some of the most frequently mentioned limitations of WP7 are described in the following:

- **Copy and paste:** This feature appears in previous Windows Mobile releases but is missing in the first release of WP7, and has been provided in a WP7 update release in March 2011. Note that iOS does not offer copy and paste until iOS 3.0. Android's full support for copy and paste starts with Android 2.3 (Gingerbread).
- **Multi-tasking for third-party applications:** While system services such as music playback and FM radio will continue to work in the background, third-party applications cannot. This is similar to the first version of iPhone. On Android, developers have the flexibility to build a service. On WP7, developers cannot create a service, and thus multi-tasking is not possible for third-party applications. It's possible that Microsoft will enable multi-tasking of a selected set of tasks, such as music playback, VoIP, and GPS, for third-party applications in a future version. However, it's unlikely that it will provide a way to write a service on WP7.

- **Native application development:** Only managed code can run on WP7. This means that third-party applications can use only those .NET classes provided by Silverlight and XNA. As a consequence, a software vendor won't be able to implement mission-critical modules in native code and access them from a managed application unless they work with a Windows Phone OEM partner directly to build the native module as an OEM extension to the operating system. (Developers can write native code modules on Android using the Native Development Kit). Microsoft is unlikely to change this design in future Windows Phone releases. The main reason for this limitation is that Microsoft wants to maintain a consistent user experience and also to make sure applications don't impact system performance and battery life.
- **HTML5 support:** The Internet Explorer web browser in WP7 does not support HTML5, whereas the browsers in Android and iOS already have solid HTML5 support because they are based on the WebKit rendering engine. Microsoft already has demonstrated Internet Explorer 9 running on WP7, which supports HTML5 in large part. It is likely that HTML5 support will be added in an update release in 2011.

Despite the limitations described in this list, there is some good news for developers: Microsoft plans to provide more frequent firmware updates to WP7 than to previous Windows Mobile releases. This means some much-needed features and flexibilities may be added to the updates.

SIDE-BY-SIDE COMPARISONS WITH ANDROID AND IPHONE

Mobile developers coming from iOS and Android will be interested in the similarities and differences between WP7, iOS, and Android. The following sections survey the three mobile operating systems from an architectural point of view.

Operating System

There are four generations of iPhones at the time of writing. Earlier iPhone models use Samsung processors, whereas iPhone 4 uses Apple's own processor, A4. All iPhones use a 3.5-inch display. The iPhone 4 has 512Mb of internal memory and either 16GB or 32GB of storage. All iPhones are equipped with GPS, Wi-Fi, and Bluetooth connectivity.

Apple derived iOS from the Mac OS X, which is Apple's desktop OS for Macintosh systems. Mac OS X is a Unix flavor combining a Mach kernel and numerous components from FreeBSD (where BSD stands for Berkeley Software Distribution) and NetBSD. iOS bears the same core OS components as Mac OS X, and Apple has optimized it for mobile devices such as iPhone, iPod Touch, and iPad with respect to performance, stability, and battery life. The iOS consists of four layers that include (from the bottom to the top): core OS layer, core service layer, media layer, and Cocoa Touch layer. iOS applications are native applications developed using the iOS SDK and Xcode development environment, and are written in Objective-C.

Android device hardware is more versatile as Google doesn't define a hardware specification for it (Microsoft does this for WP7). Generally Android devices, just like other smartphones, have cellular voice and data capability, camera, GPS, Wi-Fi, Bluetooth, and a few sensors such as an accelerometer and a proximity sensor. Screen size varies on different devices, although 3.5-inch and 4-inch screens are becoming popular.

At the core of Android device software is the Linux kernel. The drivers and the hardware abstraction layer modules are all Linux-based. On top of the Linux kernel is the Dalvik virtual machine and its runtime environment, along with native libraries and services. Dalvik is a special Java virtual machine that's optimized for running Java applications on a mobile device. The majority of the Android operating system is the application framework, which consists of Java services, APIs, and libraries, as well as native libraries and native services. Figure 1-3 shows the details (source: `http://developer.android.com/guide/basics/what-is-android.html`).

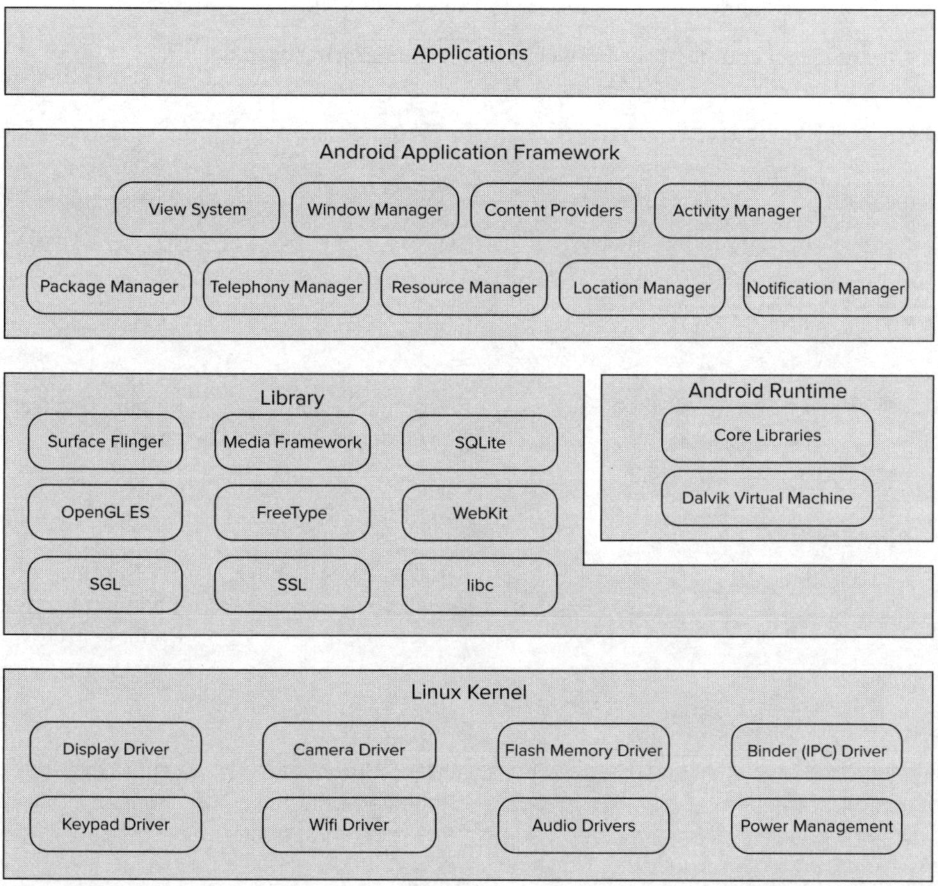

FIGURE 1-3: Android architecture

There are two application development options on the Android: Java applications that run entirely in the Dalvik virtual machine and Java applications that use JNI (Java Native Invocation) to call into native libraries on the system. You might wonder why Android developers would use a native library. Performance is the main reason, since native code generally offers better performance than Java bytecode (or dex code produced by the Dalvik virtual machine). This is particularly important for CPU intensive code and media processing code. In addition, when someone ports C/C++ code from another platform, it's natural to wrap the existing code into native libraries instead of porting it to Java code.

On WP7, developers aren't allowed to write native code; only managed code is allowed. In addition, only subsets of the general Silverlight APIs and XNA APIs are supported on WP7. However, phone makers have the means to inject some native code into the system and use it in their applications by using COM (Component Object Model) interop. Microsoft provided a special SDK (not to be confused with the publicly available SDK for WP7 application development) for phone makers, so that they can develop native COM DLLs with a limited set of native Windows CE APIs, and use them in their applications. Those COM DLLs are treated as part of the operating system. Unfortunately, mobile developers don't have access to the special SDK.

Table 1-1 shows system level comparisons between the three operating systems.

TABLE 1-1: System Level Comparisons

ITEMS	IOS	ANDROID	WP7
Kernel	Mach kernel in Mac OS X	Linux kernel	Windows CE 6
Supported CPU architecture	ARM	ARM, X86	ARM
Memory model	Paging, no disk-backed backing store	Paging, SD-card swap space optional	Paging, no swap space
OS low-memory management	Yes	Yes	Yes
Managed run time	No	Dalvik Java run time	.NET Framework Common Language Runtime (CLR)
Application sandbox	Yes	Yes	Yes
Cross-application communication	Limited, using custom URL	Intent, Broadcast Receiver, Content Provider, and Service	Launchers and chooser

Application Framework

Understanding the differences between the application frameworks of these three major operating systems is critical to cross-platform application development.

iOS applications are built on top of the Cocoa Touch layer, which consists of Objective-C application frameworks like UIKit, Game Kit, Map Kit, and iAd. The Cocoa Touch layer provides a rich set of UI objects and event-handling mechanisms for applications and windows, as well as core system services and media services enabled by the media layer and the core service layer. Developers will normally use the functions and methods provided by the Cocoa Touch frameworks, although it's also possible to call directly into the media and core service layers for some fine-grained control of specific services.

Similarly, Android Java applications are built on top of the Android framework that consists of an array of Java packages such as Activity Manager, View System, Windows Manager, Telephony Manager, Content Provider, Notification Manager, and so on. The UI is composed of a variety of "view" components in a combination of layouts. The framework's Java packages are the interface to the underlying system capabilities, as they rely on corresponding native libraries to perform low-level tasks. Developers can also inject native libraries into a Java application. Interaction between the Java code and the native code occurs within JNI. Android introduces some distinct programming concepts such as Activity and Intents, which makes it easier to extend system applications and services.

The WP7 application framework is .NET-based, and you can't use native code (as discussed in the "Limitations and Road Map" section of this chapter). Silverlight and XNA both provide a set of .NET Framework types that serve as the interface to the underlying operating system. In addition, applications can interact with media, local and remote data, sensors, location, and phone-specific information using the standard managed Windows Phone API. The .NET run time and libraries rely on native system services and libraries. Each application is isolated within a .NET runtime sandbox. The use of a sandbox means that cross-application communication isn't allowed on WP7, which means developers have to resort to web services as a bridge for cross-app communication.

Table 1-2 outlines application frameworks on the three platforms.

TABLE 1-2: Application Framework Comparisons

ITEM	IOS	ANDROID	WP7
Available application frameworks	Cocoa Touch	Android framework	Silverlight and XNA
Programming languages	Object-C	Java	.NET languages (C# and Visual Basic.NET)
Native API access	Yes	Yes, via JNI using NDK	No
SDK	iOS SDK on Mac OS X	Android SDK on Windows, Linux and Mac OS X	Windows Phone SDK on Windows 7
IDE	Xcode	Eclipse (with ADT plug-in)	Visual Studio 2010 or Visual Studio 2010 Express
UI definition	NIB resource file	Widgets and layouts	XAML
UI event mechanism	Delegates	Event listeners in View classes	Event handlers in CLR (Common Language Runtime)

Some developers want to be able to use C# and Visual Studio tools to develop mobile applications running on Android and iOS. This will make cross-platform design easier and enable code reuse across these platforms. Mono-Android (`http://mono-android.net/`) and Monotouch (`http://monotouch.net/`) are such tools for this purpose.

Application Store Process

The Apple App Store was the first mobile application store, and had the largest number of applications at the time of writing. Nonetheless, many mobile developers have questioned Apple's application review process because Apple has imposed so many restrictions and is known to hand out mysterious rejections. Each iOS app that a developer submits is verified against a functionality policy, content policy, and legal policy, which may take days to even a few weeks.

The Android Market grew rapidly as the number of shipped Android devices increased dramatically between 2009 and 2010. For instance, in December 2010 Google announced that it was activating 300,000 Android devices a day (`http://mashable.com/2010/12/09/android-device-stats-2010`). The Android Market enforces content policies that disallow hate speech, nudity, sexually explicit material, and copyright infringements. In contrast to the Apple App Store, the basic idea of the Android Market is openness — the Android Market doesn't perform any verification of a submitted application. Essentially, developers simply register an account, upload an application, and describe it, and the application appears in the Android Market. Google has a team that handles any reported violation against the content policy and may remove the application from the Market if necessary.

> *The Apple iOS content policy can be found at* `developer.apple.com/appstore/guidelines.html`. *Google Android content policy is available at* `www.android.com/us/developer-content-policy.html`.

Microsoft started to offer a mobile application store during the Windows Mobile 6.5 years. But that effort didn't take off, mainly because the accompanying OS release failed to earn significant market share. As part of the new Windows Phone project release, Microsoft has redesigned the Windows Phone Marketplace to compare with the other two competitors. Furthermore, Microsoft seems to have learned quite a bit from the problems of the Apple App Store and the Android Market, which has resulted in a fairly comprehensive application certification process that aims at verifying a submitted application against policies and technical requirements to ensure the application is reliable, makes efficient use of resources, doesn't interfere with device functionality, and is free of malicious software. The Marketplace certification process includes both static and automatic testing of the application on API usage, stability, performance, and so on. The whole process can take as long as a few weeks.

Table 1-3 outlines the three application store processes.

TABLE 1-3: Comparison of Application Store Processes

ITEMS	APPLE APP STORE	ANDROID MARKET	WINDOWS PHONE MARKETPLACE
Registration fee	$99 annually, $299 annually for enterprise	One-time $25	$99 annually, include submissions of five free apps
App submission fee	None	None	$19.99 for free apps; $99 for paid
Revenue sharing (developer/company)	70/30	70/30 (Carrier takes the 30%)	70/30
Regional availability	Worldwide with country-specific stores	Free apps are accessible everywhere, but paid app access varies.	17 countries, including some EU countries, Australia, Hong Kong, India, Singapore, and U.S.
Content policy	Yes	Yes	Yes
Testing on real devices	Yes	No	Yes
Performance requirement	No	No	Yes
App return for refund	Yes	Yes, within 15 minutes after purchase	No
Try and buy	No	No	Yes

SUMMARY

After it realized that Windows Mobile wasn't the product the consumer smartphone market wanted, Microsoft decided to change its enterprise-focused strategy to a consumer-focused one. The goal of WP7 is to provide a consistent "life-in-motion" mobile experience across devices made by various handset makers.

WP7 is a completely new mobile operating system that isn't compatible with any previous Windows Mobile releases in most cases. Microsoft redesigned the GUI to provide a finger-friendly interface with multi-touch support on capacitive touch screens. The live tile Home screen is one of a kind in the industry. The integration with Xbox LIVE, Zune media service, Office Mobile, and SharePoint gives users easy access to Microsoft technologies on the go. The application framework enables developers to build Silverlight applications and XNA-based games quickly. The Windows Phone Marketplace provides an opportunity for developers to monetize their applications and games.

WP7 compares favorably with iOS and Android from various technical perspectives. The operating system kernel is Windows CE 6, which can provide 2GB of virtual address space for each process, and has significant performance and stability improvements over Windows CE 5. The application framework enables creation of a unique UI pattern and interactions with the phone functionality in a managed environment. Even if you aren't familiar with .NET technologies, you can still ramp up quickly with Windows Phone application development.

After reading this chapter, you should have the big picture of the WP7 operating system from a technical architectural perspective. In the remaining chapters, you'll learn how to leverage these innovations to build appealing applications.

2

The Development Environment

WHAT'S IN THIS CHAPTER

- How to build your applications
- Methods for testing applications
- Using the Windows Phone Emulator
- How to publish your applications

Before exploring the details of programming for Windows Phone 7 (WP7), you need to become familiar with the WP7 development environment. The development environment for a mobile platform usually includes an Integrated Development Environment (IDE) to write your applications, build tools to create the executables, source-level debugging tools to debug the application and tune performance, device emulators to simulate the device environment, and much more.

WP7 provides a developer with a friendly development environment by relying on the features of Visual Studio. In short, you don't have to learn a new development tool to work with WP7. The WP7 development tools are almost as powerful as those that Visual Studio provides for developing desktop applications and games. At first sight these tools may even seem to be overkill for mobile development. However, by using these powerful tools to reduce the work of coding, developers can spend more time creatively thinking about application design and features. The following sections describe the WP7 development environment in detail.

OVERVIEW

Windows Phone Developer Tools provide the fundamental functionality required for WP7 development. You can obtain these tools and the corresponding release notes at `http://create.msdn.com/home/getting_started` The full installation package for Windows Phone Development Tools includes:

- Visual Studio 2010 Express for Windows Phone

- Windows Phone Emulator
- Microsoft Expression Blend for Windows Phone
- XNA Game Studio 4.0
- Silverlight and .NET Framework 4
- Samples and other software development kit (SDK) components

To download the installation package, you first need to download a smart installer. The installer can detect the current software setups on the development PC and then download a 357 MB file for the full installation. If Visual Studio 2010 Professional or higher is already installed, an add-in is also installed along with the Visual Studio 2010 Express for Windows Phone. Because of the limited localization support, the Windows Phone Developer Tools are currently available in only English, German, French, Italian, and Spanish.

To run the development environment, you need the following system requirements:

- **OS:** All editions (except Starter Edition) of Windows Vista (x86 and x64) Service Pack 2 or Windows 7.
- **Hardware:** 3GB disk space for installation and 2GB RAM to run the development environment. The emulator also requires a DirectX 10-capable graphics card with a Windows Display Driver Model (WDDM) 1.1 driver.

Compared with WP7 development, Android and iOS have different system requirements, and these are summarized in the accompanying sidebar.

ANDROID

- **OS:** Windows (XP, Vista, or Windows 7), Mac OS X 10.5.8 or later (x86 only), Linux.
- **Hardware:** 1GB disk space for installation (Java, Android SDK, Eclipse).
- **Others:** JDK 5 or JDK 6, Eclipse 3.4 IDE or later, Android SDK.

iOS

- **OS:** Mac OS X 10.5.x or later.
- **Hardware:** Intel-based Mac, 3GB disk space for installation, and 1GM RAM to run the development environment. If an Intel-based Mac is not available, Airplay SDK (`http://www.airplaysdk.com/`) provides an alternative way to develop iOS applications on a Windows-based PC.
- **Others:** Xcode IDE, Interface Builder, Instruments, and iOS SDK.

USING WINDOWS PHONE 7 DEVELOPER TOOLS

Now that you've configured the development environment, you can create a "Hello World!" application as your first Windows Phone 7 coding example. Using this simple example demonstrates how to use the Windows Phone Developer Tools.

> *This book includes source code for all downloadable examples. To open these example projects, please use File ➪ Open command to open the solution file in the IDE.*

Creating WP7 Applications with Visual Studio

Visual Studio provides easy-to-use wizards to create WP7 applications. The resulting solution includes a generated project structure and code skeleton. Use these steps to create a solution:

1. Open Visual Studio 2010 Express for Windows Phone. Windows starts the IDE for you.
2. Choose File ➪ New Project. You'll see the New Project dialog box shown in Figure 2-1. As shown in the figure, the two types of applications available are Silverlight application for Windows Phone and XNA game for Windows Phone. This example uses a Silverlight application.

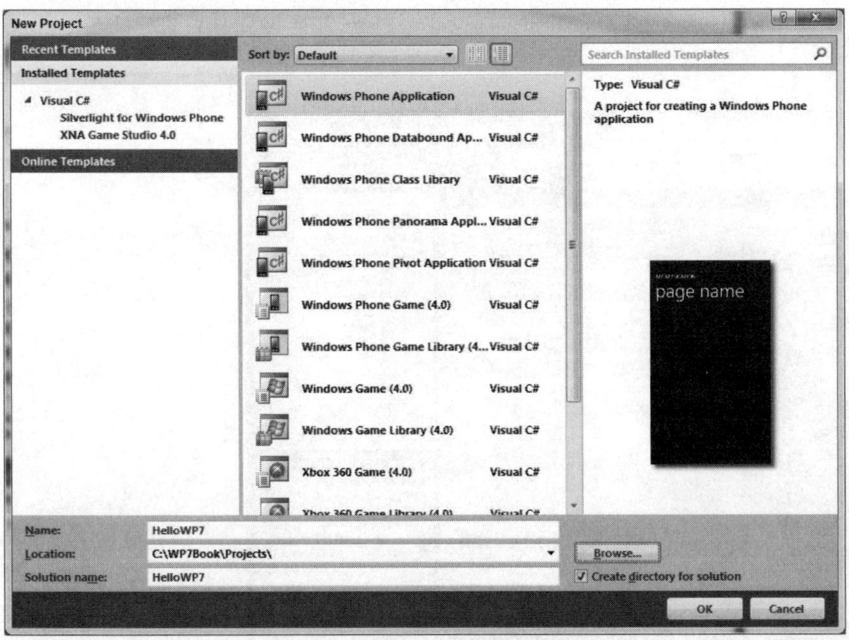

FIGURE 2-1: Creating a new Windows Phone project

3. Select Windows Phone Application from the template list. You'll see a description of this project in the right pane of the New Project dialog box as shown in Figure 2-1.

4. Type a name for the solution in the Name field. The example uses a name of HelloWP7.
5. Change the Location and Solution Name field entries if needed. The example uses the default location and solution name (which is the same as the project name found in the Name field).
6. Click OK. Visual Studio creates the project structure and generates a code skeleton with default settings and values as shown in Figure 2-2. Solution Explorer shows the files that the IDE creates for you based on the Windows Phone Application template. You'll discover later in the chapter how these files are used. The Properties window lets you view and change properties and events of selected objects.

As in other Silverlight applications, eXtensible Application Markup Language (XAML, pronounced "zammel") is used extensively in WP7 Silverlight applications. XAML files (.xaml) are normally used to define user interface (UI) elements, data binding, event handling, application properties, and other application features. In Solution Explorer, you can right-click a XAML file and choose View Designer from the context menu to open the XAML file in a designer window.

A XAML file is associated with a separate C# code-behind file (.cs), which contains the code used to handle events and manipulate the objects declared in XAML. In Solution Explorer you can right-click a XAML file and choose View Code from the context menu to open its code-behind file in an editor window.

At some point you'll want to open the `MainPage.xaml` file so that you can create a UI for your application. When you open a XAML file, the designer shows the XAML file content in split mode by default (as shown in Figure 2-2). XAML View (the right editor pane) displays the XAML markup, and Design View (the left editor pane) shows a WYSIWYG representation of the UI elements and layouts. There is a thin toolbar between Design View and XAML View. You can use the controls in this toolbar to manipulate the viewing area of each view.

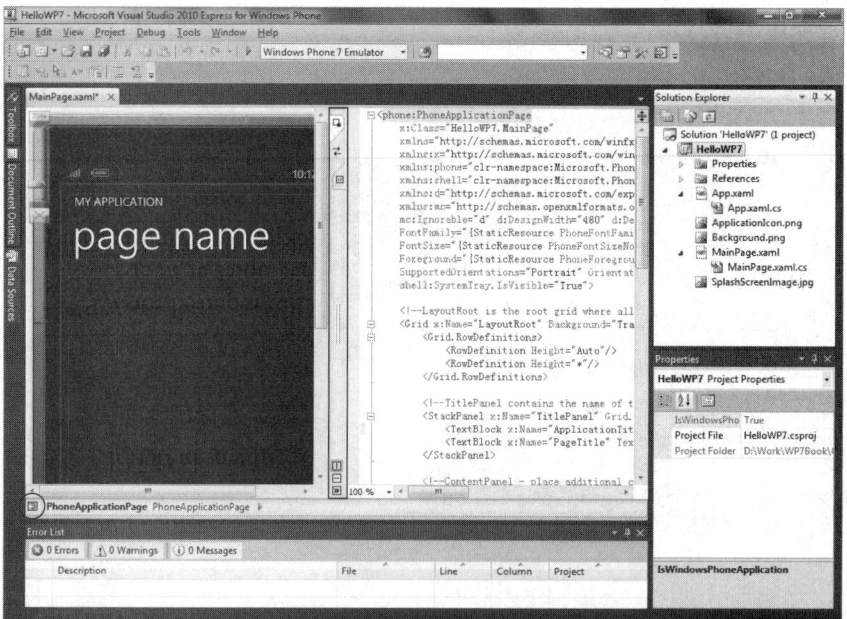

FIGURE 2-2: Visual Studio for Windows Phone 7

WP7 applications have controls the user interacts with to perform tasks, so you'll want to add controls to your application. You can drag and drop common Windows Phone controls from the Toolbox to the Design View. In this way you can manipulate UI elements and layouts graphically. You can also manually edit the tags in the XAML file, and all changes will show in the Design View immediately for your review. You can use the Document Outline window to view the hierarchical relationship among UI elements and navigate to specific elements.

It's time to customize the UI for the example a bit and add a small function to become comfortable with Visual Studio for Windows Phone. First, in the XAML View of `MainPage.xaml` file, locate the `StackPanel` element named `TitlePanel`. Set the `Text` property of the first `TextBlock` element, `ApplicationTitle`, to the string `MY FIRST WP7 APPLICATION`. You may notice that your changes appear immediately in the Design View while you are editing in the XAML View.


```
<!--TitlePanel contains the name of the application and page title-->
<StackPanel x:Name="TitlePanel" Grid.Row="0" Margin="12,17,0,28">
   <TextBlock x:Name="ApplicationTitle" Text="MY FIRST WP7 APPLICATION"
       Style="{StaticResource PhoneTextNormalStyle}"/>
   <TextBlock x:Name="PageTitle" Text="page name" Margin="9,-7,0,0"
       Style="{StaticResource PhoneTextTitle1Style}"/>
</StackPanel>
```

Code snippet HelloWP7\HelloWP7\MainPage.xaml

It's time to add the first control to the application. The following steps tell you how:

1. Drag and drop a Button control from the Toolbox onto the Design View as shown in Figure 2-3. You can adjust the button size and position it as needed.
2. Select the button and set the `Content` property in the Properties Window to `Say`.
3. Set the `Name` property to `SayButton`.
4. Click the Events tab in the Properties window to display a list of available events.
5. Locate the Click event. Press Enter in the textbox next to the Click event. The IDE generates an event handler in the code-behind file `MainPage.xaml.cs`.
6. Type the following event handler code in the `SayButton_Click()` event handler. Of course, a production application would perform a more useful task — this code is just for example purposes.


```
private void SayButton_Click(object sender, RoutedEventArgs e)
{
   PageTitle.Text = "Hello WP7";
}
```

HelloWP7\HelloWP7\MainPage.xaml.cs

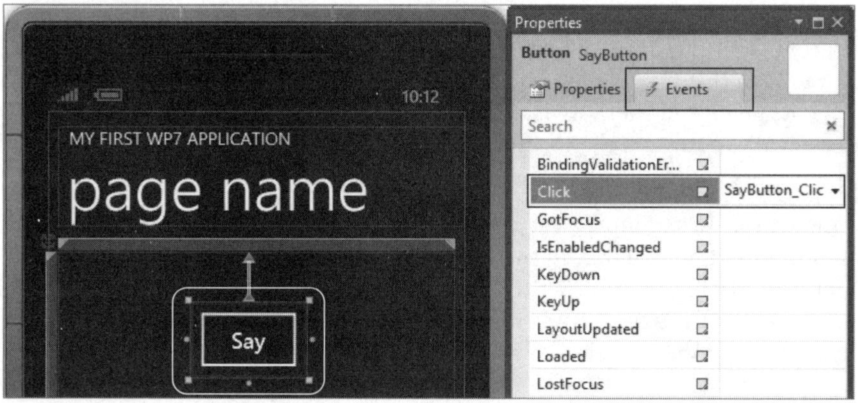

FIGURE 2-3: Adding UI elements and event handlers

For more information about using Visual Studio for Windows Phone development, please refer to Application Development in Visual Studio (`http://msdn.microsoft.com/library/h8w79z10.aspx`).

DIFFERENCES BETWEEN WP7 AND ANDROID DEVELOPMENT

If you're used to creating applications for Android, you might find WP7 development a little confusing. Knowing how the two development environments compare can make the transition easier. The following list provides an overview of this comparison:

- **IDE:** Android does not have its own dedicated IDE; instead, Android provides the Android Development Tools (ADT) plug-in for Eclipse.
- **Code generation:** The code generation approach is the same as Visual Studio uses — ADT generates basic application project structure and code skeletons for the developer.
- **UI design:** Android developers may envy Windows Phone and iOS developers for the powerful UI design tools natively provided by the IDE. ADT has limited WYSIWYG UI design tools. In addition, Android developers have to check the UI implementation results by actually deploying the applications in the emulators or devices.
- **UI and code-behind files:** The UI and code-behind files are similar to those used for Windows Phone development. Android uses an XML file to structure the UI elements and layouts. The code-behind relies on a Java file to manage the application logic.

DIFFERENCES BETWEEN WP7 AND IOS DEVELOPMENT

Someone moving from iOS development to WP7 development will definitely see differences in the way the two platforms work. However, there are also some similarities that you should know about to make the transition easier. The following list provides an overview of how the two development environments contrast:

- **IDE:** iOS developers use Xcode and Interface Builder to program iOS applications. Unlike Windows Phone and Android, the iOS UI design tool, Interface Builder, is not part of the iOS coding IDE, which is Xcode. Interface Builder compares to an alternative Windows Phone UI development tool, Expression Blend, which serves a similar purpose.
- **Code generation:** Xcode uses the same approach as Visual Studio to generate code — it generates basic application project structure and code skeletons for the developer, using a template.
- **Cross-platform development:** iOS cross-platform development is similar to Windows Phone development, which follows the generic cross-platform Silverlight or Xbox/DirectX New generation Architecture (XNA) development. Developing iOS applications with Xcode adopts the same programming paradigms used in developing other Mac OS X applications.
- **MVC pattern:** Windows Phone, Android, and iOS use a similar approach for application design. iOS applications employ the MVC (Model-View-Controller) pattern. Using the MVC pattern decouples the application UI and the way it behaves so that they can be developed, tested, and maintained independently.

Testing WP7 Applications in the Windows Phone Emulator

The Windows Phone Emulator provides an environment in which you can test and debug your applications using the same actions and functions that you'd use when working with an actual Windows Phone device. However, the performance and user experience on the emulator differ from that of the physical device. The main limitations are:

- The graphics display more slowly because the emulator doesn't benefit from hardware acceleration.
- The emulator doesn't provide multi-touch operations. A multi-touch monitor used with Windows 7 can overcome this limitation, but not every multi-touch monitor is compatible with Windows 7 (learn more at `http://www.microsoft.com/windows/windows-7/features/windows-touch.aspx`).

Unlike emulators provided for the Android and iOS, the Windows Phone Emulator doesn't include the full set of WP7 applications. However, a few applications are available, such as IE and Bing search. In addition, you can change the orientation or zooming settings of the emulator.

> *If you want to get a fully "unlocked" emulator, you can search "unlock windows phone emulator RTM" in your preferred search engine.*

Now that you have a very simple "Hello World!" Windows Phone application ready, let's use the Windows Phone emulator to test it. The following steps show how you'll typically perform this task:

1. Select Windows Phone 7 Emulator in the Device List field in Visual Studio, as shown in Figure 2-4. (You also have the option of deploying your applications to an actual device, and you'll learn how to perform this task in the next section.)

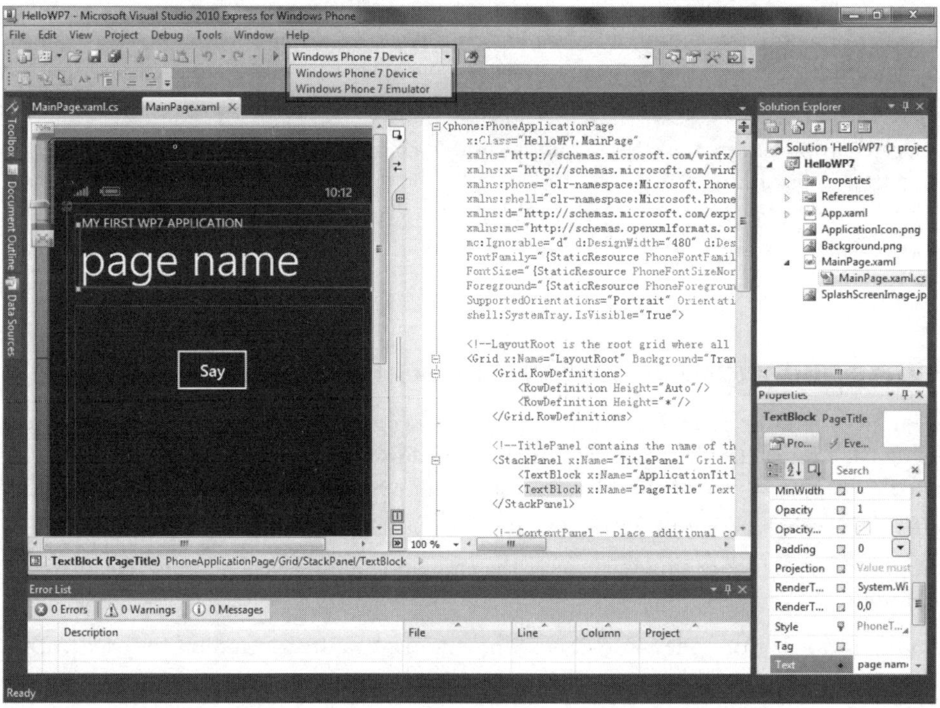

FIGURE 2-4: Selecting Windows Phone 7 Emulator in Visual Studio

2. Choose Debug ➪ Start Debugging (or press F5). Visual Studio starts the emulator. After a short wait, you see an emulator window similar to the one shown on the left side of Figure 2-5.

> *After the emulator window appears, it takes a while for the system to set up the emulator environment and deploy the application into the emulator. You can save time by leaving the emulator window open when you need to make changes to your application and then retest it.*

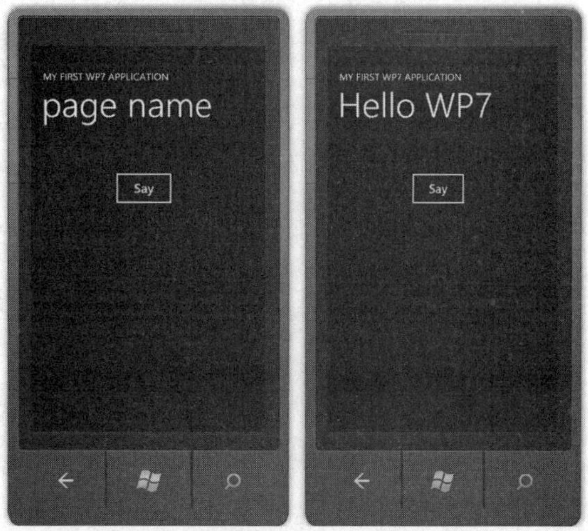

FIGURE 2-5: Deploying and running applications in Windows Phone 7 Emulator

Once your application appears in the emulator window, you can interact with it using the mouse and keyboard to simulate the touch inputs. If you have a touch-enabled monitor, you can directly touch the UI as you would if you were using an actual device. Click the Say button and the "page name" title will change to "Hello WP7" as shown on the right side of Figure 2-5.

You can also launch the Windows Phone Emulator from outside Visual Studio. The WP7 SDK installer creates a `Windows Phone 7 Emulator` shortcut in the `Windows Phone Developer Tools` folder in the Start menu of your PC.

Another way to launch the Windows Phone Emulator is through the command line. Open a command line window, type **`%ProgramFiles%\Microsoft XDE\1.0\XDE`** (in a 64-bit PC, it should be **`%ProgramFiles(x86)%\Microsoft XDE\1.0\XDE`**) and press Enter. A help window like the one shown in Figure 2-6 pops up to explain the options you can use to start a Windows Phone Emulator from the command line. Here is a typical command used to start a customized emulator (even though this command appears on multiple lines in the book, you type it on a single line at the command prompt):

```
"C:\Program Files\Microsoft XDE\1.0\XDE.exe" "C:\Program Files\Microsoft
SDKs\Windows Phone\v7.0\Emulation\Images\WM70C1.en-US.unlocked.bin"
/VMID {E575DA31-FC47-4766-853F-018D823B9EE6}
```

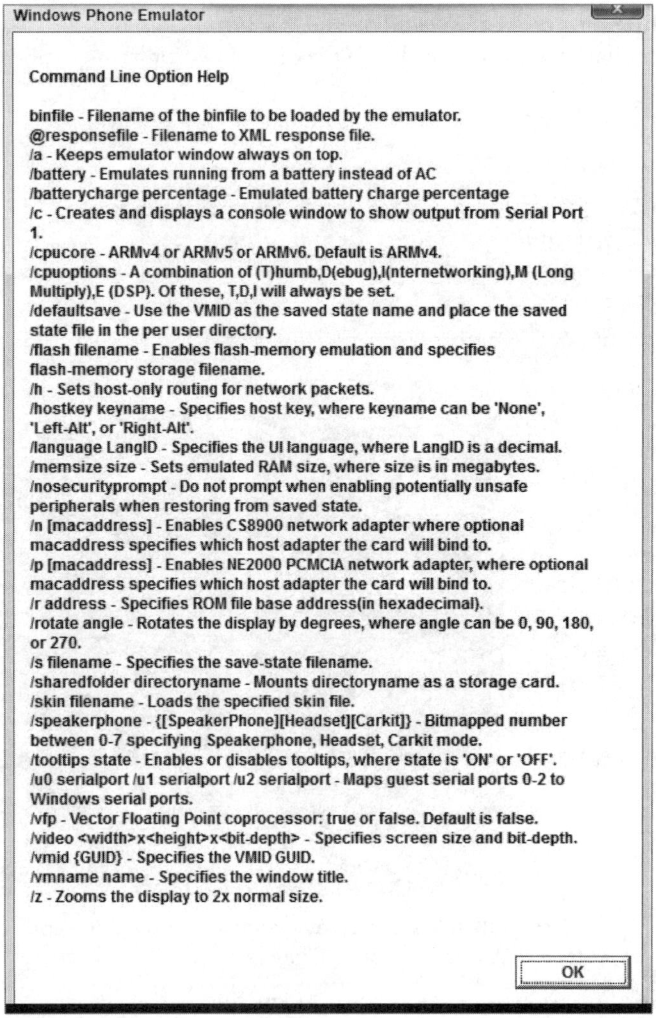

FIGURE 2-6: Command line option help for Windows Phone 7 Emulator

The Android and iOS IDEs both provide debuggers. Likewise, WP7 developers can debug applications using the debugger provided by Visual Studio. In addition to common debugging tasks such as setting break points and examining variable values, you can write run-time messages to the Output window to trace application behaviors. Choose View ➪ Output to display the Output window in Visual Studio. The following example shows how to add debug output information to the `SayButton_Click()` event handler — when you run the application you see "Button is clicked" in the Output window:

Available for download on Wrox.com

```
private void SayButton_Click(object sender, RoutedEventArgs e)
{
    Debug.WriteLine("Button is clicked.");
    PageTitle.Text = "Hello WP7";
}
```

Code snippet HelloWP7\HelloWP7\MainPage.xaml.cs

To see this code in action, set a break point at the `Debug.WriteLine()` function as shown in Figure 2-7. When you step over that line of code using the debugger, the debug output message "Button is clicked" appears in the Output window. You can use the Back button on the emulator or the Stop Debugging command in Visual Studio to stop the application and cancel the current debugging session.

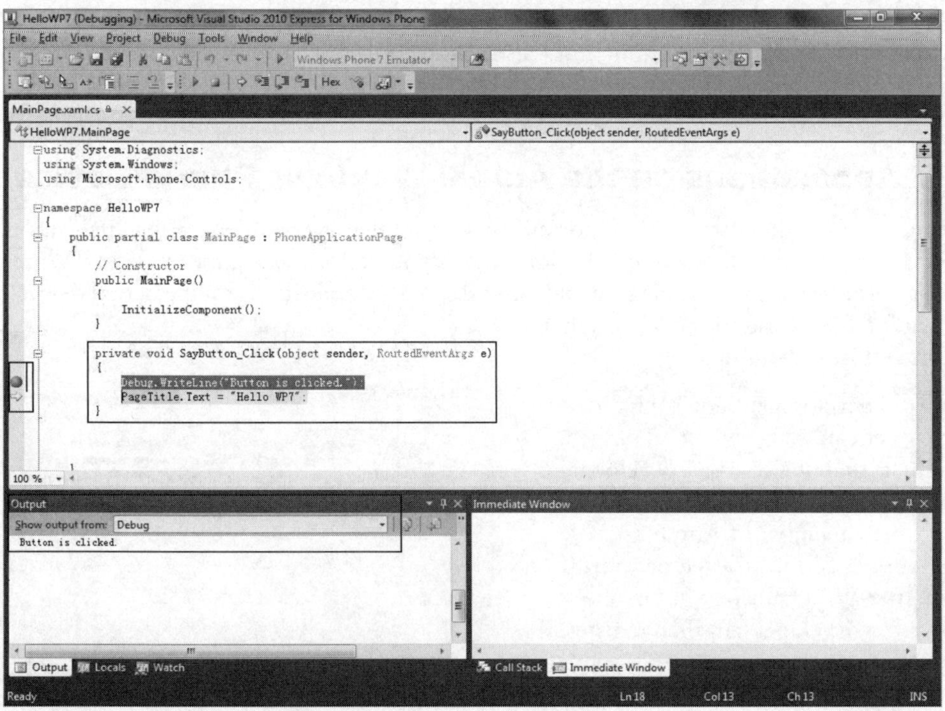

FIGURE 2-7: Debugging applications in Windows Phone 7 Emulator

WP7 EMULATOR CONSIDERATIONS FOR THE ANDROID DEVELOPER

Emulators vary considerably in their capabilities. You get used to certain features when working with the Android emulator that don't appear in the WP7 emulator. The following list describes some of these differences so you don't make assumptions that could cause problems as you develop your application.

- The official WP7 emulator does not contain a full set of basic applications that come with the Windows Phone 7 OS. You can overcome some of these issues by using the "unlock" version emulator provided by a third-party developer.
- Because isolated storage is available only while the emulator is running, any data you create while testing applications cannot persist on the WP7 emulator — the data is lost once the emulator closes.
- The telephony status, telephony actions, and location controls can't be simulated and controlled in the WP7 emulator.
- Unlike the powerful debugging tool, Dalvik Debug Monitor Server (DDMS), shipped with the Android SDK, the WP7 debugging environment cannot provide functionality such as the port-forwarding service, screen capture, thread and heap information, and logcat.

Testing WP7 Applications on the Actual Windows Phone Device

When developing a mobile application, it's important to confirm that every aspect of the application is working properly, using the actual device in addition to tests you perform using the emulator. Window Phone 7 Developer Tools make it possible to deploy and test WP7 applications on the actual device before publishing it to the marketplace, although it takes a little more effort to set it up.

Unlike the Android development environment, in which a developer can easily side-load an application to the actual device, WP7 developers follow a different course. The first step is to register and become a member of App Hub (`http://create.msdn.com`), a developer portal website created after WP7 launched in October 2010. The registered developer can then register a Windows Phone device for use in the development and debugging of WP7 applications. To register a development phone, launch the Windows Phone Developer Registration Tool that you'll find in the Windows Phone Developer Tools folder of the Start Menu. Figure 2-8 shows the initial Windows Phone Developer Registration dialog box.

FIGURE 2-8: Using the Windows Phone Developer Registration Tool to register Development Windows Phone

> *You need to install and launch the Zune software (the development phone will appear in Zune) while registering the development phone. When Zune is ready, you can connect your development phone to the PC. Zune software should keep running to maintain communications between the Windows Phone and the development PC. This requirement also applies to deploying applications to the phone and debugging on the phone.*

Make sure the phone screen is unlocked and the Status label displays "Phone ready." Enter the ID and password for your App Hub membership and then click Register. You can confirm whether the registration is successful or not by checking the Status label. Also, if the development phone has been registered successfully it should appear in the My Registered Devices list under your App Hub dashboard profile page, as shown in Figure 2-9. An individual developer can register a maximum of three devices for each App Hub account.

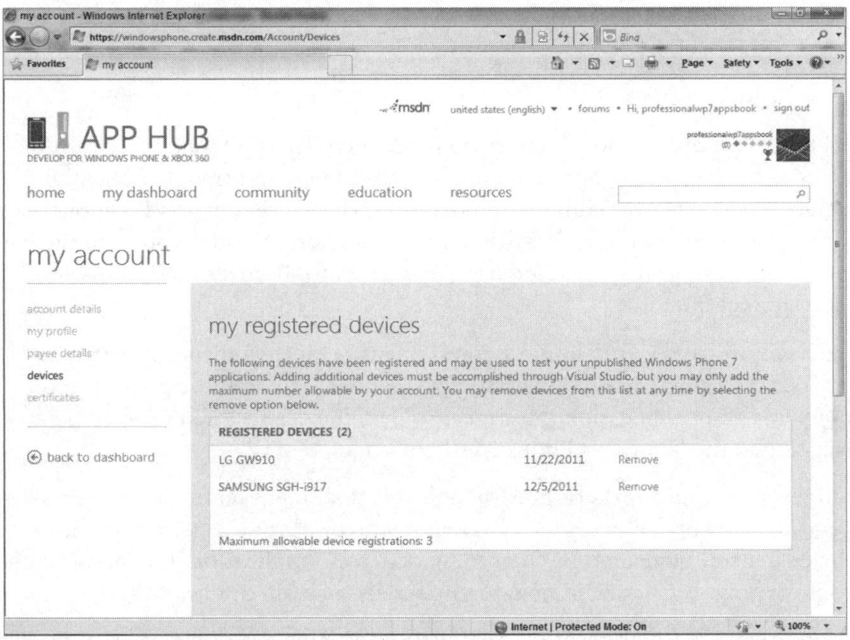

FIGURE 2-9: Registered devices list at App Hub profile page

Now you can test your WP7 applications on the actual device. To perform this task, select "Windows Phone 7 Device" in the Device list in Visual Studio and choose Debug ➪ Start Debugging (or press F5). The IDE will deploy the application to the development phone. The debugging process is the same as when you use the Windows Phone 7 Emulator. Because the application has been deployed and persisted on the phone, you can use and test it as you would any other WP7 application, unless you unregister your development phone.

PUBLISHING WINDOWS PHONE 7 APPLICATIONS TO THE MARKETPLACE

Windows Phone 7 doesn't currently allow you to side-load applications to the phone from a PC or use other methods to install applications on the phone. The only way for developers to distribute WP7 applications is to publish them to the Windows Phone Marketplace. In general, the job of building and publishing WP7 applications consists of the following steps:

1. Research the market and identify requirements.
2. Design, develop, and test application.
3. Prepare certification prerequisites.
4. Submit application to App Hub for certification.
5. Publicize application.
6. Maintain and update application.

In the sections that follow, you'll walk through the Windows Phone Marketplace application certification submission process.

Preparation

The Windows Phone Marketplace application submission is handled by App Hub. You can submit up to five applications/games for free with your App Hub account. Each additional application/game costs $19.99. Two documents at App Hub are important to the success of a WP7 application submission and even that of the application itself. Although this section provides you with the basics of submitting your application, you need these reference documents to discover the submission details for your particular application.

- **Windows Phone 7 Application Certification Requirements:** This document (at `http://go.microsoft.com/fwlink/?LinkId=183220`) provides details about the policies and technical requirements your application must meet to pass the certification process and qualify for listing in the Windows Phone Marketplace.
- **Best Practices for Application Marketing:** Although this document (at `http://create.msdn.com/home/about/app_submission_walkthrough_application_marketing`) is intended to provide general guidelines on how to market your application, it contains many useful suggestions on how to prepare submission materials for your application.

The Windows Phone 7 requirements for application certification provide a significant amount of information, but the essentials can be summarized into four fundamental rules every WP7 application should follow:

- **Applications should be reliable:** No one wants to use an application that doesn't work well. Of course, you want to ensure your application contains as few bugs as possible (the bug-free application is a myth). You also want to be sure that the application contains appropriate exception handling code so that, if it does experience an error, it can recover (or at least fail gracefully). It's important to test all the things you expect the user to do as well as things the user shouldn't do, but will probably try anyway.

- **Applications should make efficient use of resources:** Although Microsoft has specified the hardware requirements, and the Windows Phone devices are considered as high-end devices with powerful processing capability, developers should keep in mind that the mobile devices are still constrained in resources, especially those of battery and memory. So making efficient use of resources will not only provide a good user experience but will also benefit the whole Windows Phone ecosystem.
- **Applications should not interfere with the phone functionality:** The fundamental use of a smartphone is to make phone calls. Although smartphones have many additional features and functions, every application should still respect this fundamental phone functionality. The Windows Phone certification process gives this rule a very high priority.
- **Applications should be free of malicious software:** The emergence of the mobile application market makes it much easier to distribute mobile applications. However, this also brings many problems and concerns such as malicious software that violates the user's privacy and identity. Among Windows Phone, Android, and iOS, the Windows Phone application certification process can be considered as rigorous. It is also the developer's responsibility not to make malicious software.

In addition to these four fundamental rules, you should double-check some quantitative facts before you submit your application:

- **Installation file size:** The OTA (over-the-air) installation file may not exceed 20 MB. Applications in excess of that size will be downloaded via Wi-Fi or through a tethered connection to a PC running the appropriate Microsoft software.
- **Maximum XAP size:** The maximum size of the XAP package file is 225 MB.
- **Breadth of use:** The application must run on any Windows Phone 7 device, regardless of model, screen size, keyboard hardware, and manufacturer.
- **Speed of rendering:** The application must render the first screen within five seconds after launch.
- **Responsiveness to input:** Within 20 seconds after launch, the application must be responsive to user input.
- **RAM limit:** An application must not exceed 90 MB of RAM usage, except on devices that have more than 256 MB of memory."

Again, you should check every requirement listed in the Windows Phone 7 certification requirements when your WP7 application is ready for the certification. To do so will ensure a smooth certification process and shorten the time to market.

Application artwork, which represents the application in the Windows Phone Marketplace catalog, is another engineering material you should prepare before the submission. All images should be in the PNG format. The application artwork package includes:

- **Application tile artwork (icons):** Applications will be shown in both the phone marketplace catalog and the PC marketplace catalog. For the phone marketplace catalog, there are two types of icons: small (99 × 99) and large (173 × 173). The small icon is used in the regular application list, while the large icon is used when the application is listed in the front page of the phone marketplace for promotion purposes. For the PC marketplace catalog, the icon size should be 200 × 200.

- **Panoramic background art (optional):** The size for panoramic background should be 1000 × 800. Sometimes the phone marketplace will select an application and use its panoramic background as the background image of the marketplace.
- **Application screenshots:** The size for screenshots should be 480 × 800. For each application, at least one and at most eight screenshots should be provided, so that users can check out some details about the application before they make purchases.

Submission

After you've prepared all the materials required to submit your application, you can perform the formal submission. The following steps describe the submission process:

1. Log in to your App Hub account and open the page at My Dashboard ➪ Windows Phone ➪ My Apps. You see the initial submission screen in Figure 2-10.

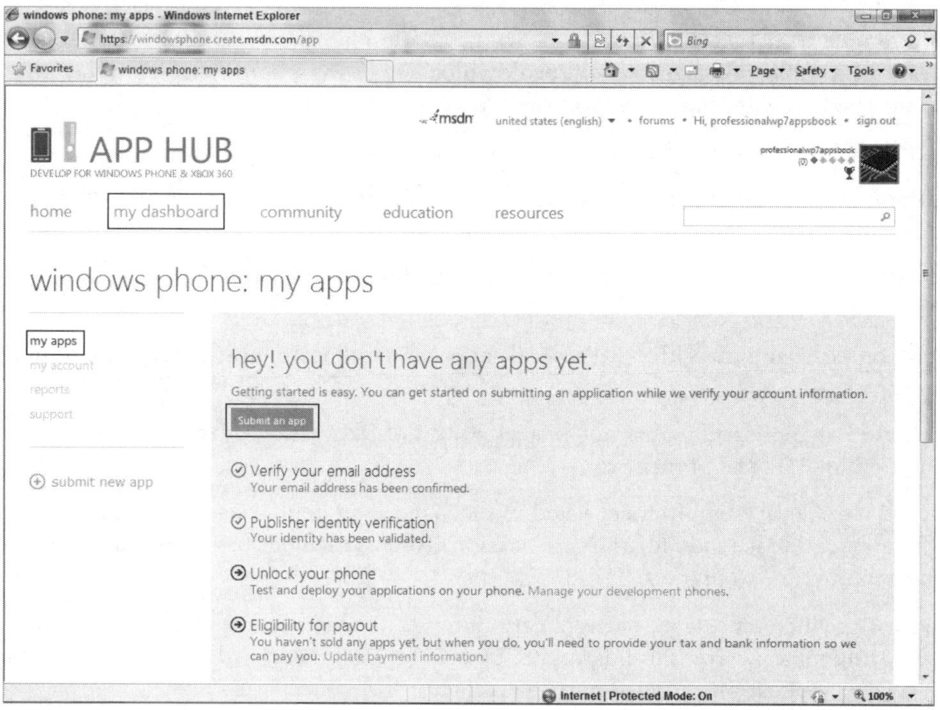

FIGURE 2-10: Login App Hub

2. Click on Submit an App. At this point, you can upload your application using the form shown in Figure 2-11. Click the plus sign in the Application Package field to upload your application's binary .XAP package. You can also find some basic information about such things as application name, application platform, default language, version, notes, and requests for exceptions. You must provide a unique application name that differs from that of any other application you've submitted. The application name is different from the application title that you enter in the next step.

> *Once you go to the next step, you can't change the application name, application platform, or default language, so be careful of your inputs. As described in the Preparation section of this chapter, the .XAP package should be less than 225 MB. The upload process may take a while, depending on the file size and your Internet connection speed.*

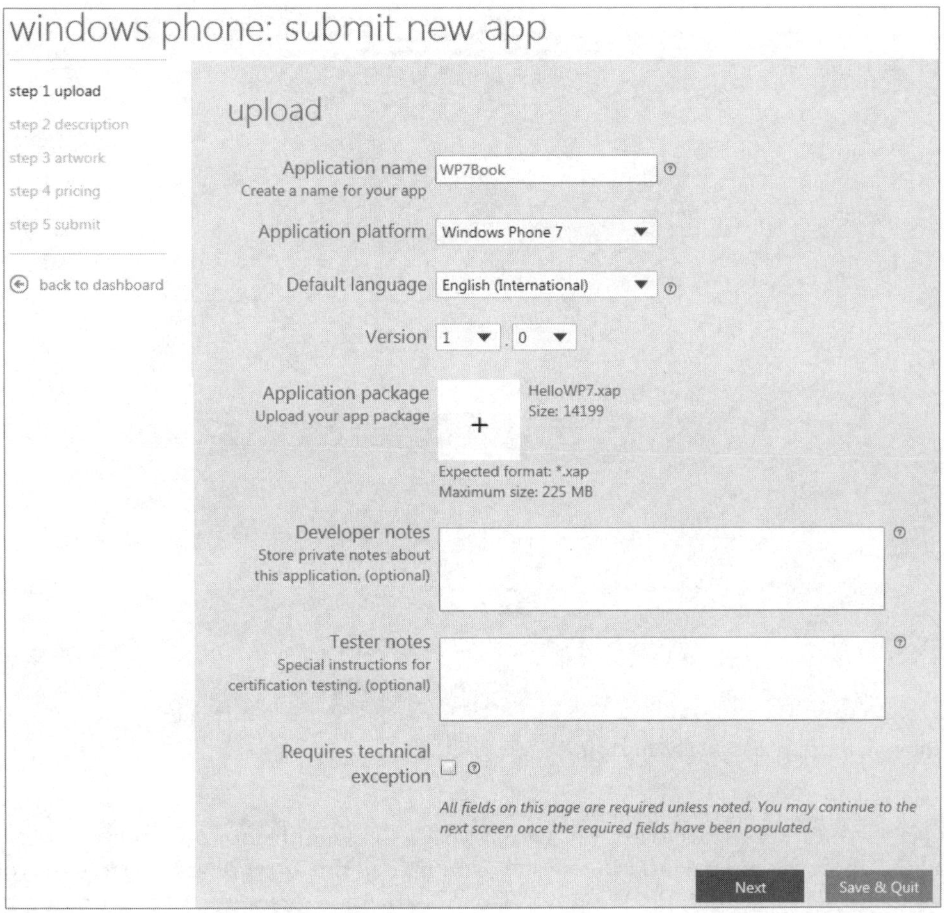

FIGURE 2-11: Uploading your application

3. Click Next. You'll see the description page shown in Figure 2-12 where you enter the application description. In this step, you provide a detailed description of your application. You can select from a total of 41 application categories, which are grouped into 16 top-level categories. You need to provide both the top-level category and sub-category of your application. If your application is a game and has been rated by the Entertainment Software Rating Board (ESRB), Pan-European Game Indicator (PEGI), or Unterhaltungssoftware Selbstkontrolle (USK), you must submit the rating certificate. If your application uses push notification, you should also provide the push notification certificates, which can be found at App Hub.

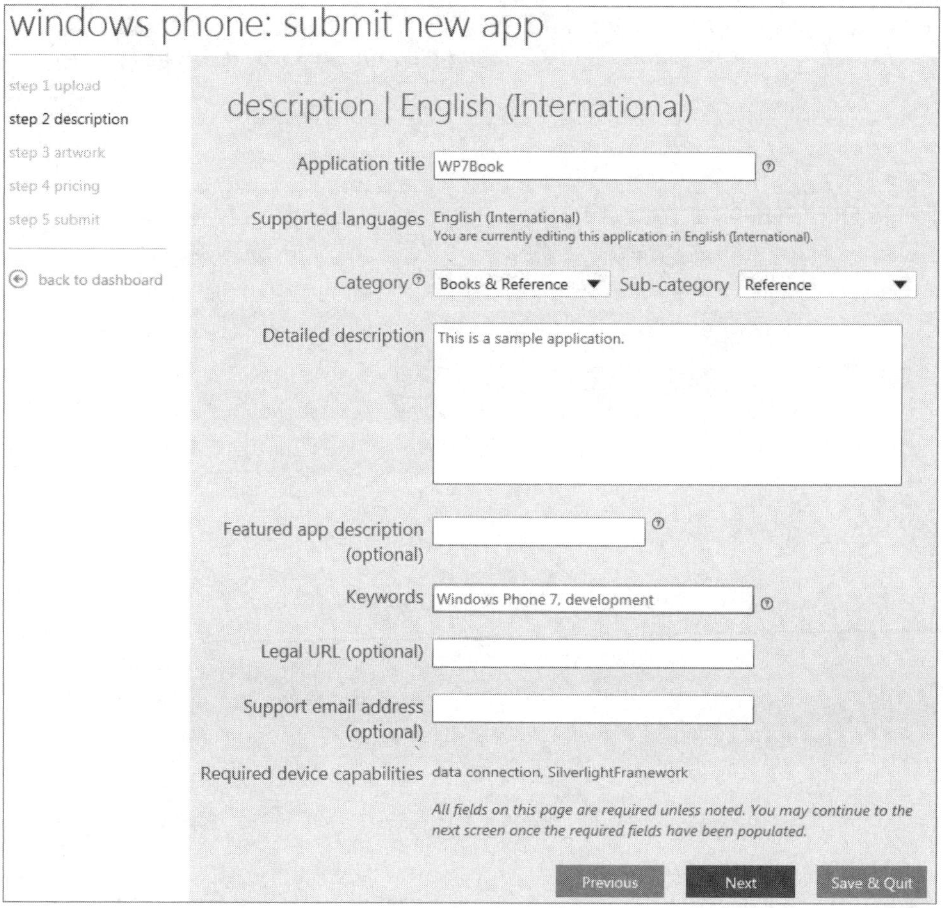

FIGURE 2-12: Entering the application description

4. Click Next. You'll see the Upload App Artwork page shown in Figure 2-13, where you upload the application's artwork. In this step, you upload the artwork associated with your application, which is used in the Windows Phone Marketplace catalog. All artwork must be

in the PNG format and customized to the resolutions specified in the Preparation section of this chapter.

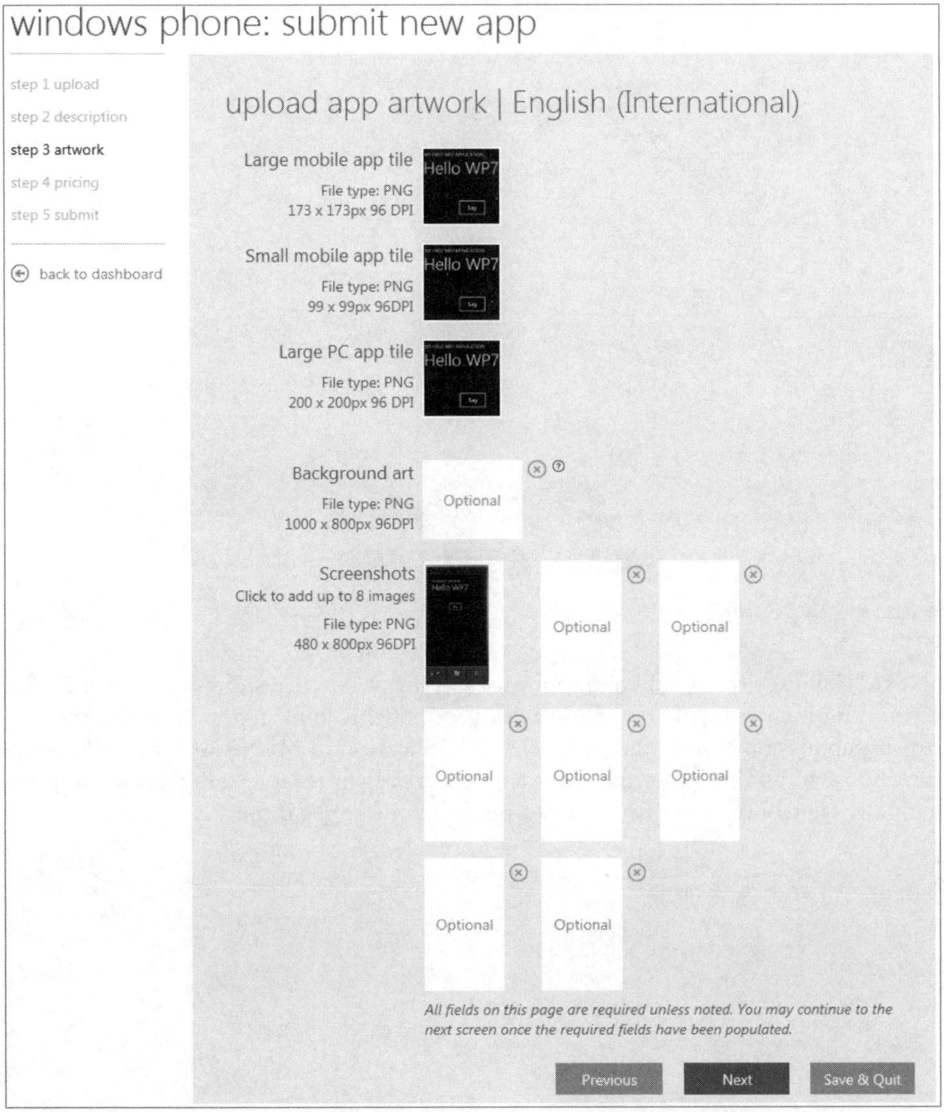

FIGURE 2-13: Uploading the artwork

5. Click Next. You'll see the Price Your App page shown in Figure 2-14, where you supply a price for your application. In this step, you also specify the market for your application. You should set payee details in App Hub in order to receive payment when someone buys your application. By default, your application is sold to all markets; however, you have options to select specific markets.

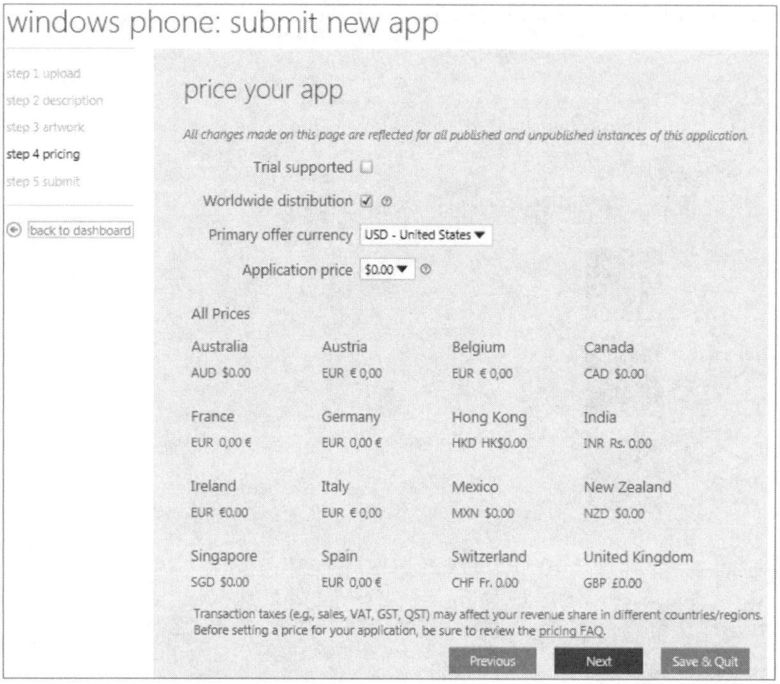

FIGURE 2-14: Pricing the application

6. Click Next. You'll see the Ready to Submit for Certification page shown in Figure 2-15, where you submit your application for certification. In this final step, you review the application submission details and submit your application to Microsoft for certification. You can choose to publish your application automatically after it passes certification, or you can manually control the time you want to publish your application.

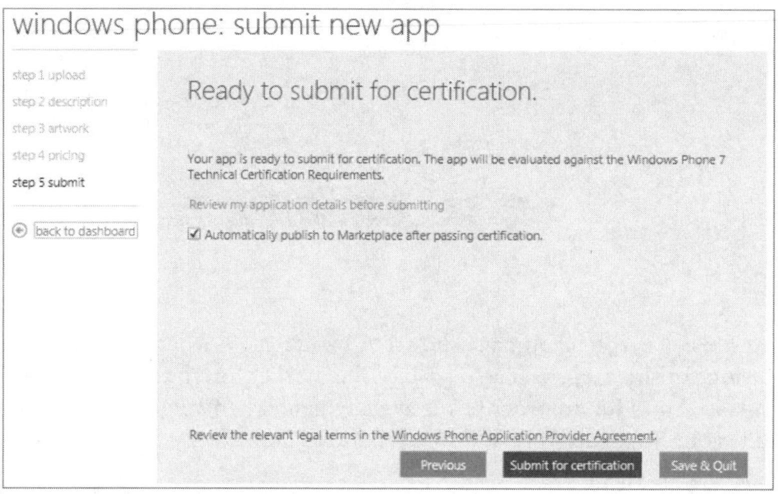

FIGURE 2-15: Submitting for certification

7. Click Submit for Certification. Your application is sent to Microsoft for certification. The Post-Submission Actions section of this chapter that follows describes what to expect during this part of the process.

Post-Submission Actions

After submitting your application to Microsoft for certification, you can check its status in My Dashboard at App Hub. The possible certification status levels are:

- **Submission in progress:** You are in the middle of entering submission metadata and have not yet clicked the Submit for Certification button shown in Figure 2-15.
- **Ready for testing:** Your application has been submitted and is waiting to enter the test queue.
- **Testing in progress:** Your application is being tested. Microsoft tries to finish application testing within five business days, but it doesn't make any guarantees.
- **Ready for signing:** Your application has passed the certification process and is waiting for Microsoft to digitally sign it before publication.
- **Ready for publishing:** You won't see this status if you elect to automatically publish after passing certification. This status notifies you that you must publish your application manually.
- **Published to marketplace:** Your application has been published and is being listed in the Windows Phone Marketplace.
- **Certification failed:** Your application failed certification testing. In this case, you'll receive an e-mail notification and test summary report to explain why the application failed. You can make the changes that Microsoft requests for certification and re-submit your application.

If your application has been published to Windows Phone Marketplace successfully, you can use Zune Redirection Service to create links to your application. After submitting your application, you can find the product ID of your application by clicking View Details at My Dashboard. If, for example, the product ID is 34432d66-ba01-e011-9264-00237de2db9e, then the link may look like: `http://social.zune.net/redirect?type=phoneApp&id=34432d66-ba01-e011-9264-00237de2db9e`. You can then use the links in your marketing materials such as e-mails, Short Message Service (SMS), web pages, and in-application links so that users can go directly to your application description in the Windows Phone Marketplace catalog.

As a responsible WP7 application vendor, you should keep updating your application in response to user feedback, bug reports, and Windows Phone updates.

Comparisons with Android and iPhone

Android Market and iPhone App Store are the two largest mobile application stores. Although Windows Phone Marketplace was launched in October 2009, it was a complete failure because it was rushed by the Windows Mobile 6.5 launch. However, Microsoft made a lot of improvements to Windows Phone Marketplace as part of the Windows Phone 7 launch and integration with Zune software. Table 2-1 compares these three mobile application stores side by side, not only from the developer point of view, but also from that of the user.

TABLE 2-1: Comparisons among Windows Phone Marketplace, Android Market, and iPhone App Store

ELEMENT	WP MARKETPLACE	ANDROID MARKET	IPHONE APP STORE
Size	20,000+	300,000+	500,000+
Developer Fee	$99/year for 5 free submissions	$25 onetime fee and $20 more for submitting paid applications	$99/year
Developer Takes	70%	70%	70%
PC Sync	Yes	No	Yes
Multimedia Content	Yes	No	Yes
Application Discovery	Device, Zune software, web	Device, web	Device, iTunes
Review Process	Mild	Loose	Rigid

Besides these basic comparisons, Windows Phone 7 has several advantages over iPhone and Android in terms of application store as described in the following list:

- A relatively small number of applications makes it is easy for the early birds in Windows Phone 7 development to stand out from the crowd.
- Both Windows Phone 7 and Microsoft's popular game console, Xbox, support XNA-based game development, which potentially means that you can easily port your XNA-based games from one platform to another. In sum, you can get your game quickly onto the larger user base of two platforms instead of one.
- The new Windows Phone 7 Metro UI provides a whole new user experience to organize mobile information. Some existing applications come in three versions for these three different platforms and have proven that the WP7 version has a better UI and is easier to use.
- The Try and Buy model makes the application store purchase experience more pleasurable and benefits both end users and developers.

SUMMARY

In this chapter, you discovered the basic Windows Phone 7 development experience and learned how to publish your applications on the Windows Phone Marketplace. This chapter also compared Windows Phone 7 with Android and iOS to show the pros and cons of Windows Phone 7 development.

Developing mobile applications in Windows Phone 7, Android, and iOS follows a similar model: IDE + emulator + application store. Silverlight, .NET, and XNA support in Windows Phone 7 makes it relatively easier to port the existing PC/Web/game applications to the Windows Phone 7 platform. Visual Studio, the best IDE so far, makes the development process more efficient. However, the Windows Phone 7 emulator definitely needs to provide more system-level controls and debugging functions. In addition, there is big potential for developers to jump into the new Windows Phone 7 platform by porting their Android and iOS applications to Windows Phone Marketplace.

In the next chapter, you'll be introduced to the basic programming concepts of building a WP7 application.

3

Fundamentals

WHAT'S IN THIS CHAPTER

- Understanding application life cycles
- Handling application state transitions
- Using launchers and choosers
- Learning work-around solutions in WP7

For beginning developers, the first question is where to start. On the other hand, experienced Android/iPhone developers might want to know how to achieve the same functionality as their current applications in Windows Phone 7 (WP7). In this chapter, we'll introduce concepts that are fundamental to WP7 application development.

This chapter starts with an introduction to the application execution model. This model includes these topics:

- Starting and stopping an application
- Managing application states
- Saving state data when an application stops
- Resuming the previous state

In addition to describing the application execution model, this chapter shows how to leverage the WP7 build-in functions to complete some common tasks, such as placing phone calls or sending e-mail.

Using the techniques in this chapter, beginning developers can familiarize themselves with WP7 technology and start writing their first WP7 applications. Experienced developers can start thinking about how to port their existing applications on other platforms to WP7.

BASIC APPLICATION PROJECT STRUCTURE

In the last chapter, you learned about the WP7 development environment. Like other mobile application development platforms, Windows Phone Developer Tools provide templates to help developers structure their WP7 application projects and generate the basic code skeletons. It's necessary to look closely at the initial application pieces generated by the SDK to better understand how a WP7 application is organized. This discussion also helps you understand the major differences between WP7 applications and Android or iPhone Operating System (iOS) applications.

Application Project Structure for Windows Phone 7

Windows Phone Developer Tools provide several project templates that are similar to those used for iOS application development. You use these templates to create the basic WP7 application skeletons. Figure 3-1 shows the New Project dialog box that is used to select one of the application templates.

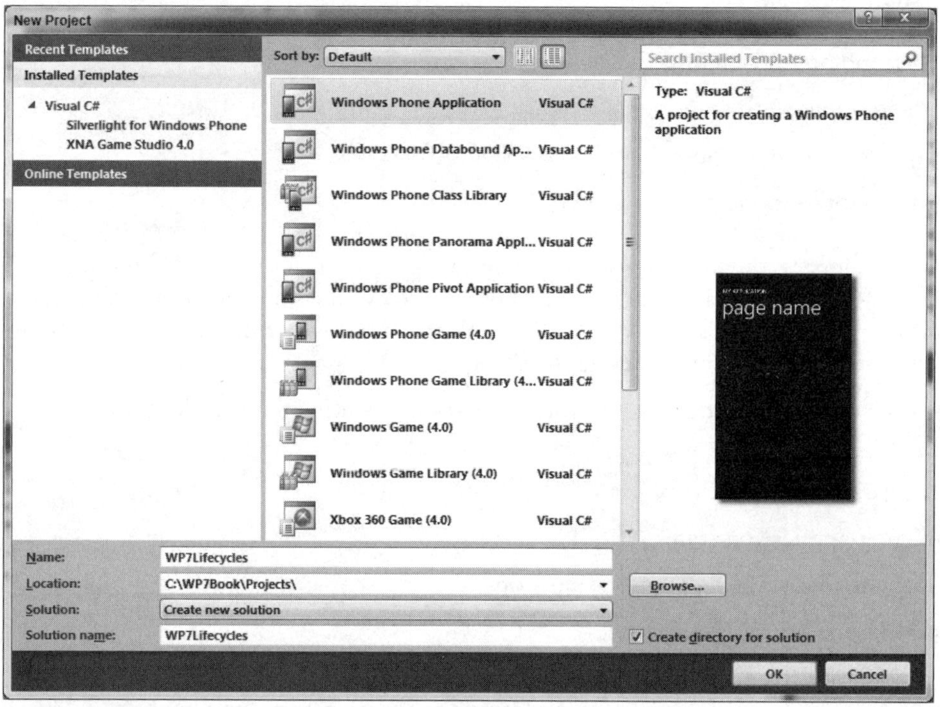

FIGURE 3-1: WP7 Application project templates

Two categories of templates are available: Silverlight for Windows Phone and the Xbox/DirectX New Generation Architecture (XNA) Game for Windows Phone. The IDE uses the templates to create code skeletons and default content for the specified application type. Developers can also create applications using the basic Windows Phone project template. In this case, the developer later modifies the application project structure and content manually. The following list describes several types of WP7 application templates:

- **Windows Phone Application:** The basic application template used as a starting point for any WP7 Silverlight application. This template provides a simple one-page application by default.
- **Windows Phone Databound Application:** This template can be used to create applications that display data. By default, it provides a two-page (main page and detail page) user interface and the basic corresponding `ViewModels` based on Silverlight Model-View-ViewModel (MVVM) pattern. For general information about MVVM, please refer to `http://msdn.microsoft.com/magazine/dd419663.aspx`.
- **Windows Phone Class Library:** The basic class library project template. This template has no UI elements by default and any number of applications can share the output DLL file.
- **Windows Phone Panorama Application:** The application template that uses a `Panorama` control. *Panorama* is a unique native Windows Phone look-and-feel. It's designed to break the confines of the phone screen by providing a long horizontal canvas to display contents.
- **Windows Phone Pivot Application:** This application template uses a `Pivot` control. *Pivot* applications provide a quick way to manage different pages within the application. The `Pivot` control serves as a container for other `PivotItem` controls. Each `PivotItem` control contains individual page content.
- **Windows Phone Game:** The basic application template used as a starting point for any WP7 XNA game.
- **Windows Phone Game Library:** The basic class library project template for WP7 XNA games.

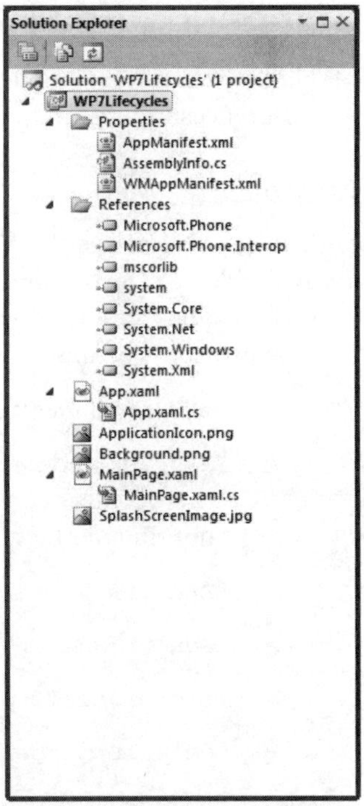

FIGURE 3-2: WP7 Silverlight Application Structure

The Visual Studio IDE generates several folders and files based on the selected template, which is similar to Android and iOS application development. Figure 3-2 shows a typical application structure.

The following list describes many of the entries displayed in Figure 3-2:

- `Properties\AppManifest.xml`: Specifies the manifest used to build the executable package.
- `Properties\AssemblyInfo.cs`: Specifies the metadata embedded in the generated assembly, such as name and version. This information is used by the system and is not generally visible to application users (except through the file's Properties dialog box or by using special utilities). It can be modified in the Assembly Information window on the Application tab of the project Properties page.
- `Properties\WMAppManifest.xml`: Contains specific metadata related to the application itself, such as title, description, author name, icon, and background image. It can be modified on the Application tab of the project Properties page.

> *Always use the project Properties page to make changes to application metadata. The IDE may overwrite any manual changes in the XML file itself with changes in the Properties page.*

- `References` folder: Contains the default library assemblies required for this application. If the application requires other library assemblies, you should add them to this folder.
- `App.xaml/App.xaml.cs`: Provides the entry point to the application. This file also controls the life cycle of the application.
- `MainPage.xaml/MainPage.xaml.cs`: Defines the first user interface users would see when the application launches. Silverlight uses Extensible Application Markup Language (XAML) to declare the UI look-and-feel. Thus files such as `MainPage.xaml` are similar to the XML layout files used in Android; `MainPage.xaml.cs` is the actual C# code used to handle the UI events.

Comparing Application Project Structure for Android and iOS

Having considered application project structure for WP7, the following short sections will explore differences in project structure for Android and iOS.

Android

Developers can use the New Project Wizard provided by Android Development Tools (ADT), available at `http://developer.android.com/sdk/eclipse-adt.html`, to create a new Android project quickly. The New Project Wizard creates several folders and files when you define a new project. The newly created folders and files include:

- `src/`: Includes all Java source code files for the application.
- `assets/` and `res/`: Contains all the resource files needed for the application. For example, these folders contain layout files, drawable files, string values, and raw assets, such as multimedia files.
- *`Android Version/`*: Includes the core Android library file `android.jar`.
- `Gen/`: Contains the generated Java files by ADT.
- `AndroidManifest.xml`: Provides the Android manifest file for the application.
- `default.properties`: Provides the Eclipse project property file for project settings.

`AndroidManifest.xml` plays an important role in Android application development — it's the configuration file that specifies components, structure, functionality, permissions, and requirements of the Android application. The following example shows typical `AndroidManifest.xml` file content:

```
<?xml version="1.0" encoding="utf-8"?>
<manifest xmlns:android="http://schemas.android.com/apk/res/android"
   package="com.wp7book"
```

```
    android:versionCode="1"
    android:versionName="1.0">
    <application android:icon="@drawable/icon" android:label="@string/app_name">
      <activity android:name=".HelloAndroid"
                android:label="@string/app_name">
          <intent-filter>
              <action android:name="android.intent.action.MAIN" />
              <category android:name="android.intent.category.LAUNCHER" />
          </intent-filter>
      </activity>
    </application>
</manifest>
```

In this example `AndroidManifest.xml` file, the `HelloAndroid` activity is defined as the entry point for this Android application. The intent filter (the combination of the `android.intent.action.MAIN` action and the `android.intent.category.LAUNCHER` category) specifies the entry point details. Thus `HelloAndroid` is the first UI users see after launching this application.

iOS

The iOS SDK and the Xcode IDE provide several project templates for iOS application development, so developers can create applications based on the generated iOS application skeletons. Like Android, the IDE automatically generates several folders and files to help structure the application project. These organization aids include source code files, resource files, user interface objects (nib files), and the application configuration file (`Info.plist`).

The `Info.plist` file is similar to the `AndroidManifest.xml` and contains the basic configuration information about the iOS application. The first UI that users see when the application launches is also defined in the `Info.plist` file. Here is an example of launching information within the `Info.plist` file:

```
<key>NSMainNibFile</key>
<string>MainWindow</string>
```

This file tells the application to display the user interface defined in the `MainWindow` nib file when the application launches.

APPLICATION EXECUTION MODEL AND LIFE CYCLES

Providing users with a fast and responsive experience is critical to mobile systems and applications. Different systems use different execution models to manage the *life cycle* of applications. The life cycle extends from the time the application is launched until it is terminated. Android and iOS support multitasking operations, while WP7 doesn't. Thus it's even more important for WP7 applications to maintain a runtime status and deal appropriately with state transitions.

Application Execution Model and Life Cycles in Windows Phone 7

Unlike Android and iOS, WP7 doesn't run multiple applications concurrently, and only one application can be active, running, visible, or focused at any given time. Like Android and iOS, the developer can manage application life cycles in WP7 through events triggered by state transitions, as shown in Figure 3-3. (Information is taken from `http://msdn.microsoft.com/library/ff817008.aspx`.)

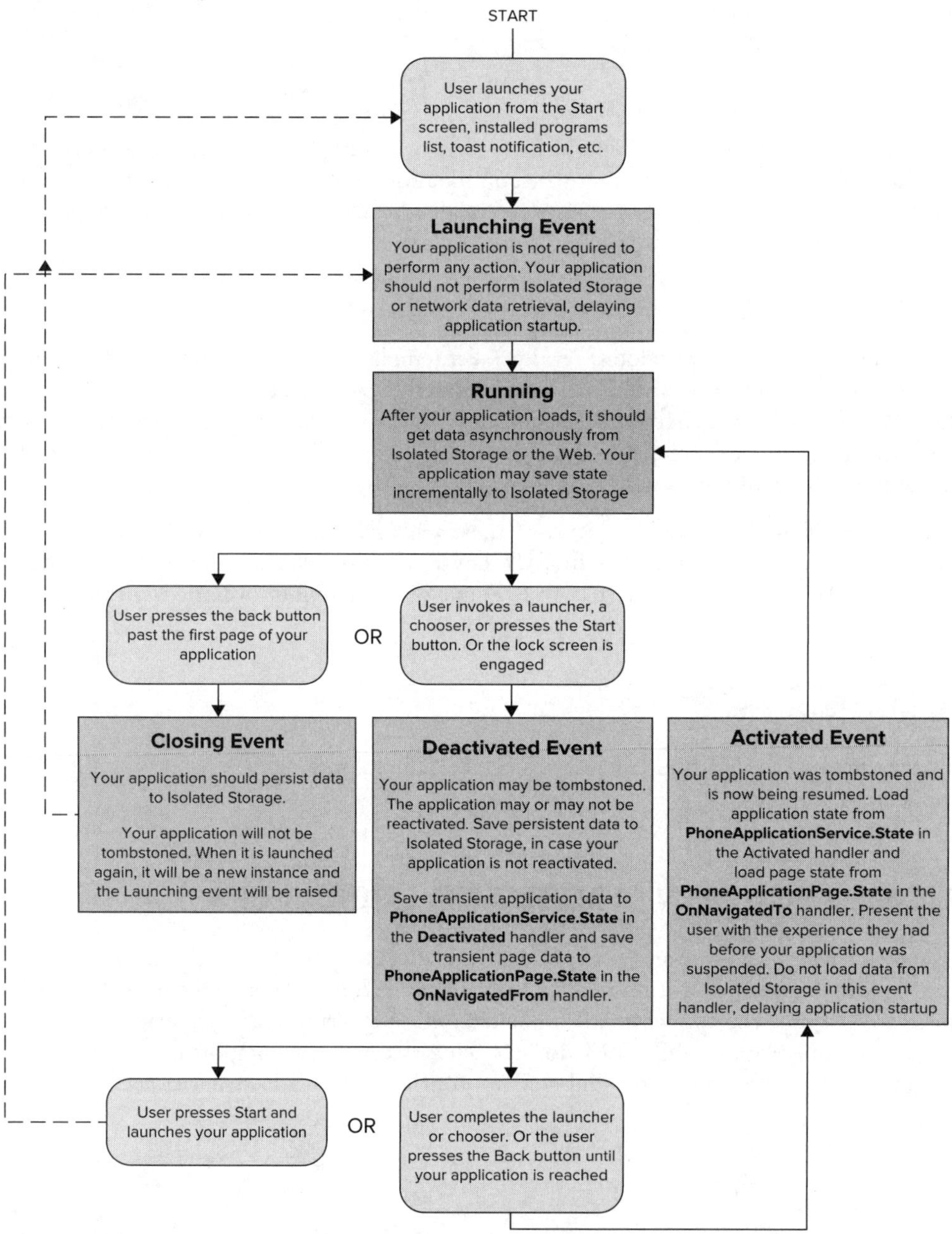

FIGURE 3-3: WP7 application life cycles

As depicted in Figure 3-3, the system triggers four events to notify the application of state transitions: launching, activated, deactivated, and closing. The notification process differs from both Android and iOS — WP7 doesn't raise these events sequentially. For example, the launching event isn't followed by the activated event as you might think. The launching and activated events are mutually exclusive (similarly, the closing and deactivating events are also mutually exclusive).

One reason for the difference between platforms is that there are two ways to bring a WP7 application to the foreground in this single task system. In the first method, the user launches applications from the Start screen, installed application list, toast notification, or other launching spot. In the second method, the user presses the Back button to return to a previously launched application, or the user completes the launcher and chooser and returns to the working application. (The launcher and chooser are introduced in the next section. They allow WP7 applications to enable a common set of tasks for users, such as selecting a contact's phone number or launching the camera.)

COMPARING THE WP7 BACK BEHAVIOR TO ANDROID AND IOS

Android devices have a Back button similar to the one provided with WP7. In Android, the Home screen is the last application the Back button can reach, which means pressing the Back button would not bring any application to the foreground once the user has reached the Home screen.

This approach is different from WP7, in which the user can bring applications to the foreground even though the user has reached the Home screen but not reached the limit of the application stack.

Unlike WP7, the iPhone does not have a physical Back button.

The first method always creates a new instance of the application, and the application is presented to the user from its root page. The second method tries to resume an application as if the user had just left it for a while and now wants to return to the application's previous state. From a user experience perspective, WP7 creates a faked multi-tasking experience. The following list describes the WP7 states:

- **Launching:** This event is raised when the user launches applications from the Start screen, installed application list, toast notification, or other launching spot. The application always appears as a new instance when the user launches the application using these methods. The corresponding event handler for this event is `Application_Launching()`. You can place some application initiation tasks in the event handler. However, you shouldn't perform some time-consuming tasks such as accessing isolated storage or the network, because the application calls the event handler before the application is visible and active to the user. Performing time-consuming tasks in this event handler will provide a bad user experience. Furthermore, the application shouldn't attempt to restore the transient state from a previous application instance.

- **Closing:** This event is raised only when the user presses the Back button to navigate backward through the application pages and past the application's first page. The corresponding event handler for this event is `Application_Closing()`. After this event returns, the system terminates the application, and the application should save persistent data to isolated storage if needed. In this case, the only way to return to the application is by launching it again as a new instance (pressing the Back button wouldn't bring it to the foreground in this case); it's unnecessary to save transient state data which is related only to the current instance. The next screen the user sees after the `Application_Closing()` event handler returns is the system Start screen (Home screen).
- **Deactivated:** This event occurs when the application loses focus, moves to the background, and becomes invisible to the user. The system raises this event in the following three situations: the user presses the Start button; the device timeout causes the system to engage the lock screen; or the application invokes a launcher or chooser.

  The corresponding event handler for this event is `Application_Deactivated()`, which is mutually exclusive with `Application_Closing()`. The system *tombstones* a deactivated application (which means the system adds the application record to the application stack so the user can bring the application back to the foreground by pressing the Back button).

  Note that this feature is available only to a deactivated application, not to a closed application. In addition, the application should save both transient state data (the application could be reactivated and resume to the previous state) and persistent data. (You must save the persistent data because there isn't any guarantee the user will reactivate the application later.) To ensure the application remains responsive, all deactivating tasks should be completed in 10 seconds; otherwise, the system will terminate the application. To avoid this problem, use incremental saving for large quantities of persisted data. Take time to compare the closing and deactivating events to see the differences between them.
- **Activated:** This event corresponds to deactivating, and it's raised in the following three situations: the user presses the Back button to reach a tombstoned application; the user unlocks the screen and the application is in the foreground; or the launcher or chooser invoked by the application returns.

  However, firing the activating event isn't guaranteed because of the following exceptions: the user launches a new application instance from Start, or the tombstoned application is knocked off the back of the application stack due to the limited size of the application stack. The corresponding handling function for this event is `Application_Activated()`. As with the launching event, applications shouldn't perform time-consuming tasks in this function, such as accessing isolated storage or a network resource. The system also shows a small progress message, "Resuming," on the screen during this event.

When you create a new application project in Visual Studio using one of the provided templates, the IDE automatically generates the application life cycle handling functions in `App.xaml.cs`. You can use following code to test the application life cycles:

Available for download on Wrox.com

```
private void Application_Launching(object sender, LaunchingEventArgs e)
{
   Debug.WriteLine("in Application_Launching:" +
                   DateTime.Now.ToLongTimeString());
}
```

```
private void Application_Activated(object sender, ActivatedEventArgs e)
{
   Debug.WriteLine("in Application_Activated:" +
                    DateTime.Now.ToLongTimeString());
}
private void Application_Deactivated(object sender, DeactivatedEventArgs e)
{
   Debug.WriteLine("in Application_Deactivated:" +
                    DateTime.Now.ToLongTimeString());
}
private void Application_Closing(object sender, ClosingEventArgs e)
{
   Debug.WriteLine("in Application_Closing:" +
                    DateTime.Now.ToLongTimeString());
}
```

Code snippet WP7Lifecycles\WP7Lifecycles\App.xaml.cs

When the IDE executes the code in debug mode, the invoke sequence of each handling function is displayed in the Output window of Visual Studio. In summary, `Application_Launching()` could be followed by `Application_Closing()` or `Application_Deactivated()`, depending exclusively on the user's actions. `Application_Launching()` is called after `Application_Closing()` when the user launches the application again. The application could call `Application_Activated()` after `Application_Deactivated()` when the application is tombstoned and remains in the application stack.

Comparing Application Model and Life Cycles in Android and iOS

Because Android and iOS support multitasking while WP7 does not, Android and iOS have their own peculiarities in terms of application models and life cycles. The following sections will explore these differences.

Android

Every Android application runs in its own process, and each process has its own virtual machine, so every Android application runs in its own isolated environment. Android allows multiple application processes to run simultaneously after it launches the applications. The application process isn't shut down until it's no longer needed or until other system or application processes need more system resources. For this reason, it's important to understand how the system manages application behavior when the application is visible to users, rather than running in the background.

You can use four types of components to compose an Android application: activity, service, broadcast receiver, and content provider. Among these, activity is the most commonly used. An activity presents a visual user interface that Android displays and the user interacts with. In Android, one application can contain multiple activities, and one activity can start another activity, even when the activities belong to different applications and run in separate processes. Activities from different applications can work together as if they are parts of the same application to achieve a certain level of functionality. Android maintains this type of user experience by keeping these activities in the same task. This sort of cross-application communication makes the definition of application a little ambiguous, because a task is actually what the user experiences as a logical application, but an application is a physical bundle of activities.

An Android activity has three states:

- **Active:** The activity is in the foreground and visible, and the user is interacting with it.
- **Paused:** The activity has lost focus, but is still visible to the user. For example, an activity could pause when another transparent activity is on top of it, or the foreground activity doesn't cover the full screen.
- **Stopped:** The activity is completely invisible to the user. It could be running in the background or the system may have killed it.

When an application is launched for the first time from the launcher, the Android system starts managing the application life cycle and the transitions among these states by notifying the underlying activities about changes. Figure 3-4 shows a flowchart of the Android life cycle. (This information is available at `http://developer.android.com/guide/topics/fundamentals.html`)

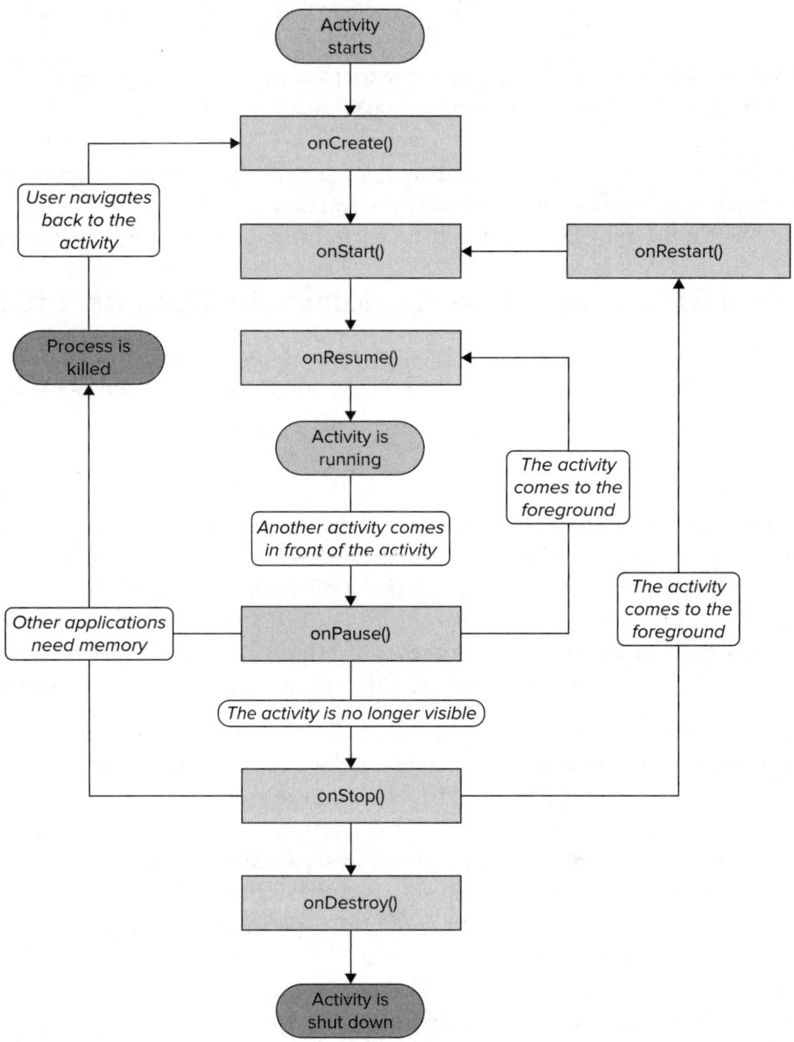

FIGURE 3-4: Android activity life cycles

As illustrated in Figure 3-4, Android developers can implement several callback functions to perform operations when the activity transitions between states. The entire lifetime of an activity is from the first call to `onCreate()` through to the final call to `onDestroy()`. The time from the call to `onStart()` until a corresponding call to `onStop()` is the visible lifetime, while the foreground lifetime refers to the time from the call to `onResume()` until `onPause()`. Note that the system may kill an activity if more resources are needed by the system or other applications. Therefore a well-designed Android application should save the current application state so that the user can return to the activity later. To implement this, the save and restore operations can be put in `onSaveInstanceState()` and `onRestoreInstanceState()`, and the system will call these two functions if needed. Because they aren't life cycle functions, the system doesn't always call them, and you shouldn't use them to save and restore the transient state of the activity. For persistent data related to the activity state, you should use `onResume()` and `onPause()` instead.

iOS

iOS also supports multitasking. Figure 3-5 shows the sequence of events between the application launch and termination. This information is available at `http://developer.apple.com/library/ios/#documentation/iphone/conceptual/iphoneosprogrammingguide/CoreApplication/CoreApplication.html`.

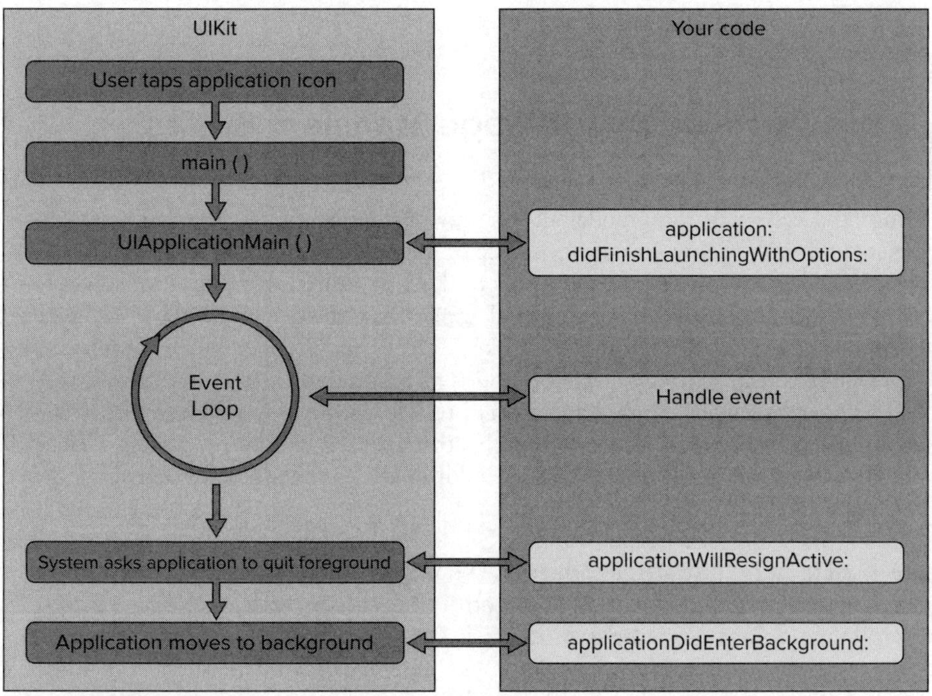

FIGURE 3-5: iOS application life cycles

Applications running in iOS can be in one of the following states at any given time:

- **Not running:** The application hasn't been launched or has been terminated by the system.
- **Inactive:** The application is running in the foreground but the user isn't interacting with it (similar to "paused" in Android).
- **Active:** The application is running and the user can interact with it (similar to "active" in Android).
- **Background:** The application is running in the background and executing code.
- **Suspended:** The application is in the background but isn't executing code (similar to "stopped" in Android).

The system calls several callback functions to manage the transitions between these states. Developers can put the transition handling logic in the following callback functions:

- `application:didFinishLaunchingWithOptions()`
- `applicationDidBecomeActive()`
- `applicationWillResignActive()`
- `applicationDidEnterBackground()`
- `applicationWillEnterForeground()`
- `applicationWillTerminate()`

Preserve/Restore Application and Page Transient States for Windows Phone 7

As mentioned before, the WP7 execution model allows only one application to run at a time. However, because of the concept of tombstoning and the existence of a Back button, there are two different experiences when a WP7 application becomes invisible to the user from the foreground. When the user presses the Back button to reactivate a deactivated application, the system displays the last page the user was viewing of the application. However, unlike Android and iPhone, which support multitasking, it's the application's responsibility to restore the application and UI state; otherwise, the application will appear as if it has just started and won't provide a seamless user experience (as it would have if the application had remained running). It's true that this limitation creates a lot of extra work for the developer, but it's worth trading the additional effort for a better user experience.

Developers use two types of state to display the application after a deactivation:

- **Page state:** Used to store UI-related information about a WP7 application page. For example, the page state affects the content of `TextBox` controls, the checked state of `CheckBox` controls, and the scroll position of a `ScrollViewer` control. The application can save and restore the UI information using the `State` property of the `PhoneApplicationPage` class (from which every Windows Phone page should inherit). In order to save and restore the page UI state, the application needs to override two methods of the `PhoneApplicationPage` class in the page class:
    - `OnNavigatedTo(NavigationEventArgs)`
    - `OnNavigatedFrom(NavigationEventArgs)`

Generally, the application uses `OnNavigatedTo()` to restore the UI state and calls it after the call to `Application_Launching()` or `Application_Activated()`. The application uses `OnNavigatedFrom()` to save the UI state and calls it before the call to `Application_Closing()` or `Application_Deactivated()`.

- **Application state:** Used to store the global application state that isn't associated with a specific page. The application saves and restores this transient state information using the `State` property of the `PhoneApplicationService` class. The application can save the state in the `Application_Deactivated()` handler and restore it in the `Application_Activated()` handler.

Creating the WP7 Life Cycles Application

It's time to create an example to demonstrate how to preserve and restore the application and page state. The purpose of this example is simple: count the number of times the user clicks a button. Use the following procedure to create the example:

1. Create a new WP7 application project that has two pages: `MainPage` and `SettingPage`. The `MainPage` is created automatically by the IDE when you create a new project. To create the `SettingPage`, you can right-click the project and choose Add ➪ New Item from the pop-up menu. You will see the Add New Item dialog box with several page templates. Select the Windows Phone Portrait page and name it `SettingPage`.
2. Add two `TextBlock` controls to the `MainPage` and name them `tbNewClicks` (used to display the number of clicks every time the system restarts the application) and `tbTotalClicks` (used to display the total number of clicks each time the system launches the application).
3. Add a `Button` control to the `MainPage` and name it `btnClick`. Every time the user clicks this button, the `tbNewClicks` and `tbTotalClicks` display will change accordingly.
4. Add a `Button` control to the `MainPage` and name it `btnSettings`. Clicking this button navigates the application to the `SettingPage`. `MainPage` should appear like the example shown on the left side of Figure 3-6.
5. Add a `CheckBox` control to `SettingPage` and name it `cbReset`. Checking `cbReset` is supposed to clear the total clicks of this application after the application quits. `SettingPage` should appear like the example shown on the right side of Figure 3-6.

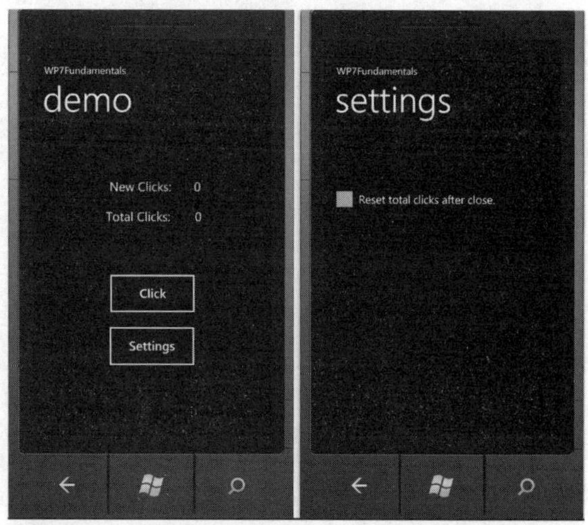

FIGURE 3-6: UI layouts of the sample application

Adding Code to the Main Page

Now that you have the example defined, you need to create some code for the various buttons. Listing 3-1 shows the code used to handle the application launching event.

Available for download on Wrox.com

LISTING 3-1: Handling the application launching event, WP7Lifecycles\WP7Lifecycles\App.xaml.cs

```
public int NewClicks { get; set; }
public int TotalClicks { get; set; }
private void Application_Launching(object sender, LaunchingEventArgs e)
{
   NewClicks = 0;
   TotalClicks = 0;
}
```

The code begins by creating two properties, `NewClicks` and `TotalClicks`, which are used to track the number of clicks the user has made with `btnClick`. Because you aren't doing anything special, you can use the short form of the property definition.

The `Application_Launching()` event handler simply sets the value of the two properties to 0.

Now that you have the properties initialized, you'll want to fill them with some information. Listing 3-2 shows the event handlers for the two buttons on `MainPage`.

Available for download on Wrox.com

LISTING 3-2: Handling the user's button clicks on MainPage, WP7Lifecycles\WP7Lifecycles\MainPage.xaml.cs

```
private void btnClick_Click(object sender, RoutedEventArgs e)
{
   int newClicks = ++(Application.Current as WP7Lifecycles.App).NewClicks;
   int totalClicks = ++(Application.Current as WP7Lifecycles.App).TotalClicks;
   tbNewClicks.Text = newClicks.ToString();
   tbTotalClicks.Text = totalClicks.ToString();
}
private void btnSettings_Click(object sender, RoutedEventArgs e)
{
   NavigationService.Navigate(new Uri("/SettingPage.xaml", UriKind.Relative));
}
```

In the event handler `btnClick_Click()`, which will be called when the user clicks the `btnClick` button, the value of `NewClicks` and `TotalClicks` is increased and the new result will be updated on the screen. In order to access `NewClicks` and `TotalClicks` (application properties defined in the `App` class) from the page class, we can use `Application.Current` property to get the `Application` object and then get the access to the properties. For more details about managing applications in .NET Framework, please refer to `http://msdn.microsoft.com/library/ms743714.aspx`.

In the event handler `btnSettings_Click()`, which will be called when the user clicks the `btnSettings` button, the application will navigate to the `SettingPage`. The page navigation is carried out by the `NavigationService` provided by the .NET Framework. For more details, please refer to `http://msdn.microsoft.com/library/system.windows.navigation.navigationservice.aspx`.

Working with Application and Page State

After testing this application, we find that it does count the button clicks. However, there are several issues related to the application and page state. The first issue is that when the application is tombstoned and reactivated, the application clears and resets the content of `tbNewClicks` and `tbTotalClicks`. This isn't what the user expects of the reactivated tombstoned application. The user expects that the application will preserve the page UI state. To fix this problem, you need to make the changes shown in Listings 3-3 and 3-4 to save and restore the page state if needed.

Available for download on Wrox.com

LISTING 3-3: Saving and restoring the page state on MainPage, WP7Lifecycles\WP7Lifecycles\MainPage.xaml.cs

```
bool newPageInstance = false;
public MainPage()
{
   InitializeComponent();
   newPageInstance = true;
}
protected override void OnNavigatedFrom(
   System.Windows.Navigation.NavigationEventArgs e)
{
   base.OnNavigatedFrom(e);
   StateUtils.PreserveState(this.State, tbNewClicks);
   StateUtils.PreserveState(this.State, tbTotalClicks);
   newPageInstance = false;
   this.State["PreservedPageState"] = true;
}
protected override void OnNavigatedTo(
   System.Windows.Navigation.NavigationEventArgs e)
{
   base.OnNavigatedTo(e);
   if (newPageInstance && this.State.ContainsKey("PreservedPageState"))
   {
      StateUtils.RestoreState(this.State, tbNewClicks, "");
      StateUtils.RestoreState(this.State, tbTotalClicks, "");
   }
   newPageInstance = false;
}
```

The `Boolean` variable `newPageInstance` determines whether the application needs to restore the page UI state. `OnNavigatedFrom()` and `OnNavigatedTo()` are two protected methods of the `PhoneApplicationPage` class and can be overridden by its subclasses. `OnNavigatedFrom()` is called when a page is no longer the active page, and `OnNavigatedTo()` is called when a page becomes the active page. Thus the next step is to save the page UI state (the contents of `tbNewClicks` and `tbTotalClicks`), using the `OnNavigatedFrom()` event handler. Then restore the page states, using the `OnNavigatedTo()` event handler. The application saves the page UI state in the `State` property of the `PhoneApplicationPage` class. `State` is a generic collection of key/value pairs with limited storage of 2MB for each page and 4MB for the entire application. Saving and restoring the page state can be achieved in a helping class listed in Listing 3-4.

Available for download on Wrox.com

LISTING 3-4: Defining helping class to save and restore states, WP7Lifecycles\WP7Lifecycles\StateUtils.cs

```
public static void PreserveState(IDictionary<string, object> state,
                                 TextBlock textBlock)
{
   state[textBlock.Name + "_Text"] = textBlock.Text;
}
public static void RestoreState(IDictionary<string, object> state,
                                TextBlock textBlock, string defaultValue)
{
   textBlock.Text = TryGetValue<string>(state, textBlock.Name + "_Text",
                                        defaultValue);
}
public static T TryGetValue<T>(IDictionary<string, object> state, string name,
                               T defaultValue)
{
   if (state.ContainsKey(name))
   {
      if (state[name] != null)
      {
         return (T)state[name];
      }
   }
   return defaultValue;
}
```

Listing 3-4 gives an example of how to save and restore the UI state of a `TextBlock` control, using the control name as the key and the actual text as the value, and then saving or restoring them from the `State` property. You can easily implement similar functions for other controls, and the full implementation of this helping class can be found in the sample code of this book.

Now when the user reactivates the application from the tombstoned state, `tbNewClicks` and `tbTotalClicks` display the same content that appeared before the application became tombstoned. This action seems correct, but we have a new problem. Even though the application restores the page UI state, the click count number will be reset to 0 and the count will start from the beginning when the user clicks `btnClick` button. This is a logic error and the reason is that the application only saves and restores the UI part of the state information; the actual values, which can be considered application states, are not saved and restored. To correct this error, you add the code shown in Listing 3-5.

Available for download on Wrox.com

LISTING 3-5: Saving and restoring the application state, WP7Lifecycles\WP7Lifecycles\App.xaml.cs

```
private void Application_Activated(object sender, ActivatedEventArgs e)
{
   Debug.WriteLine("in Application_Activated:" +
                   DateTime.Now.ToLongTimeString());
   NewClicks = StateUtils.RestoreState(PhoneApplicationService.Current.State,
                                       "NewClicks", 0);
```

```
        TotalClicks = StateUtils.RestoreState(PhoneApplicationService.Current.State,
                                              "TotalClicks", 0);
    }
    private void Application_Deactivated(object sender, DeactivatedEventArgs e)
    {
        Debug.WriteLine("in Application_Deactivated:" +
                        DateTime.Now.ToLongTimeString());
        StateUtils.PreserveState(PhoneApplicationService.Current.State,
                                 "NewClicks", NewClicks);
        StateUtils.PreserveState(PhoneApplicationService.Current.State,
                                 "TotalClicks", TotalClicks);
    }
```

The `PhoneApplicationService` class provides access to various aspects of the application's life cycles, including the management of the application state when it becomes active or inactive. The application state is a global property. When the application is deactivated, the system keeps this state information for the application and passes it back to the application when the application is reactivated. In Listing 3-5, the application retains the state information in the `State` property of `PhoneApplicationService` to save (in the handler for the deactivated event) and restore (in the handler for the activated event) the application state. Unlike the code in Listing 3-4, saving and restoring the application state is handled in the application class instead of the page class.

Adding Code for the Page Navigations

This part of the example demonstrates different uses of application state and page state. You should note that the page state is only associated with the page object, which has a narrower scope than the application state. To demonstrate the differences, you can look at the behavior that occurs when navigating to the settings page. The user can navigate to the settings page by clicking `btnSettings` button. In the settings page, the user can check the `cbReset` checkbox. However, when the user presses the device's Back button to navigate back to the main page, and then clicks `btnSettings` to access the settings page again, `cbReset` is unchecked. This isn't the expected behavior. Furthermore, when the user navigates to the settings page and presses the device's Start button, the application becomes tombstoned. If the application is later reactivated, you'll find that `cbReset` is also unchecked. To fix this problem, you might consider using the page state to save and restore the page UI state. Listing 3-6 shows how to perform this task.

Available for download on Wrox.com

LISTING 3-6: Using page state to handle the state changes for page navigations, WP7Lifecycles\WP7Lifecycles\SettingPage.xaml.cs

```
protected override void
OnNavigatedFrom(System.Windows.Navigation.NavigationEventArgs e)
{
   base.OnNavigatedFrom(e);
   StateUtils.PreserveState(this.State, cbReset);
   newPageInstance = false;
   this.State["PreservedPageState"] = true;
}
protected override void
```

continues

LISTING 3-6 *(continued)*

```
OnNavigatedTo(System.Windows.Navigation.NavigationEventArgs e)
{
   base.OnNavigatedTo(e);
   if (newPageInstance && this.State.ContainsKey("PreservedPageState"))
   {
      StateUtils.RestoreState(this.State, cbReset, false);
   }
}
```

Similar to Listing 3-3, we use `OnNavigatedFrom()` and `OnNavigatedTo()` to save and restore the UI page state for the settings page. This change solves the problem of restoring the page state when reactivating a tombstoned application. The system also maintains the page state. When the application is tombstoned, the system keeps the page state of the last page before tombstoning and passes the page state back to the last page when the application is reactivated. Thus, the UI state can be restored through the page state maintained by the system.

However, the example code doesn't solve the problem of restoring state when navigating between the main page and settings page. The reason is that the page `State` property is limited to the page object. When navigating away from the settings page and navigating back to it by using `NavigationService.Navigate()`, the system creates a new `SettingPage` object; thus the saved page state (which is in the `State` property of the previous page object) is not available anymore. Application state is different from page state; it's global to the application. So the example uses the application state approach to solve this problem by handling the changes as shown in Listing 3-7.

Available for download on Wrox.com

LISTING 3-7: Using application state to handle the state changes for page navigations, WP7Lifecycles\WP7Lifecycles\SettingPage.xaml.cs

```
protected override void
OnNavigatedFrom(System.Windows.Navigation.NavigationEventArgs e)
{
   base.OnNavigatedFrom(e);
   StateUtils.PreserveState(PhoneApplicationService.Current.State, cbReset);
   newPageInstance = false;
   PhoneApplicationService.Current.State["PreservedPageState"] = true;
}
protected override void
OnNavigatedTo(System.Windows.Navigation.NavigationEventArgs e)
{
   base.OnNavigatedTo(e);
   if (newPageInstance &&
       PhoneApplicationService.Current.State.ContainsKey("PreservedPageState"))
   {
      StateUtils.RestoreState(PhoneApplicationService.Current.State,
                              cbReset,
                              false);
   }
}
```

Persisting State Data in Isolated Storage

This chapter has discussed how to use the application and page transient state to provide a seamless user experience in the tombstoning case. The transient state isn't persisted in isolated storage, so when the application is completely closed, the state information is lost. In the example application, the application needs to keep the total number of clicks in isolated storage after the application is completely closed and loaded back after a fresh start, in order to maintain a consistent record of total clicks. Because you can't guarantee that the user will reactivate the application after the system tombstones it, the code must save the data in both the `Application_Deactivated()` event handler and the `Application_Closing()` event handler. On the other hand, because the application should perform tasks in `Application_Activated()` and `Application_Launching()`in 10 seconds or less, the code should load the saved persistent data asynchronously after the system loads the application, as shown in Listings 3-8 and 3-9.

LISTING 3-8: Persisting state data in isolated storage , WP7Lifecycles\WP7Lifecycles\App.xaml.cs

Available for download on Wrox.com

```
private void Application_Deactivated(object sender, DeactivatedEventArgs e)
{
    StateUtils.PreserveState(PhoneApplicationService.Current.State,
                             "NewClicks", NewClicks);
    StateUtils.PreserveState(PhoneApplicationService.Current.State,
                             "TotalClicks", TotalClicks);
    SaveDataToIsolatedStorage("myDataFile.txt", TotalClicks);
}
private void Application_Closing(object sender, ClosingEventArgs e)
{
    SaveDataToIsolatedStorage("myDataFile.txt", TotalClicks);
}
private void SaveDataToIsolatedStorage(string isoFileName, int value)
{
    IsolatedStorageFile isoStore =
       IsolatedStorageFile.GetUserStoreForApplication();
    StreamWriter sw = new StreamWriter(isoStore.OpenFile(isoFileName,
                                FileMode.OpenOrCreate));
    int totalClicks = value;
    bool shouldReset = StateUtils.TryGetValue(
                          PhoneApplicationService.Current.State,
                          "cbReset_IsChecked", false);
    if (PhoneApplicationService.Current.State.ContainsKey("PreservedAppState")
        && shouldReset)
    {
        totalClicks = 0;
    }
    sw.WriteLine(totalClicks);
    sw.Flush();
    sw.Close();
}
```

In this code, we define a helping function `SaveDataToIsolatedStorage()` to persist the total clicks in a data file. If the checkbox for resetting the total clicks is checked, the click number is reset to 0; otherwise, the number of current total clicks is saved. As mentioned before, `SaveDataToIsolatedStorage()` should be called in both `Application_Deactivated()` and `Application_Closing()`.

Available for download on Wrox.com

LISTING 3-9: Restoring state data from isolated storage, WP7Lifecycles\WP7Lifecycles\MainPage.xaml.cs

```
protected override void
OnNavigatedTo(System.Windows.Navigation.NavigationEventArgs e)
{
    base.OnNavigatedTo(e);
    if (newPageInstance)
    {
        if ((Application.Current as WP7Lifecycles.App).TotalClicks == -1)
        {
            GetDataAsync();
        }
        else if (this.State.ContainsKey("PreservedPageState"))
        {
            StateUtils.RestoreState(this.State, tbNewClicks, "");
            StateUtils.RestoreState(this.State, tbTotalClicks, "");
        }
    }
    newPageInstance = false;
}
public void GetDataAsync()
{
    Thread t = new Thread(new ThreadStart(GetData));
    t.Start();
}
public void GetData()
{
    string data = "0";
    IsolatedStorageFile isoStore =
       IsolatedStorageFile.GetUserStoreForApplication();
    if (isoStore.FileExists("myDataFile.txt"))
    {
        StreamReader sr =
           new StreamReader(isoStore.OpenFile("myDataFile.txt", FileMode.Open));
        data = sr.ReadToEnd();
        sr.Close();
    }
    Dispatcher.BeginInvoke(() => { SetData(data); });
}
public void SetData(string data)
{
    tbTotalClicks.Text = data;
    int totalClicks;
```

```
        bool result = int.TryParse(data, out totalClicks);
        if (false == result)
            totalClicks = 0;
        (Application.Current as WP7Lifecycles.App).TotalClicks = totalClicks;
    }
```

When the Main page is loaded, the number of total clicks must be retrieved in some way. If this number has not been initialized, it should be loaded from the isolated storage; otherwise, the page is navigated from other pages and the number of total clicks can be set from the preserved page state, as was described before.

USING LAUNCHERS AND CHOOSERS

Mobile platforms are getting more and more powerful, and mobile devices can perform many tasks in addition to making phone calls. Mobile applications also need a way to leverage the existing platform capabilities (such as placing phone calls, surfing webs, and taking pictures) to complete their own tasks. WP7 exposes a set of APIs referred to as *launchers* and *choosers* to allow applications to access some commonly used phone features.

Launchers and choosers start one of the built-in system applications to complete a task. The biggest difference between launchers and choosers is that launchers do not return a value back to the calling application, while choosers do.

You use a similar approach when working with any of the launcher APIs. First, you create an instance of the target launcher class. Next, you set the required properties, which determine the launcher behaviors. Finally, the code calls the `Show()` method of the launcher class to start the handling application. For example, to make a phone call you can use the following code:

```
PhoneCallTask phoneCallTask = new PhoneCallTask();
phoneCallTask.PhoneNumber = "1234567890";
phoneCallTask.DisplayName = "WP7";
phoneCallTask.Show();
```

Using a chooser is more complicated than working with a launcher, but choosers still follow the launcher pattern. First, you create an instance of the target chooser class. Next, you create a delegate function for the chooser's `Completed` event and assign it an event handler in the `PhoneApplicationPage()` constructor. When the chooser completes and returns, the system calls the `delegate` function. Finally, the code calls the chooser's `Show()` method to launch the handling application. For example, you can use the following code to choose a phone number:

```
PhoneNumberChooserTask phoneNumberChooserTask;
// Constructor
public MainPage()
{
  InitializeComponent();
  // Initialize the PhoneNumberChooserTask and
  // assign the Completed handler in the page constructor.
```

```
    phoneNumberChooserTask = new PhoneNumberChooserTask();
    phoneNumberChooserTask.Completed +=
       new EventHandler<PhoneNumberResult>(
          phoneNumberChooserTask_Completed);
  }
  // Invoke the camera to take a picture.
  private void button1_Click(object sender, RoutedEventArgs e)
  {
    try
    {
        phoneNumberChooserTask.Show();
    }
    catch (System.InvalidOperationException ex)
    {
    }
  }
  // The Completed event handler.
  void phoneNumberChooserTask_Completed(object sender, PhoneNumberResult e)
  {
    if (e.TaskResult == TaskResult.OK)
    {
        string s = e.PhoneNumber;
    }
  }
```

When an application starts a launcher or chooser, the system generally deactivates the calling application. So it's important to save and restore the calling application state as described in the "Preserve/Restore Application and Page Transient States for Windows Phone 7" section of this chapter to maintain a seamless user experience. The event handler calling sequence when using choosers is: `OnNavigatedFrom()` ➪ `Application_Deactivated()` ➪ `Application_Activated()` ➪ chooser's completed function ➪ `OnNavigatedTo()`.

WORK-AROUND SOLUTIONS IN WINDOWS PHONE 7

Some fundamental Android and iOS features aren't available in WP7. This section discusses these problems and highlights potential work-arounds.

Multitasking

WP7 doesn't support multitasking. However, it turns out that there is a misconception about this.

In fact, the platform itself and some system applications clearly multitask. For example, there are many system-level processes that run in the background. The user can listen to music while browsing the Web, make a phone call while taking notes, and perform other tasks that require multitasking support.

What WP7 doesn't support is multitasking between third-party applications. Android supports real multitasking — any given application continues running as normal even if it's in the background. iOS supports limited multitasking — a set of service APIs allows the application to continue accessing some system functionality when it's in the background. The background services include: audio, location, Voice over Internet Protocol (VoIP), task completion, and local notifications. What's missing in WP7 are the APIs or other mechanisms for the third-party applications to leverage the system's multitasking capability as other system applications do. Currently the best way to bring the multitasking experience to users is to handle the application life cycle state changes and save/restore the state status correctly as has been described in this chapter.

Even in the current WP7 platform, it's quite possible that users won't notice that they're actually single-tasking. As mentioned before, the system tombstones an application when it is navigated away from the foreground. However, when this application uses launchers and choosers to leverage the existing functionality, the launchers and choosers are actually other individual applications. To ensure a common flow between the application and the function it has called, the system treats them both as the same application. The system doesn't tombstone the calling application automatically. The launchers and choosers that support this experience include: `PhotoChooserTask`, `CameraCaptureTask`, `MediaPlayerLauncher`, `EmailAddressChooserTask`, `PhoneNumberChooserTask`, and `GameInviteTask`.

Background Service

Services are an important concept in modern mobile systems. Services usually run in the background and provide some commonly used functions to other applications. In Android, developers can even develop new services that other applications can use.

WP7 applications can't run in the background, and there isn't any way to develop a background service like the existing system services in the current release. However, Windows Phone provides a push notification mechanism through which back-end services can send messages to the phone to notify the user of any changes. For example, developers can develop a stock-watching service in Android to run in the background and notify users of interesting changes. To support a similar feature using WP7, developers can use a push notification. For details about using push notifications, please refer to Chapter 6.

Data Sharing

Every application has its own isolated storage space to store application-specific data in WP7, which is similar to the storage used in Android and iPhone. Other applications can't access the data (except that system-wide information such as contacts and pictures can be accessed by choosers).

In Android, an application can share its data with other applications through Content Provider. There is no such concept in WP7. However, if a WP7 application really needs to share its data with other applications, it can achieve this through cloud service. For details, please refer to Chapters 5 and 6.

System Event Hooks

One important feature found on other platforms is the ability to monitor or receive system change events, such as information that the battery is low or that the time zone has changed. Currently the application can receive only a few limited and predefined system-change event notifications as described in Chapter 4. One frequently used event is the device orientation change. Currently there isn't a good work-around solution for other system changes.

SUMMARY

This chapter introduced you to the fundamental concepts of WP7 development, including application project structure, execution model, and application life cycles, as well as how to use launchers and choosers to perform common tasks.

Because WP7 does not support real multitasking, you need to save and restore application states when applications switch states. Using this approach provides a user experience that's similar to multitasking systems such as Android and iOS. Even though it isn't mandatory to maintain application state, you need to handle application life cycles and states to ensure a better user experience.

After understanding these fundamental concepts, in the next chapter you'll be introduced to building the user interface for WP7 applications.

4

User Interfaces

WHAT'S IN THIS CHAPTER?

- Understanding WP7 UI principles
- Building a UI
- Customizing a UI

The touch user interface (UI), one in which the user interacts with the device and applications by using the touch gesture on the display screen, now is *the* UI for modern smartphones. Windows Phone 7 (WP7) introduces its touch UI based on a design system codenamed Metro, which is completely different from its predecessors. While the Metro UI concept is new to every mobile application developer, it's based on Silverlight, a powerful development platform for creating interactive user experiences for web, desktop, and mobile applications. Thus it is quite straightforward to port existing Silverlight applications to the WP7 platform. The new UI is one of the most exciting changes in WP7.

This chapter is all about creating UI on WP7. Three topics are discussed:

- UI design guidelines
- Basic steps of generating the UI
- Device UI customization

We start with UI design guidelines, followed by detailed information about UI elements and controls, UI system behaviors, and the interaction model for the touch interface. This chapter also covers common tasks to customize the UI for WP7 applications. Because the UI itself could be a very broad topic and WP7 is based on Silverlight, this chapter doesn't discuss every detail about each control or about Silverlight programming; instead, the chapter focuses on the key WP7 UI concepts and comparisons with Android and iOS.

UI DESIGN PRINCIPLES

Metro is the design system for the new WP7 UI, and the concept is inspired by the visual language of airport and metro system signage.

"Keep things simple, direct, and relevant" is the fundamental principle of the Metro design and it means:

- **Clean, light, open, and fast:** Focus on the primary tasks, make delightful use of white space, and do not fill every pixel.
- **Content, not chrome:** Reduce visuals that aren't content.
- **Integrated hardware and software:** Create a seamless user experience.
- **World-class motion:** Provide responsive and animated transitions.
- **Soulful and alive:** Serve personalized, automatically updated views of the information that matters most.

Although these design principles are already enforced in the WP7 platform design and implementation, third-party WP7 developers are also encouraged to follow these principles in order to develop harmonious, functional, and attractive WP7 applications.

UI Design Resources

Several documents are available for WP7 developers to understand and master the WP7 UI principles, and they can be downloaded from `http://msdn.microsoft.com/library/ff637515.aspx`.

- **UI Design and Interaction Guide:** This guide provides detailed information about UI principles, usage, guidelines of UI elements and controls, and UI system behaviors. Designers and developers should read this guide to learn the dos and don'ts of UI design and implementation for WP7 applications. The rest of this chapter also presents the key points from this guide.
- **Windows Phone Design System–Codename Metro:** This is a visual explanation of the inspiration behind the Metro concept. It helps designers and developers to better understand the Metro design language and align their applications with the whole system design.
- **Design Templates for WP7:** This is a collection of 28 layered Photoshop template files that can be used to create pixel-perfect application mockups.

Platform Characteristics

Before exploring the details of WP7 application UI design and implementation, this section briefly highlights some basic facts that directly impact UI matters.

Currently, all WP7 phones have Wide Video Graphic Array (WVGA) screens that provide 480 × 800 pixel resolution, no matter the physical screen size. Even so, all the controls and UI elements within Windows Phone Developer Tools are sized to support all possible screen sizes.

COMPARING THE WP7 DISPLAY TO ANDROID AND IOS

Currently WP7 supports only one resolution, but support for other resolutions might be added in future updates. It's unclear how supporting multi-resolutions will impact the WP7 application implementation; thus developers do not need to consider this issue for the current release.

Android applications run on devices with different screen sizes and resolutions. Android's support for multiple screen characteristics is achieved through a set of built-in compatibility features that manage the rendering of application resources for the current device screen. Developers can create screen-specific resources for precise UI control if needed; otherwise the platform can handle most of the work of adapting applications to the current device screen.

iOS currently supports three different resolutions: 320 × 480 for the original iPhone screen, 640 × 960 for the iPhone 4 screen, and 768 × 1024 for the iPad screen. Similar to Android, most of the work of handling the different screens is done by the system framework. Developers can bundle different image resources to take advantage of extra pixels.

After users power on their phones, the Start screen is the first (and also the most viewed) interface. The Start screen displays application Tiles that users have pinned for quick launch; thus you should carefully consider the potential that users may pin and display your application Tile in the Start screen. No matter what application is running, pressing the Start button on the phone always brings the user to the Start screen.

COMPARING THE WP7 START SCREEN TO ANDROID AND IOS

In a similar way to WP7, users can pin an application's icon to the Android or iOS Home screen for quick launch.

Unlike WP7, Android also supports App Widgets in the Home screen. App Widgets are small application views that can be placed in the Home screen and receive periodic updates. App Widgets provide users access to some of the corresponding application features directly from the Home screen, without launching the application.

There are two display areas beyond each WP7 application content space: Status Bar (on the top) and Application Bar (at the bottom). The Status Bar is an indicator that displays system-level status information such as signal strength, battery level, and time. The Status Bar is system-reserved and cannot be modified. The Application Bar provides a place to display icon buttons with text hints and an optional context menu for the common application tasks. The Application Bar is customizable.

WP7 uses Segoe WP as the system font, which is a Unicode font. You can embed your own fonts for use within your application, but these embedded fonts will be available only to your application and cannot be shared with others. The recommended font size is bigger than 15 points, because fonts smaller than 15 points are hard to read and too small to touch. Currently WP7 supports only five languages: English, French, Italian, German, and Spanish. A standard set of East Asian reading fonts that supports Chinese standard, Japanese, and Korean is also included.

On WP7, a touch target is an area that's defined to accept touch input and isn't visible to the user, and a touch element is the visual indicator of the touch target that's visible to the user. Usually the touch element size is 60 percent to 100 percent of the touch target size. The recommended touch target size is 9mm or 34 pixels square, while the minimum touch target size is 7mm or 26 pixels square. The minimum spacing between each touch control (a touch target that is combined with a touch element that the user touches) is 2mm or 8 pixels.

BUILDING THE WP7 UI

The examples in previous chapters demonstrate how to develop a simple WP7 application, including how to build the UI. Visual Studio 2010 Express for Windows Phone helps the developer construct the base page layout and then add more controls to the page through the Toolbox or create new pages using templates. Now it's time to take a close look at the UI implementation code under the hood.

Defining WP7 UI with XAML

WP7 applications are based on the Silverlight 3.0 platform, and Silverlight applications generally use Extensible Application Markup Language (XAML) for defining the application's visual compositions and layouts. For example, in the HelloWP7 example, the IDE generates `MainPage.xaml` as the base UI layout. This approach is similar to what Android does. Android uses an XML layout file to define the application layout. On WP7, XAML is the primary format for declaring a WP7 UI and elements in that UI. A WP7 difference from Android is that the XAML file is also associated with a code-behind C# file to define the interaction logic. The HelloWP7 project shown in Chapter 2 uses a simple XAML file, `MainPage.xaml`, which can be used to explain the XAML basics for the WP7 UI. The first part of this XAML file is the `PhoneApplicationPage` element and its attributes.

```
<phone:PhoneApplicationPage
    x:Class="HelloWP7.MainPage"
    xmlns="http://schemas.microsoft.com/winfx/2006/xaml/presentation"
    xmlns:x="http://schemas.microsoft.com/winfx/2006/xaml"
    xmlns:phone="clr-namespace:Microsoft.Phone.Controls;
       assembly=Microsoft.Phone"
    xmlns:shell="clr-namespace:Microsoft.Phone.Shell;assembly=Microsoft.Phone"
    xmlns:d="http://schemas.microsoft.com/expression/blend/2008"
    xmlns:mc="http://schemas.openxmlformats.org/markup-compatibility/2006"
    mc:Ignorable="d" d:DesignWidth="480" d:DesignHeight="768"
    FontFamily="{StaticResource PhoneFontFamilyNormal}"
    FontSize="{StaticResource PhoneFontSizeNormal}"
    Foreground="{StaticResource PhoneForegroundBrush}"
    SupportedOrientations="Portrait" Orientation="Portrait"
    shell:SystemTray.IsVisible="True">
```

The `PhoneApplicationPage` element defines the general settings for the whole page, such as font, foreground color, and screen orientation. To get the full list of attributes you can manipulate, place the cursor anywhere in the PhoneApplicationPage keyword and press F1 to display the PhoneApplicationPage class help document. Another approach is to place the cursor anywhere in the PhoneApplicationPage element block so that the IDE displays all the configurable attributes in the Properties window. The following code defines the basic container layout inside the phone page as a Grid:

```
<Grid x:Name="LayoutRoot" Background="Transparent">
    <Grid.RowDefinitions>
        <RowDefinition Height="Auto"/>
        <RowDefinition Height="*"/>
    </Grid.RowDefinitions>
</Grid>
```

A Grid is a table-like control where you can create rows and columns with absolute or relative sizes; it's similar to TableLayout on Android. In order to access this grid later in the code, you should assign a name to it. In this example, it's named LayoutRoot. Giving the grid a name also declares a Grid object in the XAML file at design time. This object is instantiated and used at run time. In the RowDefinitions element (similarly for ColumnDefinitions), you can set attributes for each row. The most common attribute is the size of each row or column. You can use three types of values: double value (floating-point value for pixel count, typically specified as an integer); start sizing (a weighted factor for the size of rows or columns to take the remaining available space, for example, *, 2*, 3*, etc.); and auto (the size to contain controls inside, which must be described by the literal Auto). Inside the Grid, you can define the content of each row:

```
<StackPanel x:Name="TitlePanel" Grid.Row="0" Margin="12,17,0,28">
    <TextBlock x:Name="ApplicationTitle" Text="MY FIRST WP7 APPLICATION"
       Style="{StaticResource PhoneTextNormalStyle}"/>
    <TextBlock x:Name="PageTitle" Text="page name" Margin="9,-7,0,0"
       Style="{StaticResource PhoneTextTitle1Style}"/>
</StackPanel>
<Grid x:Name="ContentPanel" Grid.Row="1" Margin="12,0,12,0">
    <Button Content="Say" Height="85" HorizontalAlignment="Center"
       Margin="151,59,171,0" Name="SayButton" VerticalAlignment="Top"
       Width="134" Click="SayButton_Click" />
</Grid>
```

The first row is a StackPanel control which can align objects vertically or horizontally. The placement of this StackPanel is set through Grid.Row="0". Like the Grid control, StackPanel also serves as a container, which means you can place controls inside it. In this example, two TextBlock controls are used to show the application tile and page title. Another control you can use as a container is the Canvas control, in which you can explicitly position the child elements by coordinates. The second row in the LayoutRoot is another Grid called ContentPanel in which you can place your main content.

Defining WP7 UI Programmatically

Normally you define the WP7 UI using XAML at design time, which is good for UI verification and prototyping. However, like Android, you can also define the UI programmatically if needed. For example, if you want to add a button to the ContentPanel, you can create the button object dynamically and then add it to the `ContentPanel`. You can implement this in the constructor of the `MainPage` class:

```
public MainPage()
{
    InitializeComponent();

    var SayButton = new Button
    {
        Name = "SayButton",
        Content = "Say",
        HorizontalAlignment =
            HorizontalAlignment.Center,
        VerticalAlignment =
             VerticalAlignment.Top,
        Height = 85,
        Width = 134,
        Margin = new Thickness (151,59,171,0)
    };
    SayButton.Click += new RoutedEventHandler(SayButton_Click);

    ContentPanel.Children.Add(SayButton);
    Grid.SetRow(SayButton, 0);
}
```

If you want to remove the button definition from the `MainPage.xaml` and use this code in the `MainPage.xaml.cs` instead, the UI and the function remain the same. First, you create a button object, and then set its properties and event handler individually. Finally, you add it as a child of the `ContentPanel` and set its row number.

Pages and Navigation Among Pages

Like PC or web applications, a mobile application usually contains a collection of data and needs to present this data in different views. Sometimes, different functions in an application also have different UI presentations. Navigations and transitions among these different views are one of the primary tasks that mobile developers need to deal with.

Activity is normally used to provide a screen with which users can interact to achieve certain functions on Android. Thus navigation from one screen to another in Android is achieved when one activity starts another activity. Regarding the UI implementation, the screen navigations in Android are usually triggered by a touch on the UI widgets/views and are handled through *intents* (messages exchanged between Android components) in the touch event handlers.

A UI element called the *navigation bar* can be used to enable screen navigation on iOS. A navigation bar is at the upper edge of an application and just below the status bar. You can add content-specific controls to the navigation bar so that users can tap these controls to navigate to the different content views. The iPhone doesn't have a hardware Back button, which differs from the Android and WP7. Thus a Back button on the left edge of the navigation bar is used to return to the previous screen.

WP7 Silverlight applications are based on the Silverlight page model where users can navigate forward through different pages of content via links and backward using the Back button. The user experience of this page model is similar to web page browsing. In addition, WP7 also provides a new user experience for navigating through data and information, which is carried out through the *Pivot* and *Panorama* controls. We will introduce these two controls later and focus on the basic page-based navigation described in this section.

In a WP7 application, a *frame* is the top-level container control. It contains the *page* control and other system elements such as a system tray and application bar. Every application can have only one frame, and the frame exposes properties from a hosted page such as screen orientation. The `PhoneApplicationFrame` type is the reference object for the application frame. When you create a WP7 application project, access to the application's root frame is also created in the `App.xaml.cs` file.

```
// The root frame of the Phone Application
public PhoneApplicationFrame RootFrame {get; private set;}
```

A page fills the entire content region of the frame. If the status bar and the optional application bar are visible, the drawable area of the page is smaller. The status bar is 32 pixels high and the application bar is 72 pixels high, so the minimal height of a page is 696 pixels and the maximal height is 800 pixels. The relationship between frame, page, and other UI elements is depicted in Figure 4-1. On WP7, every page you define is derived from the `PhoneApplicationPage` class.

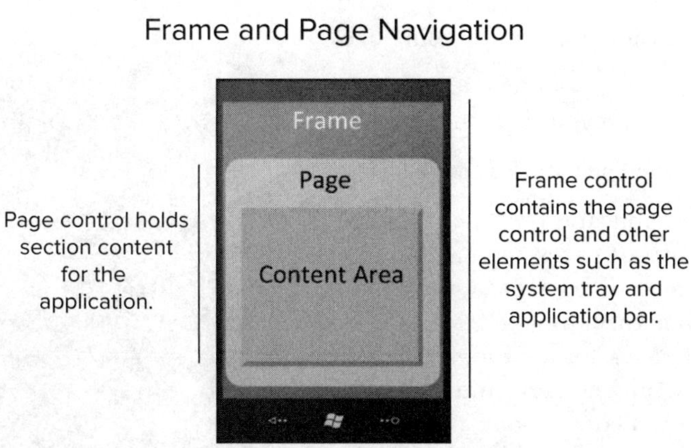

USED WITH PERMISSION OF MICROSOFT
(SOURCE: HTTP://MSDN.MICROSOFT.COM/LIBRARY/FF402536.ASPX)

FIGURE 4-1: Frame and page for WP7 applications

Basic Page Navigation

Navigation on WP7 means a transition between pages. As you may have seen in the previous chapter samples, the page navigation can be achieved through the `NavigationSevice` property of each `PhoneApplicationPage`. The most frequently used properties/methods to control the navigation include:

- `CanGoBack`: Gets a value that indicates whether there is at least one entry in back navigation history.
- `CanGoForward`: Gets a value that indicates whether there is at least one entry in forward navigation history.

- `Navigate()`: Navigates to the supplied URI, and each page can be identified by a URI.
- `GoBack()`: Navigates to the most recent entry in back navigation history, if there is one. You should check `CanGoBack` before calling this method.
- `GoForward()`: Navigates to the most recent entry in forward navigation history, if there is one.

> *The `GoForward()` method will always throw an exception because there isn't any forward navigation stack for the current WP7 version.*

It's time to create a sample application to demonstrate the usage of these key properties and methods. The sample application includes the following features:

- There are three pages with "page 1," "page 2," and "page 3" as the page title. All three pages have the same layout and functions, and page title is the only difference among them.
- In each page, users can navigate to other pages by clicking the listed page number. The current page number is grayed out and disabled.
- In each page, there are two navigation buttons for going back and going forward if possible.
- In each page, the source page is the previously displayed page.
- In each page, the entry count in the back stack is displayed.

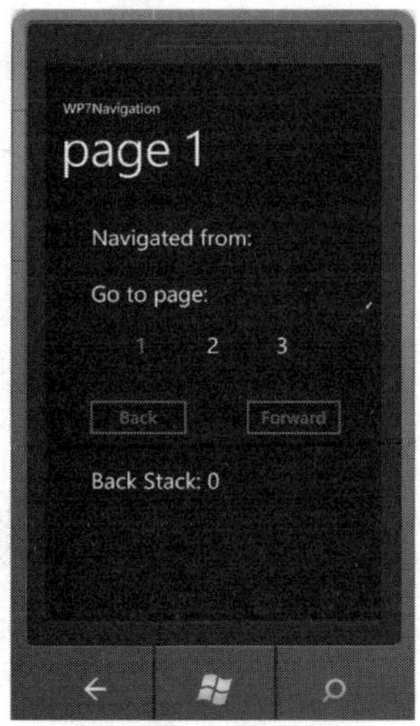

FIGURE 4-2: A sample application to demonstrate page navigation

Figure 4-2 is a screenshot of this sample application, `WP7Navigation`. You can create three Windows Phone portrait pages and name them `Page1.xaml`, `Page2.xaml`, and `Page3.xaml`. In each page, use a `TextBlock` for the page number and name them as `textPage1`, `textPage2`, and `textPage3`. Use a `TextBlock` for the forward and back buttons, naming them as `buttonFwd` and `buttonBack`, and then install the event handlers `buttonFwd_Click()` and `buttonBack_Click()` respectively. Use a `TextBlock` for the source page and back stack, and name them `textFrom` and `textBack`. You also need to define a local property `currentPage` to create a record of the current page number. You can obtain this information from the page title.

You can make `Page1.xaml` as the default page when the application starts by modifying the `WMAppManifest.xml` file. After doing this, you can safely delete the `MainPage.xaml`.

```
<Tasks>
   <DefaultTask Name = "_default" NavigationPage="Page1.xaml"/>
</Tasks>
```

As just mentioned, you can use `NavigationService` to manage the page navigation in the corresponding event handlers shown in Listing 4-1.

Available for download on Wrox.com

LISTING 4-1: Managing page navigation, \WP7Navigation\WP7Navigation\Page1.xaml.cs

```
private void buttonFwd_Click(object sender, RoutedEventArgs e)
{
    if (NavigationService.CanGoForward)
        NavigationService.GoForward();
}
private void buttonBack_Click(object sender, RoutedEventArgs e)
{
    if (NavigationService.CanGoBack)
        NavigationService.GoBack();
}
protected override void OnManipulationStarted(
    ManipulationStartedEventArgs args)
{
    if (args.OriginalSource == textPage1 ||
        args.OriginalSource == textPage2 ||
        args.OriginalSource == textPage3)
    {
        string text = (args.OriginalSource as TextBlock).Text;
        string uriString = "/Page" + text + ".xaml";
        int targetPage = Convert.ToInt32(text);
        if (currentPage != targetPage)
        {
            NavigationService.Navigate(
               new Uri(uriString, UriKind.RelativeOrAbsolute));
        }
    }
}
```

When users touch the screen, the system calls the `OnManipulationStarted()` event handler of the `Page` class. If the user touches a page number and current page isn't the target page, then the `Navigate()` method of `NavigationService` is used to navigate from the current page to the target page. You should note that when you navigate to a page with the `Navigate()` method, WP7 always creates a new page instance. A good practice is to first check `CanGoForward` and `CanGoBack` when users click the forward or back button, before using `GoForward()` and `GoBack()` to navigate along the navigation history.

Sharing Data among Pages

Beside the UI connections among pages, pages also need to share data in many cases. When working with Android you can bundle the data in an intent object and pass it to the target activity because the screen navigation is handled through intents among activities. In iOS, the Model-View-Controller (MVC) pattern is adopted to share data among different views.

There are several ways for pages to share data on WP7. The first way is to add query arguments in the URI when you use `NavigationService.Navigate()` to pass simple types of data to the target page. This operation is similar to the method used to pass parameters between web pages. In the sample application, you need to display the source page, which is the previous page. You can achieve this goal by passing the current number from the source page to the target page. It's possible to change the URI used in Listing 4-1 to:

```
string uriString = "/Page" + text + ".xaml" + "?from=" + currentPage;
```

To retrieve the passing arguments, you can use the `QueryString()` method provided by `NavigationContext`, a property of `PhoneApplicationPage`, which contains information about the navigation request. The code would look like this:

```
string from = "";
if (NavigationContext.QueryString.TryGetValue("from", out from))
   textFrom.Text = "Navigated from: " + from;
```

As we mentioned in the last chapter, the `Application.Current` property provides global access to the `Application` object associated with the program. Thus you can also store data you want to share among different pages in the `App` class, which derives from the `Application` class. In the sample application, you need to track the entry count in the back stack. You can define a property called `BackStack` in the `App` class to keep the entry count updated. Then you can make minor modifications to the code in the Listing 4-1 to achieve this.

```
protected override void OnManipulationStarted(
    ManipulationStartedEventArgs args)
{
    if (args.OriginalSource == textPage1 ||
        args.OriginalSource == textPage2 ||
        args.OriginalSource == textPage3)
    {
        string text = (args.OriginalSource as TextBlock).Text;
        string uriString = "/Page" + text + ".xaml" + "?from=" + currentPage;
        int targetPage = Convert.ToInt32(text);
        if (currentPage != targetPage)
        {
            (Application.Current as App).BackStack++;
            NavigationService.Navigate(
               new Uri(uriString, UriKind.RelativeOrAbsolute));
        }
    }
}
private void buttonBack_Click(object sender, RoutedEventArgs e)
{
    (Application.Current as App).BackStack--;
    if (NavigationService.CanGoBack)
        NavigationService.GoBack();
}
```

Before the application navigates to a new page the code increases the count number, and decreases the count number every time the application returns to the previous page.

Navigation Events

While navigating from/to a page, the `PhoneApplicationPage` exposes several methods you can override to take actions (such as initializing the page, managing the page states, canceling navigation, etc.) against navigation-related events. These methods include:

- `OnNavigatingFrom()`: Called just before the user is about to navigate from the current page. This is the place you can cancel the navigation if needed.

- `OnNavigatedFrom()`: Called when the current page is no longer the active page. This is the place you can preserve the page state.
- `OnNavigatedTo()`: Called after the user navigates to the current page. This is the place you can initialize the page or restore the page state.
- `OnBackKeyPress()`: Called when the hardware Back button is pressed. This is the place you can override the Back button's default behavior when going to the previous page isn't the logical behavior.

In the sample application, you need to gray out the current page and disable the Back or Forward button if there is no entry in the back or forward history stack. These initializations can be done in `OnNavigatedTo()` method as shown in Listing 4-2.

LISTING 4-2: Handling navigation events, \WP7Navigation\WP7Navigation\Page1.xaml.cs

Available for download on Wrox.com

```
protected override void OnNavigatedTo(NavigationEventArgs args)
{
    base.OnNavigatedTo(args);

    if (!NavigationService.CanGoBack)
        buttonBack.IsEnabled = false;
    if (!NavigationService.CanGoForward)
        buttonFwd.IsEnabled = false;

    textPage1.Opacity = 1;
    textPage2.Opacity = 1;
    textPage3.Opacity = 1;
    string pageNumber = PageTitle.Text.Substring(PageTitle.Text.Length - 1);
    currentPage = Convert.ToInt32(pageNumber);
    switch (currentPage)
    {
        case 1:
            textPage1.Opacity = 0.5;
            break;
        case 2:
            textPage2.Opacity = 0.5;
            break;
        case 3:
            textPage3.Opacity = 0.5;
            break;
    }

    string from = "";
    if (NavigationContext.QueryString.TryGetValue("from", out from))
        textFrom.Text = "Navigated from: " + from;

    textBack.Text = "Back Stack: " + (Application.Current as App).BackStack;
}
```

In Listing 4-2, the code first uses `NavigationService` to check the `CanGoBack` and `CanGoForward` properties to disable the buttons if needed. Because there is no forward navigation stack, the

Forward button is always disabled. Then the code uses the page title to decide the current page number and to gray out the corresponding control. Finally, the code displays the source page and the entry count of the back stack.

The default built-in behavior of the hardware Back button is to navigate to the previous page. There is no code required for the default operation. The system maintains a back stack for navigation caching, and a call of the `Navigate()` method will always add the source page to the back stack.

In the sample application you also track the entry count of the back stack, and the code decreases the count number every time the application returns to the previous page through the soft Back button control `buttonBack`. Therefore you need to override the `OnBackKeyPress()` method to tell the system that the code will handle the hardware Back button behavior instead, using the default operation.

LISTING 4-3: Overriding the Back button, \WP7Navigation\WP7Navigation\Page1.xaml.cs

Available for download on Wrox.com

```
protected override void OnBackKeyPress(System.ComponentModel.CancelEventArgs e)
{
    if ((Application.Current as App).BackStack != 0)
    {
        e.Cancel = true;
        buttonBack_Click(null, null);
    }
}
```

In Listing 4-3, the code stops the navigation by using the `Cancel` property in the `CancelEventArgs` passed from the system. Setting the `Cancel` property to `true` will tell the system to cancel the Back button's default operation and let the application handle it.

Using Controls

Like Android and iOS, WP7 provides a comprehensive control set that developers can use to build WP7 applications and maintain a consistent user experience. Table 4-1 lists the most commonly used WP7 controls and their counterparts on Android and iOS. The full list of the built-in base controls is available at `http://msdn.microsoft.com/library/ff402549.aspx`.

TABLE 4-1 WP7, Android, and iOS UI Controls

WP7	ANDROID	IOS
Border	n/a	n/a
Button	Button	UIButton/UIBarButtonItem
Canvas	FrameLayout/AbsoluteLayout	n/a
CheckBox	CheckBox	UITableView with checkmark accessory
Grid	TableLayout	UITableView

WP7	ANDROID	IOS
HyperlinkButton	Linkify TextView	n/a
Image	ImageView	UIImageView
ListBox	ListView	UITableView
MediaElement	MediaController	Media Player
MultiScaleImage	n/a	n/a
PasswordBox	EditText password	UITextField with secureTextEntry property
ProgressBar	ProgressBar	UIProgressView
RadioButton	RadioButton	UIPickerView
ScrollViewer	ScrollView	UIScrollView
Slider	SeekBar	UISlider
StackPanel	LinearLayout	UITabeleView
TextBlock	TextView	UILabel
TextBox	EditText	UITextView/UITextField
WebBrowser	WebView	UIWebView

You can find these controls in the Toolbox of Visual Studio, so you can drag and drop them into your application design window. You can also instantiate them at run time if needed. In addition to these built-in controls, a Silverlight for the Windows Phone Toolkit can be downloaded from `http://silverlight.codeplex.com/`. This open source toolkit is designed and developed by Microsoft, and it offers additional controls that match the WP7 user experience.

> *The sample application WP7Controls introduced later in this chapter will need a control from the Silverlight for Windows Phone Toolkit. Thus you must install this toolkit in order to try the sample code.*

The most recent release of the Silverlight Toolkit includes these components: `AutoCompleteBox`, `ContextMenu`, `DatePicker`, `GestureService`/`GestureListener`, `ListPicker`, `LongListSelector`, `PerformanceProgressBar`, `TiltEffect`, `TimePicker`, `ToggleSwitch`, and `WrapPanel`. This toolkit also includes a set of components to provide page transition effects. After you download and install the toolkit, you have to manually add them to the Toolbox: right-click in the Toolbox, select Choose Items in the pop-up menu, and check the controls you want to add in the pop-up dialog. Then you'll see the selected controls appear in the Toolbox.

> *While using some controls from the Silverlight Toolkit you may notice that some icons are missing. To fix this issue, you have to add these icons into your project. You can find these icons in the toolkit installation folder, for example:* `%Program Files%\Microsoft SDKs\Windows Phone\v7.0\Toolkit\Feb11\Bin\Icons`. *Then copy the icons into your application project by putting them in a folder named* `Toolkit.Content`.

Most controls, including the built-in controls and those from the Silverlight Toolkit, behave as you expect them to from your previous experience on other platforms, so this section doesn't discuss the controls in detail. It does focus on two unique WP7 controls: `Pivot` and `Panorama`.

Overview of Pivot and Panorama Control

Compared to PCs, the screen size of a smartphone is small. To organize and display information (even different types of information) in a mobile application, you usually have to put information content on multiple pages/screens on Android and iOS and connect them through page navigation. As shown in the previous section, WP7 also provides a similar mechanism to display information. In addition, WP7 offers two special controls, `Pivot` and `Panorama`, to organize and display content.

`Pivot` is similar to the Tab Layout on Android or Segmented Control on iOS: it manages different views or pages without leaving the current context. Typical `Pivot` applications include the e-mail client and calendar application on a WP7 device. The `Pivot` control has two major usages: filtering and viewing multiple data sets, and switching application views.

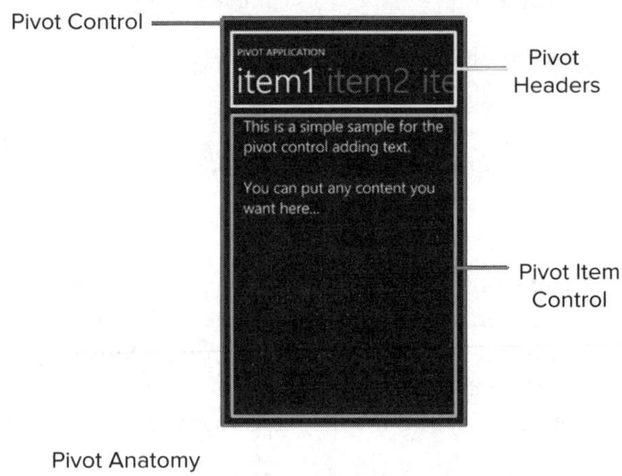

USED WITH PERMISSION OF MICROSOFT (SOURCE: HTTP://MSDN.MICROSOFT.COM/LIBRARY/FF941097.ASPX)

FIGURE 4-3: Pivot control for WP7 applications

Figure 4-3 illustrates the composition of a `Pivot` control. The base `Pivot` control is a container, and a `PivotItem` control is the actual container that hosts content and other controls. Each `PivotItem` has a `Header` property, and all headers appear in the same location and are the same size as the page titles of the normal WP7 page.

The `Pivot` control has built-in support for touch interaction and navigation. Users can navigate to each `PivotItem` by tapping its header or horizontally panning/flicking the `Pivot` control. The `Pivot` control handles all this navigation, so no code is needed. `PivotItem` control supports only portrait orientation, which means that, inside a `PivotItem`, content and controls can only scroll vertically regardless of whether the whole `Pivot` control is in portrait mode or landscape mode.

The `Panorama` experience is one of the major native Metro experiences that differentiate WP7 from other platforms. Many major applications such as People Hub, Media Hub, Picture Hub, Game Hub, Office Hub, and Marketplace Hub on WP7 are all implemented based on the `Panorama` control. By offering a long horizontal canvas, the contents and controls in a `Panorama` control can virtually spread beyond the screen boundary.

Figure 4-4 illustrates the composition of a `Panorama` control. The base `Panorama` control is similar to the `Pivot` control. It's a container, and the developer can add multiple `PanoramaItem` controls to the `Panorama` surface. Generally, each `PanoramaItem` control offers a distinct functional purpose in its own section. The `PanoramaItem` is similar to the `PivotItem` control. It's the container that hosts the content and other controls. The text size of the whole `Panorama` control title is designed to be large with a cut-off effect to motivate users to explore the contents outside the screen boundary.

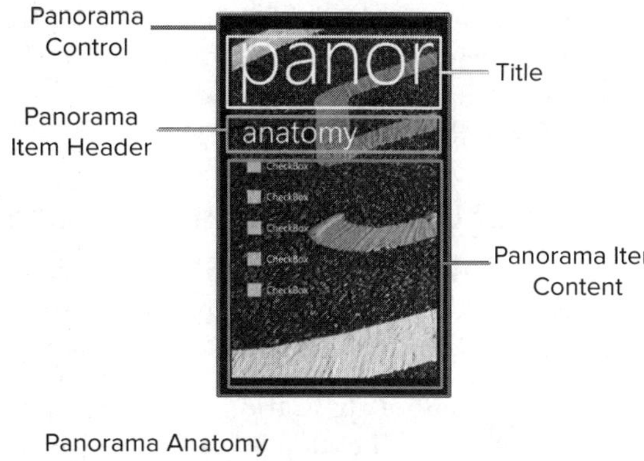

USED WITH PERMISSION OF MICROSOFT (SOURCE: HTTP://MSDN.MICROSOFT.COM/LIBRARY/FF941126.ASPX)

FIGURE 4-4: Panorama control for WP7 applications

Unlike the `Pivot` control, users can navigate to each `PanoramaItem` only by horizontally panning/flicking the `Panorama` control. Only the current `PanoramaItem` title is visible, so there isn't any way to tap the title to navigate to a new section. The `PanoramaItem` control supports both portrait and landscape orientation. When a `PanoramaItem` control is set to landscape orientation, the content inside the `PanoramaItem` is placed off the screen instead of being clipped. Thus users can pan around the contents in a wide `PanoramaItem` before snapping to another `PanoramaItem`.

Although `Pivot` and `Panorama` are conceptually similar, they have different usage scenarios in practice as listed in Table 4-2.

TABLE 4-2 Usage Comparisons between Pivot and Panorama

CHARACTERISTIC	PIVOT	PANORAMA
Data	Filters-related content	Presents unrelated data
Views	Maintains consistent behavior among different views	Offers different presentations or different experiences among different views
Application Bar	Supports	Does not support
Content Organization	Optimizes contents to the screen size in the horizontal direction	Organizes contents in a long canvas beyond the screen boundary to offer a continuously explorative navigation experience

Performance is one issue that many developers may not think about when using the `Pivot` and `Panorama` controls. Even though the contents in the `PivotItem` or `PanoramaItem` control may not be visible to users (especially with `Panorama`), by default they have been loaded to the memory when the page is displayed. This extra data will not only cause the application to start slower, but will also damage application responsiveness. Thus it is worth keeping the `Pivot` and `Panorama` control content items simple and few in number to maintain a better user experience. Generally there are no more than seven items in a `Pivot` control and four items in a `Panorama` control. An alternative solution is to implement the `SelectionChanged()` event handler for the `Pivot` and `Panorama` controls to defer the content loading.

Example of Using Pivot and Panorama Control

In order to demonstrate how to use the `Pivot` and `Panorama` controls, you create a sample application called WP7Controls. This sample application also appears in the next section to explain data binding and Model-View-ViewModel (MVVM) pattern. The sample application has the following features:

- This is an application to manage notes on the phone. There are two types of notes: text and video, and each note can be flagged as urgent or normal.
- Use a `Panorama` control for the default page. It contains three content sections which you implement using `PanoramaItem` controls: (1) links to view and change notes state; (2) list of text notes; (3) list of video notes.
- Use a `Pivot` control to show the collection of notes with detailed information. The notes are filtered by type. In addition, urgent notes appear in red, while regular notes are in green. The user can change the note's urgency flag.

FIGURE 4-5: Screenshot of the sample application WP7Controls

The completed application should look like the screens shown in Figure 4-5. In this section, we will just focus on how to build up the basic UI.

Although Visual Studio provides a project template to create a Pivot/Panorama application, it's more flexible to add Pivot/Panorama controls to a regular WP7 application manually. Use the following steps to perform this task:

1. Right click the project in the Solution Explorer and choose Add ➪ New Item from the context menu. You'll see the Add New Item dialog box.
2. Select Windows Phone Panorama Page template.
3. Type **`PanoramaPage.xaml`** in the Name field.
4. Click Add to add a new Panorama page to the project.
5. Repeat steps 1 through 4 to add a new Pivot page and name it **`PivotPage.xaml`**.

Visual Studio adds a `Panorama` control with two `PanoramaItem` controls to the page to `PanoramaPage.xaml`:

```
<Grid x:Name="LayoutRoot">
    <controls:Panorama Title="my application">
        <!--Panorama item one-->
```

```
        <controls:PanoramaItem Header="item1">
            <Grid/>
        </controls:PanoramaItem>
        <!--Panorama item two-->
        <controls:PanoramaItem Header="item2">
            <Grid/>
        </controls:PanoramaItem>
    </controls:Panorama>
</Grid>
```

Usually the `Panorama` control has a background image. The recommended background image size is 800 pixels in height and less than 2000 pixels in width. The picture is stretched to 800 pixels if its height is less than 800 pixels and it's clipped to 2000 pixels if its width is more than 2000 pixels. To apply the background image, you can set the `Background` property of the `Panorama` control:

```
<controls:Panorama Title="WP7Controls">
    <controls:Panorama.Background>
        <ImageBrush ImageSource="PanoramaBackground.jpg"/>
    </controls:Panorama.Background>
</controls:Panorama>
```

In the first `Panorama` section, you need to add several links that can be used to navigate to the note details page. Because the `PanoramaItem` control is a layout container, you can put most controls inside it.

```
<controls:PanoramaItem Header="collections">
    <Grid>
        <StackPanel>
            <TextBlock Text="all" Name ="tbAll"
               Style="{StaticResource PhoneTextExtraLargeStyle}"/>
            <TextBlock Text="text notes" Name="tbText"
               Style="{StaticResource PhoneTextExtraLargeStyle}"/>
            <TextBlock Text="video notes" Name="tbVideo"
               Style="{StaticResource PhoneTextExtraLargeStyle}"/>
        </StackPanel>
    </Grid>
</controls:PanoramaItem>
```

Now you add three `TextBlock` controls to the first `PanoramaItem` control. They're named `tbAll`, `tbText`, and `tbVideo`, respectively. The user expects to navigate to the note details page `PivotPage.xaml` after tapping any of these three controls. This page shows the corresponding type of note list. Thus in the code-behind file `Panorama.xaml.cs`, we can add code to handle the tapping event.

```
protected override void OnManipulationStarted(
   ManipulationStartedEventArgs args)
{
    string type = "";
    if (args.OriginalSource == tbAll)
        type = "all";
    else if (args.OriginalSource == tbText)
        type = "text";
    else if (args.OriginalSource == tbVideo)
```

```
            type = "video";

        if (type != "")
        {
            string uriString = "/PivotPage.xaml?type=" + type;
            NavigationService.Navigate(
               new Uri(uriString, UriKind.RelativeOrAbsolute));
        }
    }
```

After identifying the source of the tapping event, the code can decide the note type and then use the `NavigationService` to navigate to the `PivotPage` and pass the note type to it. In the second and third Panorama section, you want to list text and video notes, respectively.

```
<!--Panorama item two-->
<controls:PanoramaItem Header="text notes">
    <Grid>
        <ListBox Name="TextList" ItemsSource="{Binding}">
        </ListBox>
    </Grid>
</controls:PanoramaItem>
<!--Panorama item three-->
<controls:PanoramaItem Header="video notes"  Orientation="Horizontal">
    <ListBox Name="VideoList" ItemsSource="{Binding}">
        <ListBox.ItemsPanel>
            <ItemsPanelTemplate>
                <toolkit:WrapPanel x:Name="wrapPanel" Width="600" />
            </ItemsPanelTemplate>
        </ListBox.ItemsPanel>
    </ListBox>
</controls:PanoramaItem>
```

The example uses a `ListBox` control to host the note list. Setting the `ItemSource` property to `{Binding}` means the `ListBox` content is determined at run time. The next section discusses the details of this topic. For video notes, you set the `PanoramaItem`'s orientation to `Horizontal` and use a `WrapPanel` control to define the content layout in this `PanoramaItem`. The `WrapPanel` control comes with the Silverlight Toolkit for Windows Phone introduced earlier, and it arranges the controls inside it in from left to right and top to bottom. Set the `WrapPanel` width to 600 pixels, which is wider than the screen width. This makes the video notes section stretch across multiple screens.

Because the basic function of the `PivotPage` page is to list detailed note information, which is decided at run time, the default layout definition of `PivotPage.xaml` is sufficient after you add one more `PivotItem` control and set its titles.

```
<controls:Pivot Title="NOTES" Name="NotesPivot">
    <!--Pivot item one-->
    <controls:PivotItem Header="all">
        <Grid/>
    </controls:PivotItem>
    <!--Pivot item two-->
    <controls:PivotItem Header="text">
        <Grid/>
```

```
        </controls:PivotItem>
        <!--Pivot item three-->
        <controls:PivotItem Header="video">
            <Grid/>
        </controls:PivotItem>
    </controls:Pivot>
```

You can also connect the `PivotPage` to the `PanoramaPage` by implementing the `OnNavigatedTo()` event method in the code-behind file `PivotPage.xaml.cs`.

```
protected override void OnNavigatedTo(
   System.Windows.Navigation.NavigationEventArgs e)
{
    base.OnNavigatedTo(e);
    string type = "";
    if (NavigationContext.QueryString.TryGetValue("type", out type))
    {
        if (type == "all")
            NotesPivot.SelectedIndex = 0;
        else if (type == "text")
            NotesPivot.SelectedIndex = 1;
        else if (type == "video")
            NotesPivot.SelectedIndex = 2;
    }
}
```

When the user navigates to the `PivotPage` from the `PanoramaPage`, the application gets the note type using `QueryString()`. The note type is used to determine the focused `PivotItem` by setting its `SelectedIndex` property.

At this point, you have the basic UI set up for the sample application. The next section introduces how to integrate the data and control logic with the UI by using data binding and the MVVM pattern.

Data Binding and MVVM

Data binding is a widely used technology in Silverlight applications to enable different types of data sources to be connected with the UI elements. It makes data-UI connection and synchronization easy and automatic. The concept of data binding is also familiar to Android and iOS developers.

On Android, an *AdaptView* is a type of view whose children and contents are determined by an `Adapter` that binds to the concrete data. `ListView`, `GridView`, `Spinner`, and `Gallery` are well-known subclasses of `AdapterView`. An `Adapter` retrieves data from a data source, either a static resource or an external source that relies on code output or query results. `AdapterView` uses the `setAdapter()` method to set the adapter. Because `Adapter` provides the data and the views to present the data, when the data changes, the content displayed in `AdaptView` also changes.

On iOS, *Cocoa bindings* are a key technology to fully implement a MVC pattern. Cocoa bindings simplify programming using the MVC pattern. They provides means to keep UI presentations (views) and data (models) synchronized through a mediated connection; thus a change in one leads to the corresponding change in the other. Cocoa bindings rely on Key-Value Coding (KVC) and Key-Value Observing (KVO).

Data binding on WP7 is similar to the iOS and Android, and it provides a way to display and interact with data. Silverlight applications use XAML files to declare UI elements, layouts, and their properties. In the previous sample applications, most UI element properties are hard-coded as static values. This approach isn't sufficient for the WP7Controls sample in this section — you need to display the note list whose content isn't fixed, but may change at run time. Even though you can still use code to manually update the list UI when there is a UI data change, this method is error-prone because the greater amount of code increases the potential for bugs and development overhead.

You can create data binding in XAML using a binding expression: `<object property="{Binding propertyPath, oneOrMoreBindingProperties}" .../>`. For example, `<TextBox Text="{Binding Path=Name, Mode=OneWay}"/>` means the content displayed in the `TextBox` is bound to the `Name` property of an object, and if the value of `Name` changes, the content of `TextBox` also changes. In this way, the UI presentation is bound to the data object. The commonly used properties include:

- `Source`: Specifies a reference to an object or a collection of objects.
- `Path`: Specifies the path to the binding source property.
- `Mode`: Specifies the direction of binding.
    - `OneTime` binding updates the target when the binding is created.
    - `OneWay` binding updates the target whenever the source property changes.
    - `TwoWay` binding updates source property whenever the target property is updated and vice versa.
- `Converter`: Specifies the converter object that modifies the data as it is passed between the source and target.

The full list of binding properties is available at `http://msdn.microsoft.com/library/cc189022.aspx`.

MVVM Pattern

As an iOS developer, you must be comfortable using the MVC (Model-View-Controller) pattern to design and implement your application. The MVC pattern defines the roles of objects in an application and how they communicate with each other. In the MVC model:

- View is the UI.
- Model encapsulates the data that the view displays.
- Controller contains the business logic that modifies the model depending on the user interaction with the view and updates the UI when model changes.

The MVC pattern can help decouple models and views to reduce application complexity and increase application flexibility.

WP7 relies on the MVVM (Model-View-ViewModel) pattern, which is close and related to the MVC pattern. In the MVVM pattern:

- View defines what the user sees on the screen such as controls, layouts, styling, and data template.
- Model encapsulates business logic and data.
- ViewModel encapsulates the presentation logic and data for the view.

There is no direct communication between View and Model, and ViewModel acts as the middleman between View and Model to handle the data update and UI presentation. The implementation of the MVVM pattern in WP7 heavily leverages the data-binding technology we just introduced. Figure 4-6 illustrates the relationship between the MVVM pattern and data binding.

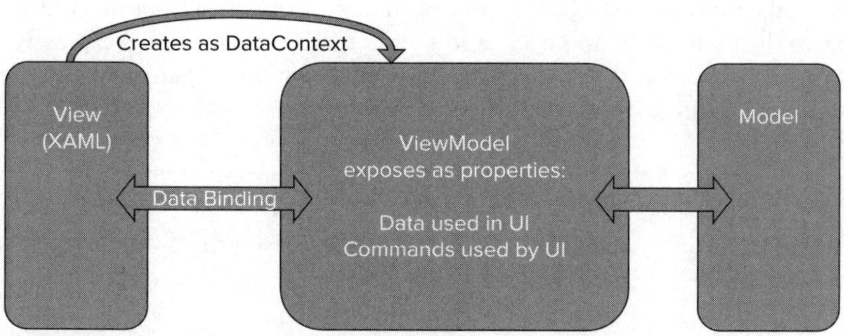

FIGURE 4-6: Data binding and MVVM pattern

The View can access the ViewModel by setting its `DataContext` property. The properties of the controls in a View are bound to the properties exposed by the ViewModel through data binding. In addition, the View may use `Converter` to format the data before it is displayed. The UI presentation, logic, and behavior can be defined and implemented in the XAML file or the code-behind file.

In contrast to the View, the Model classes are non-visual classes, and they are all about the application data and the business logic among this data. The Model classes usually don't have direct access to the View or ViewModel and don't know how they're implemented. Instead, they typically implement the change notification event interfaces `INotifyPropertyChanged` and `INotifyCollectionChanged`. In this way, changes in Model can be easily propagated to the View through data bindings between View and ViewModel.

Similar to the Model, the ViewModel classes are also non-visual classes. The ViewModel coordinates interaction between the View and the Model. As the name suggests, the ViewModel is the model of the view, and the actual data (not the data model and business logic defined in the model) is managed by the ViewModel. The ViewModel may implement additional properties that may not be defined in the Model for better presentation and consumption by the View.

To explain how to implement the MVVM pattern in a WP7 application, let's complete the sample application, WP7Controls, by binding the application data with the UI you have already created.

Creating the Model

Because the model defines the basic data structure and business logic, we can define a class named `Note` to hold the properties we need for the sample application:

- **Title:** The title of a note, which is displayed in the note list
- **Type:** The type of a note, either text or video
- **Urgent:** A flag used to define a note as urgent or not

To better organize your project files, you need to create a new folder `Model` for the project: right-click the project in Solution Explorer and choose Add ⇨ New Folder from the context menu. Type **`Model`** and press Enter. Use the same technique to create the `View` and `ViewModel` folders.

After creating the folders, you can create a new class in the `Model` folder: right-click the `Model` folder in Solution Explorer and choose Add ⇨ Class from the context menu. You'll see an Add New Item dialog. Type **`Note.cs`** in the Name field and click Add to add the class. The model typically implements the `INotifyPropertyChanged` interface so that the model can propagate its properties' changes by raising the `PropertyChanged` events, as shown in Listing 4-4.

Available for download on Wrox.com

LISTING 4-4, Model class for WP7Controls application, \WP7Controls\WP7Controls\Model\Note.cs

```
public class Note : INotifyPropertyChanged
{
    public string Title{get;set;}
    public string Type{get;set;}

    private bool _urgent;
    public bool Urgent
    {
        get{return _urgent;}
        set
        {
            _urgent = value;
            RaisePropertyChanged("Urgent");
        }
    }

    public event PropertyChangedEventHandler PropertyChanged;
    private void RaisePropertyChanged(string propertyName)
    {
        if (this.PropertyChanged != null)
        {
            this.PropertyChanged(this,
               new PropertyChangedEventArgs(propertyName));
        }
    }
}
```

As shown in Listing 4-4, the code only raises the `PropertyChanged` event in the `Urgent` property, and this means that the user can only change the `Urgent` property. Later when you bind the UI elements to the `Note` model, the bound UI elements actually subscribe to the `PropertyChanged` event. If a UI element is bound to a property that raises the event, the subscriber UI element gets the notification.

Creating the ViewModel

ViewModel is the connection between the Model and the View, so that it hosts the actual data used in the View. In the sample application, the data is a collection of `Note` objects. The `ObservableCollection` class implements the `INotifyCollectionChanged` interface in the same way as `INotifyPropertyChanged` that provides notifications when items of the collection are changed. Because we need to bind the object collection to a list type control such as `ListBox`, the `ObservableCollection` class can be used to define the collection of notes.

The ViewModel used in this example contains the code to get the collection of notes. For demo purposes, the example populates the collection with sample data as shown in Listing 4-5. For real applications, the data could be from isolated storage or remote services.

Available for download on Wrox.com

LISTING 4-5 ViewModel class for WP7Controls application, \WP7Controls\WP7Controls\ViewModel\NoteViewModel.cs

```
public class NoteViewModel
{
    public ObservableCollection<Note> Notes{get;set;}

    public void GetNotes()
    {
        ObservableCollection<Note> notes = new ObservableCollection<Note>();

        notes.Add(new Note()
        {
            Title = "note 1",
            Type = "text",
            Urgent = false
        });

        notes.Add(new Note()
        {
            Title = "note 2",
            Type = "video",
            Urgent = true
        });
        // more sample data can be added

        Notes = notes;
    }
}
```

In the ViewModel, the property `Notes` is used to hold the collection of notes, and the ViewModel provides a method called `GetNotes()` to the View to access this property.

Creating the View

In MVVM pattern implementation, View can be a window, a page, or a UI control. In the sample application, several application elements provide a View, and the application needs to bind to the ViewModel through data binding. In the `PivotPage` which is used to display the note details, there are three `PivotItem` controls defined earlier that are used to display the note list. The only difference between the contents shown in these three `PivotItem` controls is the data, and the `PivotItem` controls used in this sample share the same UI controls and layout. Thus, we can create a UI control called `NoteListView` and bind the ViewModel with this control, and then we can add it to the `PivotItem` controls. The `NoteListView` contains a `ListBox` and each item in the `ListBox` has the following columns:

- **Title:** A `TextBlock` is used to display the `Title` property. The binding mode is `OneWay`.
- **Type:** A `TextBlock` is used to display the `Type` property. The binding mode is `OneWay`.
- **Urgent:** A `CheckBox` is used to indicate whether a note is urgent or not. The user can check this item and the code updates the Model through data binding.

To create the `NoteListView` control, you can right-click the `View` folder in Solution Explorer and choose Add ➪ New Item from the context menu. You'll see the Add New Item dialog box. Select Windows Phone User Control entry and type **`NoteListView.xaml`** in the Name field. Click Add to add the control. Listing 4-6 shows the UI definition of the `NoteListView`.

Available for download on Wrox.com

LISTING 4-6: View for note list used in PivotPage, \WP7Controls\WP7Controls\View\NoteListView.xaml

```
<UserControl.Resources>
    <src:TitleColor x:Key="TitleColor">
    </src:TitleColor>
</UserControl.Resources>
<Grid x:Name="LayoutRoot" Background="{StaticResource PhoneChromeBrush}">
    <ListBox Name="NoteList" ItemsSource="{Binding}">
        <ListBox.ItemTemplate>
            <DataTemplate>
                <Grid>
                    <Grid.ColumnDefinitions>
                        <ColumnDefinition Width="200"/>
                        <ColumnDefinition Width="180"/>
                        <ColumnDefinition Width="100"/>
                    </Grid.ColumnDefinitions>
                    <TextBlock x:Name="Title"
                       Text="{Binding Path=Title, Mode=OneWay}" Grid.Column="0"
                       HorizontalAlignment="Left" VerticalAlignment="Center"
                       Style="{StaticResource PhoneTextExtraLargeStyle}"
                       Foreground="{Binding Path=Urgent,
                                   Converter={StaticResource TitleColor}}"/>
                    <TextBlock x:Name="Type"
                       Text="{Binding Path=Type, Mode=OneWay}" Grid.Column="1"
                       HorizontalAlignment="Left" VerticalAlignment="Center"
                       Style="{StaticResource PhoneTextLargeStyle}"
                       Foreground="{Binding Path=Urgent,
                                   Converter={StaticResource TitleColor}}"/>
```

```
                        <CheckBox x:Name="Urgent"
                           IsChecked="{Binding Path=Urgent, Mode=TwoWay}"
                           Grid.Column="2" HorizontalAlignment="Left"
                           VerticalAlignment="Center"/>
                    </Grid>
                </DataTemplate>
            </ListBox.ItemTemplate>
        </ListBox>
    </Grid>
```

The `ListBox` control uses `DataTemplate` to define the UI display for each note item in the bound collection of notes. Each UI control that's used to display the note details is bound to the ViewModel using binding syntax. For example, `Text="{Binding Path=Title, Mode=OneWay}"` binds the `Text` property of a `TextBlock` control to the `Title` property of a `Note` object. Because the `TextBlock` control is only for display, we can't change its `Text` property through user interaction. Thus the binding mode is `OneWay`. `IsChecked="{Binding Path=Urgent, Mode=TwoWay}"` binds the check state of a `CheckBox` to the `Urgent` property of a `Note` object. Because the user can change the check state of the `CheckBox` control, and the application requires the Model data to update its value when the user changes the `CheckBox` check state, the binding mode is `TwoWay`.

Using Converters

Data binding is used to bind the UI control property to the Model data of the same data type: the `Text` property of the `TextBlock` is bound to the `string` type `Title`, and the `IsChecked` property is bound to the `Boolean` type `Urgent`. However, sometimes an application needs to bind the UI control property to Model data with a different data type. For example, you might want to display the `Title` and `Type` in green if a note isn't urgent and in red if it is urgent. In this case, the `Foreground` property, which is a `Brush` type, needs to bind to the `Urgent` property of a `Note` object, which is a `Boolean` type. *Converters* provide the flexibility required to establish bindings between two different data types. `Converter` is a class that implements the `IValueConverter` interface. You can create a custom converter named `TileColor` that converts a `Boolean`-type `Urgent` property to a `Brush` object and then bind it with the `Foreground` property of a `TextBlock` control. Put the class definition of `TileColor` in the `NoteListView.xaml.cs` as shown in Listing 4-7.

Available for download on Wrox.com

LISTING 4-7: Converter for binding different data types, \WP7Controls\WP7Controls\View\NoteListView.xaml.cs

```
public class TitleColor : System.Windows.Data.IValueConverter
{
   public object Convert(object value, Type targetValue,
      object parameter, CultureInfo culture)
    {
        SolidColorBrush brush = new SolidColorBrush();
        bool b = (bool)value;
        if (b)
            brush.Color = Color.FromArgb(255, 255, 0, 0);
        else
            brush.Color = Color.FromArgb(255, 0, 255, 0);
        return brush;
```

continues

LISTING 4-7 *(continued)*

```
        }

        public object ConvertBack(object value, Type targetType,
            object parameter, CultureInfo culture)
        {
            throw new NotImplementedException();
        }

    }
```

A converter class should implement the `Convert()` method (mapping the source value to another data type that the target can use) and the `ConvertBack()` method (mapping the target value to a data type that the source can use). In the sample application, because the `Foreground` binding is `OneWay`, you don't need to implement `ConvertBack()`. Before you can use this converter, you need to declare it as a resource in `NoteListView.xaml`.

```
<UserControl.Resources>
    <src:TitleColor x:Key="TitleColor">
    </src:TitleColor>
</UserControl.Resources>
```

To use this converter, you need to set the `Converter` binding property to the resource you just declared: `Foreground="{Binding Path=Urgent, Converter={StaticResource TitleColor}}"/>`. In this way, the `TextBlock` color is bound to the `Urgent` property of a note object.

After you've defined the `NoteListView` custom control, you can add it the `PivotPage` as shown in the Listing 4-8.

LISTING 4-8: adding NoteListView to PivotPage, \WP7Controls\WP7Controls\PivotPage.xaml

Available for download on Wrox.com

```
<Grid x:Name="LayoutRoot" Background="Transparent">
    <!--Pivot Control-->
    <controls:Pivot Title="NOTES" Name="NotesPivot">
        <!--Pivot item one-->
        <controls:PivotItem Header="all">
            <Grid>
                <views:NoteListView x:Name="AllList">
                </views:NoteListView>
            </Grid>
        </controls:PivotItem>

        <!--Pivot item two-->
        <controls:PivotItem Header="text">
            <Grid>
                <views:NoteListView x:Name="TextList">
                </views:NoteListView>
            </Grid>
        </controls:PivotItem>

        <!--Pivot item three-->
```

```
            <controls:PivotItem Header="video">
                <Grid>
                    <views:NoteListView x:Name="VideoList">
                    </views:NoteListView>
                </Grid>
            </controls:PivotItem>
        </controls:Pivot>
    </Grid>
```

Binding View and ViewModel

Now you have the View, Model, and ViewModel ready, and the ViewModel is able to provide Model data to the View. The last piece to make them work together is to connect the View and ViewModel. As described earlier, the View accesses the ViewModel through its `DataContext` property as shown in Listing 4-9.

Available for download on Wrox.com

LISTING 4-9: binding view and view model, \WP7Controls\WP7Controls\PivotPage.xaml.cs

```
public partial class PivotPage : PhoneApplicationPage
{
    private NoteViewModel vm;
    public PivotPage()
    {
        InitializeComponent();
        vm = new NoteViewModel();
    }

   protected override void OnNavigatedTo(
      System.Windows.Navigation.NavigationEventArgs e)
    {
        base.OnNavigatedTo(e);

        vm.GetNotes();
        AllList.DataContext = vm.Notes;
        TextList.DataContext = from Note in vm.Notes
                               where Note.Type == "text"
                               select Note;
        VideoList.DataContext = from Note in vm.Notes
                                where Note.Type == "video"
                                select Note;
        string type = "";
        if (NavigationContext.QueryString.TryGetValue("type", out type))
        {
            if (type == "all")
                NotesPivot.SelectedIndex = 0;
            else if (type == "text")
                NotesPivot.SelectedIndex = 1;
            else if (type == "video")
                NotesPivot.SelectedIndex = 2;
        }
    }
}
```

In Listing 4-9, the code instantiates a ViewModel object `vm` and populates the sample Model data by calling `vm.GetNotes()`. For the all-notes list in `PivotPage`, the code displays the whole collection; thus you can set its `DataContext` to the collection of the note objects: `AllList.DataContext = vm.Notes`. On the other hand, for the text notes list and video notes list, the code displays a subset of the whole collection based on type, so the code uses Language Integrated Query (LINQ) statements to populate the data.

```
TextList.DataContext = from Note in vm.Notes
                       where Note.Type == "text"
                       select Note;
VideoList.DataContext = from Note in vm.Notes
                        where Note.Type == "video"
                        select Note;
```

Similarly, the code can connect the View and ViewModel in the `PanoramaPage` using the same technique as shown for the `PivotPage`.

Handling UI Events

Users interact with WP7 applications through touch gestures, and WP7 applications respond to the user interactions by handling UI events.

Gesture Support

WP7 UI controls are gesture-aware, and support gestures such as tap, double tap, tap and hold, pan, flick, and pinch and stretch. Some controls already have the support built-in for these gestures, along with the corresponding UI transitions. For example, the `Pivot` and `Panorama` controls already support the pan and flick gestures, so you don't need to write any code to support the switch from one section to another.

However, sometimes you want to handle the touch gesture manually by using *manipulation events*. Manipulation events are supported on controls derived from `UIElement`:

- `ManipulationStarted`: Occurs when the user places the finger(s) on the screen.
- `ManipulationDelta`: Occurs repeatedly when the finger(s) move on the screen.
- `ManipulationCompleted`: Occurs when the finger(s) are removed from the screen.

The Silverlight framework provides additional support on simple gestures, such as tap, double tap, and tap and hold, through mouse events.

- Tap: Uses the `MouseLeftButtonUp` event.
- Double tap: Starts a `TimeSpan` timer when the `MouseLeftButtonDown` event occurs. If there are two `MouseLeftButtonUp` events and one `MouseLeftButtonDown` event occurs before the timer expires and the `MouseMove` event doesn't occur between the first `MouseLeftButtonUp` event and the second `MouseLeftButtonDown` event, a double tap occurs.
- Tap and hold: Starts a `TimeSpan` timer when the `MouseLeftButtonDown` event occurs, and stores the users' touch location. If the `MouseLeftButtonUp` event doesn't occur before the timer expires and the user doesn't move the touch location more than a few pixels in the `MouseMove` event, a tap and hold occurs.

Orientation

Orientation change is another important UI event that an application may be interested in. To support orientation changes on Android you can provide different resources and layouts for different screen orientations. By default, the Android application will rotate when the screen rotates and pick the right layouts and resources to render on the screen. On iOS you need to programmatically declare the supported orientations and program the view controllers to make changes to the views in response to the system notifications.

WP7 supports three screen orientations: portrait, landscape left (the top of the page rotates to the left), and landscape right (the top of the page rotates to the right). By default the application pages support only one orientation, either portrait or landscape depending on the template you selected when you created the page. To support multiple orientations, you have to declare it in the `SupportedOrientations` property of the `Page` class. This property can take `Portrait`, `Landscape`, or `PortraitOrLandscape` as the possible value. However, you can't specify only left landscape or only right landscape if you support orientation changes — you have to support both. If it's set to `PortraitOrLandscape`, the page re-orients itself automatically when the user rotates the phone. You can't force a page to re-orient in code.

To handle the orientation change, you can install an event handler for the `OnOrientationChanged()` event, which is called after the `Orientation` property of the `Page` class changes. An `OrientationChangedEventArgs` object is passed to the event handler, and you can get the current orientation through its `Orientation` property. You might want to add transition animation effects to the orientation change as you can do on Android. Unfortunately, custom screen transition animation effects are prohibited as part of the application, so they're implemented and handled by the system framework.

Other UI Considerations

In addition to the core concepts and UI controls, some other UI considerations are worth exploring to make your WP7 applications more efficient and appealing.

Application Bar

The Application Bar is a built-in control that allows you to add a toolbar to your WP7 applications. It provides functions, such as Menu, that are similar to the Action Bar on Android and Toolbar on iOS. You should use the Application Bar instead of creating your own menu system if a menu system is needed. This design approach helps create a consistent user experience with other applications on the device.

The Application Bar is displayed as a row of up to four icon buttons to access the most commonly used application-specific tasks and views. In addition to the icon buttons, you can add additional actions to the Application Bar by adding more text-based menu items. You don't have to use all four icon buttons if the actions aren't primary, and you can put those less-used actions on the menu. When you use the menu item, you should pay attention to the text length because the title will run off the screen if it's too long. The recommended text length is 14 to 20 characters. Windows Phone Developer Tools provides a set of icons that are used by the WP7 system applications. You may want to use these icons if you implement similar functionality. You can find these icons at `%Program Files%\Microsoft SDKs\Windows Phone\v7.0\Icons`.

When you create a new `Page` in Visual Studio, the IDE generates some sample code showing Application Bar usage. You can uncomment this code if you want to add support for the Application Bar and customize it as needed.

```
<phone:PhoneApplicationPage.ApplicationBar>
    <shell:ApplicationBar IsVisible="True" IsMenuEnabled="True">
      <shell:ApplicationBarIconButton
         IconUri="/Images/appbar_button1.png" Text="Button 1"/>
      <shell:ApplicationBarIconButton
         IconUri="/Images/appbar_button2.png" Text="Button 2"/>
      <shell:ApplicationBar.MenuItems>
        <shell:ApplicationBarMenuItem Text="MenuItem 1"/>
        <shell:ApplicationBarMenuItem Text="MenuItem 2"/>
    </shell:ApplicationBar.MenuItems>
  </shell:ApplicationBar>
</phone:PhoneApplicationPage.ApplicationBar>
```

The `ApplicationBarIconButton` and `ApplicationBarMenuItem` classes are controls for the icon buttons and menu items used in the Application Bar. Both classes expose a `Click` event that can be handled like other button controls.

Splash Screen

When you create a new WP7 project, a default splash screen image, `SplashScreenImage.jpg`, is added to your project. The user will see this image immediately when the application is launched until the first page of the application is displayed.

You can replace the default splash screen image with another image you like. However, you must follow some rules to make the new image work.

- The image must be 480 × 800 pixels in size.
- The image much be named `SplashScreenImage.jpg`.
- The Build Action property for this image file must be set to `Content`.

UI CUSTOMIZATION

When you develop an application, sometimes it needs a customized look-and-feel to match the brand image of your product. This section introduces the WP7 techniques for UI customization.

Styles and themes are used to personalize an application's visual elements. This concept is widely adopted in web design. Styles such as Cascading Style Sheets (CSS) define control attributes, including height, width, font size, and color. On Android, styles are defined as XML resources and are separate from the XML that defines the layout.

Like Android, on WP7 *Style* is an object which is comprised of a collection of property setters and targets a particular control type. Styles are usually defined in XAML files. By using styles, it's easy to maintain the same look-and-feel for controls in an application and improve the application maintainability.

On WP7, you have four options to set a UI control's look-and-feel, (listed in order of precedence from low to high):

1. Use the default control property value. This option has the lowest precedence, and the control properties are set to the default value if you don't use any style or explicitly set the value.
2. Use the default style. You can find the list of the default styles at `%Program Files%\Microsoft SDKs\Windows Phone\v7.0\Design\ThemeResources.xaml`.
3. Define your own styles and set the `Style` property of the control to apply the customized styles to the controls you want to customize. This may affect all controls that use the same style.
4. Explicitly set the control's UI properties, such as font, color, and padding. This applies only to the control object you're working on, and it won't affect other control objects with the same type. This option has the highest precedence, and you can override any style or default value by explicitly setting the property value.

This section shows how to apply the customized styles to the WP7 applications. The first thing you need to do is define your own style. Let's use the WP7Controls sample application for the demonstration. If you look at `App.xaml`, you'll see a placeholder element for customized styles: `<Application.Resources/>`. You can define your own style here. For example, if you want to customize the `TextBlock` control with a red foreground color and an italic font style, you can define a style as:

```
<Style x:Key="MyTextBlockStyle"
        BasedOn="{StaticResource PhoneTextBlockBase}"
        TargetType="TextBlock">
    <Setter Property="Foreground" Value="Red"/>
    <Setter Property="FontStyle" Value="Italic"/>
</Style>
```

In the style definition, you use the `Key` property to identify the style and later reference it in control definitions in the XAML file. The `BasedOn` property specifies the style that's the basis of this style. Each style supports only one `BasedOn` value. The `TargetType` property sets the type for which this style is intended. The example defines a style named `MyTextBlockStyle`. You then use setters to apply property values that you want to use the style. In this example, the style is based on `PhoneTextBlockBase`, the default `TextBlock` style, and it should be applied to the `TextBlock` controls.

After defining a custom style, you can apply it to the `TextBlock` controls used in the WP7Controls application. For example, in `PanoramaPage.xaml`, you can use `MyTextBlockStyle` to style one `TextBlock` in the first `PanoramaItem`.

```
<TextBlock Text="all" Name ="tbAll"
   FontSize ="{StaticResource PhoneFontSizeExtraLarge}"
   Style="{StaticResource MyTextBlockStyle}"/>
```

Now the foreground color of this `TextBlock` control is red and its font style is italic.

On WP7, *theme* is a color-related concept. WP7 allows the user to customize the system theme on their phones. A WP7 system theme is a combination of a background and an accent color. In detail, the user can select light or dark background; and the user can select an accent color from a choice of 10 accent colors. You have the option of using the system theme color to color the application controls. For example, you can use `PhoneAccentBrush`, `PhoneAccentColor`, and `PhoneTextAccentStyle` in the properties that are related to colors. Then these control properties will use the same color as the color that the user selects for the system theme. In the `PanoramaPage.xaml`, you can use `PhoneTextAccentStyle` to style another `TextBlock` in the first `PanoramaItem`.

```
<TextBlock Text="text notes" Name="tbText"
   FontSize ="{StaticResource PhoneFontSizeExtraLarge}"
   Style="{StaticResource PhoneTextAccentStyle}"/>
```

Now if the user changes the system accent color, this `TextBlock` control's color will change accordingly.

SUMMARY

This chapter highlights key concepts about creating a user interface for WP7 applications. Page is the basic building block for WP7 applications. Each page is composed of controls, either built-in controls or custom controls. Developers can personalize application look-and-feel through styles and theme.

MVVM is the design pattern widely adopted in WP7 applications, and it decouples the application UI and data. Implementing MVVM relies heavily on data-binding technology which makes data changes reflect in the UI automatically.

Starting with the next chapter, the book focuses more on how to use specific WP7 platform features such as application data storage.

5

Application Data Storage

WHAT'S IN THIS CHAPTER?

- Understanding common application storage scenarios
- Using isolated storage on WP7
- Using external storage

Application data comes from various sources, such as user-generated inputs, operating system data, data received from the Web, and application-produced data. Some of the application data is transient, and you don't need to save it in persistent storage, such as a Secure Digital (SD) card or device memory, or to save it on the Web; but other data may need to be persisted either in local data storage or to the Web so the application can access it later. This chapter shows how to leverage local data storage on the device and cloud storage (data storage on the Web, accessed via web services) on WP7 to save persistent data. After reading this chapter, developers will be able to choose between local data storage and cloud storage to implement common data storage scenarios.

APPLICATION STORAGE ON MOBILE DEVICES

Generally, you have control over a few data storage design decisions when creating a mobile application:

- Where to save the data
- What format to save the data in
- How data is shared between applications

Historically, mobile applications on major software platforms usually have saved data locally on the device. For example, on Windows Mobile devices an application can use the registry to save some application-specific settings and use filesystem APIs to save other data to either device memory or an SD card. The iPhone Operating System (iOS) and Android both provide APIs to save data locally and privately — only the application can access its data. As more applications start to embrace the cloud and web services, it becomes possible to push data to the cloud while keeping a local copy in a cache. Thus data synchronization becomes a critical issue. Cloud data storage is discussed in the "Saving Data to the Cloud" section later in this chapter.

Mobile applications normally support three kinds of data format: application-specific, structural, and settings and preferences. Application-specific data can appear within plain text or XML files, or an application can store data within special binary data files. Some applications rely on a local SQL database to save structure data and enable the use of SQL-like queries and data manipulation. In fact, most of today's mobile software platforms including Android and iOS support SQLite (`www.sqlite.org`), a compact, lightweight relational-database engine.

Application data also includes settings and preferences set by a user. All major mobile software platforms, including WP7, provide some way to save these settings. This topic is discussed in the "Saving Settings in Isolated Storage" section of this chapter.

Data sharing across applications on the device varies on different software platforms. At one time there wasn't any concept of an application sandbox as a design choice, so applications simply shared data among each other using either a file or traditional operating system Interprocess Communication (IPC) methods such as sockets, messages, shared memory, and so on.

An application sandbox refers to a system mechanism that isolates an application's running process to restrict its access to the underlying system and other processes. For example, an application in a sandbox will not crash the whole operating system and cannot access data of other applications.

Today's mobile applications are mostly isolated in their own runtime sandboxes on major software platforms as a means to enhance security and reliability of the system. In addition, both Android and iOS provide a cross-application data-sharing method for applications designed to serve data to other applications. Android addresses this issue with Content Provider, a special facility to enable one application to provide data for others. On iOS, an application can register a custom URL with the system and handle data-encoded custom URLs sent from other applications. Unfortunately, at the time of writing, WP7 does not provide an on-device cross-application data-sharing method, meaning that there is no way to directly access another application's data on the same device.

On the other hand, developers can always use cloud storage and web services as a way to share data across applications on the same device and across devices. The "Saving Data to the Cloud" section of this chapter will discuss this further. In addition, all three major operating systems allow applications to access system data such as contact information, call information, calendars, and e-mails.

Local and Cloud

Mobile applications commonly combine local and cloud data as part of the data usage model. Sometimes this is the only viable approach, because the cloud data, such as RSS feeds, isn't under the developer's control.

When an application is running, it can store its data completely in memory when the data size isn't too large. If the data becomes too large to store in memory, the application relies on persistent storage on the device. In both cases, the application can communicate with cloud storage or a web service provider to download the latest data, as well as upload updates to the cloud or web service. Communication of this sort generally requires a synchronization framework that developers can use to build both the device-side and cloud-side applications to ensure data consistency without worrying about a data connection and synchronization management. Data synchronization usually relies on version numbers or timestamps for each data item.

Android provides a synchronization service and a `SyncAdapter` class that developers can use to build data synchronization with cloud storage (the server). By extending the `AbstractThreadedSyncAdapter` class and implementing the `onPerformSync(Account, Bundle, String, ContentProviderClient, SyncResult)` method, the system's sync service can perform the specified data sync operations. The device-side application must store its data in a Content Provider. On the server side, the application can store data in any structure as long as the client and the server can communicate effectively. Usually the application performs this task using a JavaScript Object Notation (JSON)–based web service (see Chapter 6 for details about using JSON).

iOS doesn't provide a synchronization framework that developers can plug in to data-processing methods. In order to enable local and cloud synchronization, the developer must build everything from scratch, but the basic idea is the same as with Android: applications poll the server, using web service calls to obtain and cache data locally, as well as to update data on the server side. Device-side data is usually saved in a SQLite database or in XML files if the data isn't too large.

On WP7, developers can leverage Microsoft's Azure synchronization framework (`www.microsoft.com/windowsazure`) to handle data synchronization. The "Using Cloud Data Services" section of this chapter discusses use of Microsoft's Azure.

Local Files and Databases

When an application has smaller data requirements (usually in the form of key-value pairs), all it really needs is a custom file or an XML file to save them. The application simply implements data serialization and de-serialization to and from the file.

iOS provides programming constructs such as `NSMutableDictionary` and `NSArray` to hold the data, and file system methods such as `NSSearchPathForDirectoriesInDomains()` and `writeToFile()` to read and write a file. You can parse XML files using the `NSXMLParser` class.

Android provides standard Java filesystem methods such as `FileOutputStream()` and `FileInputStream()` to access the application's private storage. Developers can use the `android.utils.Xml` class or Simple API for XML (SAX) in the `org.xml.sax` namespace and the `javax.xml.parsers` namespace to parse XML files.

WP7 provides similar XML-based serialization and de-serialization classes as detailed in the "Using Isolated Storage" section of this chapter. In fact, these classes not only allow serialization and de-serialization of text strings and primitive types, but also .NET objects.

In addition to using custom files or XML files, using a SQL database engine is a natural choice for applications that employ a complex data model. The advantages of using a SQL database to store and access data are:

- ➤ A database manages structured data and enables flexible access.
- ➤ A database maintains consistent data with schemes and transactions.
- ➤ Applications can leverage easy-to-use SQL commands to query and update data.
- ➤ A database provides good query and update performance.
- ➤ A database provides enhanced data security.

Both iOS and Android provide APIs to manipulate data in a SQLite database. For example, Android provides an `android.database.sqlite` package that includes such classes as `SQLiteDatabase` and `SQLiteQueryBuild`. Thus it can directly interact with a private database (accessible only from the application) using SQL commands. Note that the SQLite database created by an application is not accessible outside the application on Android.

On the other hand, iOS provides a list of SQLite functions to perform the same tasks. For example, the header file `/usr/include/sqlite3.h` includes declarations of functions such as `sqlite3_open`, `sqlite3_execute()`, and `sqlite3_close()`. In addition, `sqlite3_prepare_v2()` is the function to prepare a SQL statement, and `sqllite3_step()` is used to step through a record set.

WP7 did not provide a built-in SQL database engine at the time of this writing. Developers will have to use cloud storage that provides a database engine, or explore third-party solutions such as winphone7db (`winphone7db.codeplex.com`) or Perst Embedded Database (`www.mcobject.com/perst_eval`).

USING ISOLATED STORAGE

On WP7, each application has a private storage area in the device memory called isolated storage. Similar to common filesystem access, an application can create directories and files in its isolated storage and read data from those files, as well as delete them. The recommended approach to using isolated storage is building serialization and de-serialization capability into an application, making it easy to save and restore object collections.

> *WP7 does not allow an application to access the SD card on the device. This is different from Android, which provides APIs to enable read-and-write access to the SD card.*

Where Is Isolated Storage?

Unlike Windows Mobile, WP7 does not provide common filesystem APIs to explore and access the underlying filesystem of the device. Instead, an application can access a piece of the filesystem only through the isolated storage APIs; it has no way to access the isolated storage used by other applications. Thus applications are completely isolated from each other with respect to data access, preventing malicious applications from accessing user data as well as system data. The same design philosophy has been adopted by iOS and Android.

Isolated Storage Structure

An application's isolated storage is organized in the form of a regular directory structure, where the application can create directories and subdirectories, and files within these directories. The layout of the isolated storage is shown in Figure 5-1.

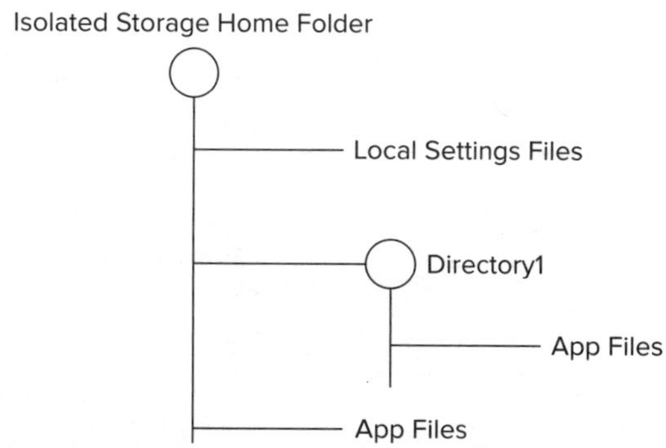

FIGURE 5-1: WP7 isolated storage

Isolated storage contains both application settings and application data. An application can save some application-specific setting data in local settings files, using APIs such as `IsolatedStorageSettings.ApplicationSettings.Add()` and `IsolatedStorageSettings.ApplicationSettings.Save()` in the `IsolatedStorage` namespace. The settings files are managed by the system, and applications do not need to open these files directly. An example of using local settings is shown later in this section.

Applications can create any directory structure in their isolated storage. The APIs in the `IsolatedStorage` namespace provide such common filesystem capabilities as:

- Creating and deleting directories
- Creating, opening, copying, and deleting files
- Enumerating directories and files in the root directory

> *If an application is uninstalled from the device, the system wipes out its isolated storage. This includes all directories, files, and application settings.*

WP7 doesn't impose a quota on the size of an application's isolated storage. However, Microsoft suggests that applications use the space only when needed and delete files when no longer needed. The system monitors the free storage space of the whole system, and prompts the user to delete files and applications when free space is below 10 percent of the available space.

Using Files in Isolated Storage

Common scenarios for using isolated storage include data access and application setting access. You gain this access by leveraging the classes in the `System.IO.IsolatedStorage` namespace. For example, the following code snippet shows how to open a file in the application's isolated storage:

```
using System.IO; //FileMode and FileAccess modes
using System.IO.IsolatedStorage;
...
try
{
    using (var store = IsolatedStorageFile.GetUserStoreForApplication())
    {
        if(store.FileExists(fileName))
        {
            IsolatedStorageFileStream fs =
            store.OpenFile(fileName, System.IO.FileMode.Open, FileAccess.READ);
            //Perform file Read operations using fs.Read()
            fs.Close();
        }

    }
}
catch (IsolatedStorageException e)
{
    //Handle the exception
}
```

To create a file in isolated storage, you set the file mode parameter in the `IsolatedStorageFile.OpenFile()` method to either `FileMode.CreateNew` or `FileMode.Create`. The difference between those two file modes is that `FileMode.CreateNew` will throw an `IsolatedStorageException` when a file of the same name exists, whereas `FileMode.Create` will open the file of the same name if it exists and truncate it to a zero length, then return to the calling method. The snippet below shows how to create a new file even if a file of the same name exists:

```
using System.IO; //FileMode and FileAccess modes
using System.IO.IsolatedStorage;
...

try
{
   using (var store = IsolatedStorageFile.GetUserStoreForApplication())
   {
      IsolatedStorageFileStream fs =
          store.OpenFile(fileName, System.IO.FileMode.Create);
      //Perform file write operations using fs.Write()
      fs.Close();
   }
}
catch (IsolatedStorageException e)
{
   //Handle the exception
}
```

Data Serialization for Isolated Storage

Developers can certainly save application data to a file in isolated storage in a custom format. In this case, developers must implement a custom serialization method and de-serialization method. A better approach is to use the data serialization APIs provided by WP7 Silverlight. The `System.Runtime.Serialization` namespace contains a number of classes that can be used to serialize and de-serialize a CLR object or an object graph (a set of related objects) to a file or a file stream in XML format. The following steps are required to use the provided serialization and de-serialization capability:

- Establish a data contract between the underlying XML serializer and the application's data class. This is done by adding the `[DataContract]` and `[DataMember]` attributes to the data class definition.
- Instantiate a `DataContractSerializer` object with the data class and file stream associated with the file, and call its `ReadObject()` and `WriteObject()` methods to perform serialization and de-serialization.

The sample application in this chapter uses a `NoteEntry` class to model the user's data. The `NoteEntry` class consists of an `int Id` field, a `string Text` field, and a `LastUpdateTime` field of type `DateTime`. Listing 5-1 shows the `NoteEntry` class used for serialization.

The chapter provides the `AppDataSample` application that demonstrates the use of local data storage and cloud storage on WP7. Because the application requires a web service reference (which is provided by another sample project, `SampleCloudStorage`), its data model class is actually automatically generated by the IDE when you add the web service reference to the project.

Note that because the AppDataSample application is a bit complicated, this section will not be able to cover the complete step-by-step creation of this project. Readers are strongly encouraged to download and compile the application code from the book's website and go through the section with the project loaded in IDE.

Available for download on Wrox.com

LISTING 5-1: An example of data class used for data serialization, AppDataSample\AppDataSample\NotesData.cs

```
using System.Collections.Generic;
using System.Runtime.Serialization;
using System.IO.IsolatedStorage;
...
[DataContract]
public class NoteEntry
{
    //"id" field
    private int id;
    [DataMember(Name = "id")]
```

continues

LISTING 5-1 *(continued)*

```
    public int Id
    {
        get { return id; }
        set { id = value; }
    }

    //"lastUpdateTime" field
    private DateTime lastUpdateTime;
    [DataMember(Name = "lastUpdateTime")]
    public DateTime LastUpdateTime
    {
        get { return lastUpdateTime; }
        set { lastUpdateTime = value; }
    }

    //"text" field
    private string text;
    [DataMember(Name = "text")]
    public string Text
{
        get { return text ?? ""; }
        set { text = value; }
    }
}
```

With the data contract in place, the application can easily serialize a list of `NoteEntry` objects into a file stream and de-serialize them back to a collection of such objects. Listing 5-2 shows an example of serialization and de-serialization for a list of `NoteEntry` objects.

LISTING 5-2: An example of data serialization, AppDataSample\AppDataSample\NotesData.cs

Available for download on Wrox.com

```
public class NotesData
{
    private static List<NoteEntry> noteList = new List<NoteEntry>();
    private static string fileName = "Notes.xml";

    public static IEnumerable<NoteEntry> GetBindingData()
    {
        return noteList;
    }

    public static void AddOneEntry(NoteEntry note)
    {
        if (note == null) return;
        note.Id = noteList.Count + 1; //Note id starts from 1
        noteList.Add(note);
    }
    public static void SaveToFile()
    {
```

```
    if (noteList.Count < 1) return;

    try
    {
       using (var store = IsolatedStorageFile.
          GetUserStoreForApplication())
       {
          IsolatedStorageFileStream fs =
              store.OpenFile(fileName, System.IO.FileMode.Create);
          DataContractSerializer ser =
              new DataContractSerializer(
                  typeof(List<NoteEntry>));
          ser.WriteObject(fs, noteList);
          fs.Close();
       }
    }
    catch (SerializationException serEx)
    {
       MessageBox.Show(serEx.Message);
    }
    catch (IsolatedStorageException e)
    {
       MessageBox.Show(e.Message);
    }
}

// Read all notes from file, refresh the in-memory list
public static void ReadFromFile()
{
    try
    {
        using (var store =
           IsolatedStorageFile.GetUserStoreForApplication())
        {
            if(store.FileExists(fileName))
            {
                IsolatedStorageFileStream fs =
                store.OpenFile(fileName,
                FileMode.Open,
                FileAccess.Read);

                IEnumerator<NoteEntry> noteEnum =
                   noteList.GetEnumerator();
                DataContractSerializer ser =
                    new DataContractSerializer(
                        typeof(List<NoteEntry>));
                noteList = (List<NoteEntry>)ser.ReadObject(fs);
                fs.Close();
            }

        }
    }
    catch (SerializationException serEx)
    {
       MessageBox.Show(serEx.Message);
    }
```

continues

LISTING 5-1 *(continued)*

```
            catch (IsolatedStorageException e)
            {
                MessageBox.Show(e.Message);
            }
        }
    }
```

The static list `List<NoteEntry> noteList` of the `NotesData` object contains both notes data retrieved from the application's isolated storage when the application starts, and the user's input data. The sample application performs de-serialization by calling `ReadFromFile()` to build the initial `noteList` from `notes.xml` in isolated storage. Once the user adds a new note, the code creates a new `NoteEntry` object and adds it to `noteList`. When the application is deactivated or stopped, `noteList` is de-serialized to the same data file in isolated storage.

Saving Settings in Isolated Storage

You can save application settings to isolated storage by using the `IsolatedStorageSettings` class. The `IsolatedStorageSettings.ApplicationSetting` property contains a dictionary object that applications can use to save key-value pairs. The following code snippet, excerpted from the AppDataSample project, shows how to save a key-value pair to isolated storage:

Available for download on Wrox.com

```
using System.IO.IsolatedStorage;
...
try
{
    var appSettings = IsolatedStorageSettings.ApplicationSettings;
    if (!appSettings.Contains(SyncSettingKey))
    {
        appSettings.Add(
            SyncSettingKey,
            SyncToCloudWhenSave);
    }
    else
    {
        appSettings[SyncSettingKey] = SyncToCloudWhenSave;
    }

    appSettings.Save();

    return true;
}
catch (Exception e)
{
    MessageBox.Show(e.Message);
    return false;
}
```

Code snippet AppDataSample\AppDataSample\AppSettings.cs

The code creates a key-value pair from *`SyncSettingKey`* (the key) and *`SyncToCloudWhenSave`* (the corresponding value). The sample application uses this key-value pair to hold a user setting. Adding the pair to isolated storage is as simple as adding it to the dictionary object. Note that the code must call the `IsolatedStorageSettings.Save()` method to save the in-memory `IsolatedStorageSettings` instance to device storage.

The following code snippet shows retrieving the same key-value pair from isolated storage:

Available for download on Wrox.com

```
using System.IO.IsolatedStorage;
...
try
{
    var appSettings = IsolatedStorageSettings.ApplicationSettings;
    if (!appSettings.Contains(SyncSettingKey))
    {
        return false;
    }
    else
    {
        return (bool)appSettings[SyncSettingKey];
    }
}
catch (Exception e)
{
    MessageBox.Show(e.Message);
    return false;
}
```

Code snippet AppDataSample\AppDataSample\AppSettings.cs

SAVING DATA TO THE CLOUD

Many mobile applications use cloud storage to save user data so the user won't lose it even after the application is uninstalled from the device. There are also applications that allow multiple types of clients, such as a desktop application, a website, and device applications on various platforms, to access and manage the same data set. In both cases, developers must design a unified data model that all client applications use, and web services that provide access to the cloud storage.

Building a Cloud Data Service

Microsoft provides Windows Azure service as a cloud storage solution, but developers are free to choose any web service provider. This section discusses using the Windows Azure development environment to build, debug, and deploy cloud web services for WP7 applications quickly. You'll also find that applications on other software platforms can use the web services you create.

> *Server-side development, such as building Windows Azure Services, requires knowledge of specialized tools to implement web service methods and related database operations. This section provides an overview of building a Windows Azure service for WP7 applications. However, interested readers should also review the documentation for the tools used to create the example. For more details, see* `http://www.microsoft.com/windowsazure/getstarted/`.

Windows Azure Service

The first step towards building a Windows Azure cloud service is to download Windows Azure SDK and Tools for Visual Studio (`http://www.microsoft.com/downloads/en/details.aspx?FamilyID=7a1089b6-4050-4307-86c4-9dadaa5ed018`). Be sure you follow the instructions on the web page because there are other software products you have to install before using the Windows Azure SDK and tools. For example, for a basic setup, you need a copy of Visual Web Developer 2010 Express (development environment, downloadable at `www.microsoft.com/express/Downloads/#Visual_Studio_2010_Express_Downloads`) and SQL Server 2008 Express (database engine, downloadable at `www.microsoft.com/express/Downloads/#SQL_Server_2008_R2_Express_Downloads`) to complete the example. You can, of course, use the full version of Visual Studio and SQL Server for the same purpose.

The Windows Azure SDK and Tools installed in Visual Web Developer 2010 Express enables developers to build and debug web services locally by packaging and deploying web services directly to a built-in web server. When you're ready to deploy the web service project, you must obtain a Windows Azure account and then use the Visual Web Developer 2010 Express tool to deploy the services to Windows Azure. (At the time of this writing, you could obtain a free trial of Windows Azure at `http://www.microsoft.com/windowsazure/free-trial/` until June 30, 2011, and Microsoft will almost certainly extend the free trial beyond this point.)

> *You might see the following error message when you open the chapter's cloud service sample in Visual Web Developer 2010 Express:*
>
> *"SampleCloudService.ccproj cannot be opened because its project type (.ccproj) is not supported by this version of the application. To open it, please use a version that supports this type of project."*
>
> *Note that this is an issue with the Windows Azure Tools. To fix this problem, simply do a "devenv /resetSkipPkgs" on the command line. The online forum* `http://social.msdn.microsoft.com/Forums/en/windowsazure` *provides help for other issues with Windows Azure Tools.*

Building a Windows Azure Service

To build a web service with cloud storage that a WP7 device application can use, create a "Windows Azure project" in Visual Web Developer 2010 Express by selecting "Cloud" under the Visual C# template list and "Windows Azure Project." Next, when prompted with a list of .NET Framework Roles for the solution, add a "Windows Communication Foundation (WCF) Service Web Role" to the solution. The web role refers to different types of services that a Windows Azure project can offer. Once you perform these two steps, the IDE automatically creates a basic web service project, and the default web role project name is `WCFServiceWebRole1`.

The key components of a Windows Azure service project appear below, along with the automatically generated filenames:

- **Service interface:** Defines the service contract. This component appears in `IService1` in `IService1.cs`. Add those web service methods that you want to provide into this interface. If a data model (a composite type) is used on both the service side and the device side, you should add a data contract class to this file as well.
- **Service class:** Implements the service methods. This appears in `Service1.svc.cs` for class `Service1` that implements the service interface `IService1`).
- **Data folder:** This is a folder named `App_Data` where data files or a SQL database can be created.

An example of a service interface, `IMyNoteService`, is shown below. This is the service interface that you'll use when working with the sample project, `SampleCloudService`, in this chapter.

Available for download on Wrox.com

```
[ServiceContract]
public interface IMyNoteService
{
    [OperationContract]
    int PushNotesData(List<NoteEntry> listNote);

    [OperationContract]
    List<NoteEntry> PullNotesData();


    [OperationContract]
    int GetNoteCount();

    [OperationContract]
    int DiscardAll();
}
```

Code snippet CloudServiceSample\CloudServiceSample\IMyNoteService.cs

The methods listed in the `IMyNoteService` interface are those that a web service client can call to perform a specific operation. As can be seen in the next section, a WP7 application project in Visual Studio 2010 Express can call those methods just as it would call methods of a regular class in the project. The application gains this access by adding a web reference to the web service into the WP7 application project. Visual Studio 2010 Express automatically generates proxy classes for the referenced web services.

Now it's time to discuss the server side in more detail. Visual Web Developer 2010 Express deploys the web service to a built-in web server running locally at 127.0.0.1 (the localhost). Once the service is successfully deployed, a web page is automatically launched to show the service URL, such as `http://127.0.0.1:84/MyNoteService.svc`. This URL is used in the device development environment to obtain a web reference and create the proxy classes.

Using Cloud Data Services

Use the following steps to add a service reference to a WP7 project:

1. Right-click the References entry in Solution Explorer and choose Add Service Reference from the context menu. The Add Service Reference dialog (Figure 5-2) will appear.
2. Enter the service URL obtained from the service provider or from the service development environment. Then choose the desired service classes to generate. For example, the sample project AppDataDemo uses services provided by another sample project CloudStorageDemo. Figure 5-3 shows the service class `MyNoteService` and its operations that the AppDataDemo project chooses to use.
3. Give the generated proxy classes a namespace name. The example uses <add specific information here> as a namespace name.

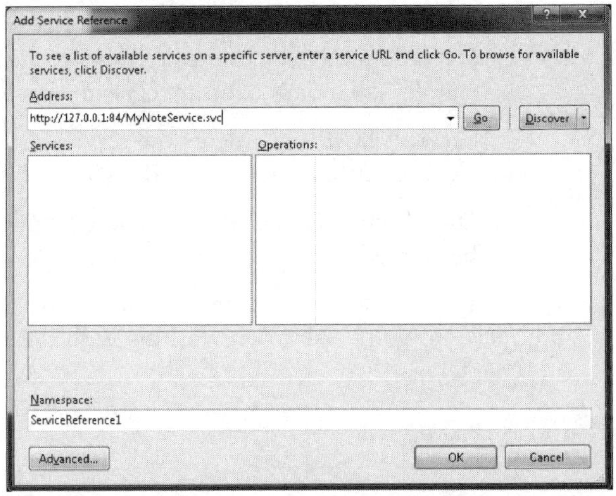

FIGURE 5-2: The Add Service Reference dialog to a project

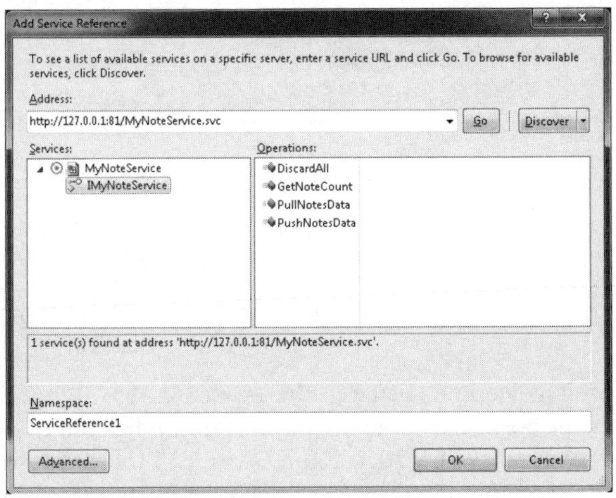

FIGURE 5-3: Select service references

Once the IDE generates the proxy classes, it adds them to the Object Browser. For example, if the server-side service interface name is `IMyNoteService` and the service class name is `MyNoteService`, the client-side proxy class name is `MyNoteServiceClient`. There are also proxy classes that you use in asynchronous callback methods as parameters. All these classes are shown in the file `reference.cs`, which is automatically generated when you add the web service reference.

This chapter uses a custom web service that is dedicated to the example WP7 application. For consuming standard web services, such as those provided by Facebook and Twitter, please see Chapter 6.

Calling Azure Service Asynchronously

Web service calls in the service proxy classes are all performed using an asynchronous pattern. For example, if a web service call method is named `PushNotesData` as shown in the service contract on the server side, then in the WP7 IDE the service proxy class has two methods for this specific web service call: the `PushNotesDataAsync()` method that initiates the call, and a custom callback method that the system calls when the web service call completes.

The following code Listing 5-3 shows an example of calling `PushNotesDataAsync()` of the proxy class `MyNoteServiceClient` from within the sample WP7 application `AppDataSample`. Note the use of an asynchronous call pattern in this example.

Available for download on Wrox.com

LISTING 5-3: An example of data class used for data serialization, AppDataSample\AppDataSample\NotesData.cs

```
public static void SaveAllToCloud()
{
    try
    {
        MyNoteServiceClient client =
                new MyNoteServiceClient();
        client.PushNotesDataCompleted +=
            new EventHandler<PushNotesDataCompletedEventArgs>
                (Client_PushCompleted);

        client.PushNotesDataAsync(
            NotesData.ToObservableCollection());

        // Always close the client.
        client.CloseAsync();
    }
    catch (Exception e)
    {
        MessageBox.Show(e.Message);
    }
}

public static void Client_PushCompleted(
    object sender,
    PushNotesDataCompletedEventArgs e)
{
    if (e.Error != null)
    {
        MessageBox.Show(e.Error.Message);
    }
    else
    {
        Debug.WriteLine("Data push succeeded");
    }
}
```

In the above example, the code doesn't use the return value of the `PushNotesData()` web method. The web service does provide a return value as the `PushNotesDataCompletedEventArgs` object `e`, which is passed back to the callback method `Client_PushCompleted()`.

Microsoft Sync Framework

One of the exciting capabilities of WP7 application development is the data synchronization framework provided with the Windows Azure cloud service. The rationale behind providing this capability is quite simple; in contrast to Apple and Google, Microsoft has a large product line that offers complete software and service for the three screens that most developers work with today: desktop, mobile, and web. Each of these screens is powered by Microsoft database products and programming facilities such as Silverlight and WCF.

For a WP7 application that requires a server data store, the Windows Azure web service platform provides SQL Azure (`http://msdn.microsoft.com/en-us/library/gg521148.aspx`), a cloud-based relational database built on top of Microsoft SQL Server technologies. In addition, Microsoft provides Sync Framework that enables easy data synchronization between a WP7 Silverlight application and a SQL Azure instance.

The Microsoft Sync Framework adopts the Open Data (OData) standard and the OData Sync protocol for data exchange between a sync client and the sync end points in the cloud. The Sync Framework provides APIs that enable data conversion between a wide range of data sources (XML, JSON, etc.) and OData. The basic model of data synchronization on WP7 consists of the following components and relationships:

- **On the device:** Application data in isolated storage
- **On the device:** WP7 sync provider
- **In the cloud:** SQL Azure sync end points and data stores

An application uses sync framework APIs to interact with the WP7 sync provider that can handle synchronization with the back-end data store. For more information about using Sync Framework for WP7, please visit MSDN site `http://msdn.microsoft.com/library/gg507824.aspx`.

This chapter's sample applications `AppDataSample` and `SampleCloudService` provide a simple local-cloud storage model where both the cloud and the device client use an XML file to store user data. You can extend the example to use a SQL database in the cloud, which allows more structural data manipulation.

DATA STORAGE DESIGN CONSIDERATIONS

A critical problem with using local data is UI performance. A general rule of thumb is to avoid any storage access when the device is busy rendering graphics and media. If blocking data access is not absolutely needed, then it should be done in a background thread with reduced thread priority whenever possible. In addition, Microsoft recommends using buffered file access to reduce I/O operations.

The problem with the combination of local data and cloud storage occurs when there are multiple copies of the same data. It becomes a challenge to manage and update these copies, while at the same time providing the user with good performance. The places where a copy can reside include:

- Data hosted in cloud storage, which can be as simple as some XML files, or as complex as a database consisting of a number of tables, views, or other data elements.
- Data on supported mobile devices, which can be a SQLite database created by an Android client or an iOS client. Or a data file created and managed by a WP7 client.

- Data on a desktop or a laptop computer, which can be created and managed by a stand-alone desktop application. This is a rare situation, as many applications choose to provide web-based access rather than a stand-alone desktop application.
- Data managed by an HTML5 web browser, as offline storage is one of the new features supported by the new HTML5 standard.

For each data set, you must exercise care to resolve synchronization conflicts between existing data and a received update. One common approach is to use timestamps to compare the freshness of the data and choose the latest update. Another approach just prompts the user to determine what to keep, and then updates local data and the cloud accordingly.

Cloud data generally serves as the central data source so other data sets on clients can keep synchronizing with it. On a mobile device, an application must handle data in at least three locations, as shown in Figure 5-4: local database or files, memory data, and cloud data.

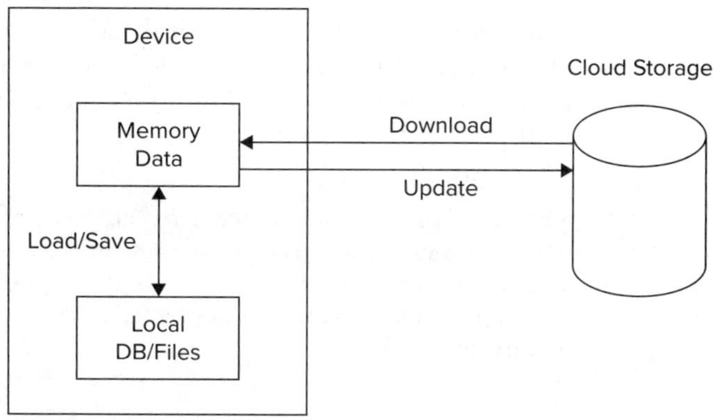

FIGURE 5-4: Application data model

The application can load data from a local database or a local file into memory, as well as save data to the database or the file. It may download data from cloud storage and update that with changes done on the device. It becomes a challenging issue to make better use of these assets to provide a good user experience and performance. There are a few synchronization strategies that you can apply to the scenario, specifically for client application design, as discussed below:

- **Real-time update:** Any update on cloud data is pushed to the client by using push notifications. If the device application is running, then device local data is updated immediately; if it's not running, then a push notification is sent to the device and gives the user a chance to launch the application and update the data. See Chapter 6 for more information on push notifications on the three platforms.

  Conversely, any client-side update is pushed to the cloud right after a change has been made on the local copy.

  This strategy is highly dependent on network connectivity. Also, frequent changes to the presented data on the UI may confuse the user. This strategy is required by those applications that offer real-time information such as flight information, game scores, and so on, since data freshness is of paramount importance to these applications.
- **Client start-stop update:** A device application downloads the latest update from the cloud when it starts to run and uploads any changes to the cloud when it stops. Unless a user initiates a data synchronization operation, applications simply use the downloaded local copy while the application runs.

Many news and magazine applications that work well in offline mode and do not require real-time updates have adopted this strategy. These applications may react to network connectivity events and try to download the update from the cloud when network connectivity resumes. The downside is that application launching performance suffers when the initial update is slow. This is a problem across WP7, iOS, and Android. A work-around is to load local data first to unblock the UI thread, and then launch a background thread (such as using .NET ThreadPool) to perform cloud synchronization, and update the UI when the synchronization is done.

- **Scheduled update:** An application can use a schedule to perform data synchronization, and a user can also specify periodical one-way or two-way updates. This is technically possible for applications on Android, where the system can create a background service to handle scheduled updates, even if the application itself isn't running. There are two variants of this approach: one is to design the client service process so it launches periodic polls to its server for updates, and the other is to maintain a long-lived heartbeat connection between the client and the server so the server can push updates at any time without clients' solicitation.

  This strategy offers a balance between data refreshes and offline access, and has been used by many e-mail and social applications. This strategy has an impact on battery life as frequent updates may drain the battery quickly. A WP7 application cannot create a service to perform scheduled updates all the time, but it can do this when the application is running.

SUMMARY

This chapter has discussed application data storage for a WP7 application. WP7 provides isolated storage and serialization classes so an application can save and restore data easily. The isolated storage local setting class is used to store application settings. Microsoft suggests using cloud storage because WP7 doesn't provide an on-device database engine. The chapter shows how to create a Windows Azure web service and use it to back up and restore data.

There are a few platform-agnostic design issues to consider when building applications on all three major platforms, such as choosing a good data synchronization strategy against performance and data freshness, while providing offline access. The chapter provides some design considerations for these issues. There are other interesting issues in this field that the chapter doesn't cover, such as the use of third-party synchronization frameworks and data providers for WP7 and Android. Interested readers are encouraged to learn more by following the links in the chapter.

This chapter has been about using a hosted web service. The next chapter is dedicated to consuming standard HTTP web services and push notifications. Both of these data sources are important topics for web-enabled mobile applications.

6

Web Services and Push Notifications

WHAT'S IN THIS CHAPTER

- Consuming web services
- Parsing data
- How to manage push notifications
- Understanding interstitial ads
- How to create and display ads

Today's smartphones come with mobile data service and the traditional voice call service. Wi-Fi is also a common component and widely used at home or in public places. This ubiquitous connectivity provides opportunities for mobile applications to access the Internet in various ways. Many websites offer web service APIs that a client can consume regardless of what software platforms the application runs on. This chapter looks into what Windows Phone 7 (WP7) provides for consuming web services. You'll learn about the push notification service, a key service that can post data to applications on devices for a web service. This chapter also explores mobile advertising and major mobile advertising providers.

USING WEB SERVICES

How do you consume popular web services on WP7? This section will answer that question. For those not familiar with web service technologies, a primer of web services is provided. Developers who have used web services on iOS and Android will find it useful to leverage their knowledge of those platforms in order to move to WP7. Therefore the section will also provide a brief coverage of using web services on iOS and Android, followed by a detailed discussion of making web service calls and parsing web service data on WP7.

A Primer of Web Services

This section will introduce key concepts in the web service domain, including the two major web service architectures: RESTful and SOAP, and the two dominating web service data formats: JSON and XML.

RESTful and SOAP Web Services

Two of the types of web service architectures used today are Representational State Transfer (REST) web services and Simple Object Access Protocol (SOAP) XML (eXtensible Markup Language) web services. People in the industry always refer to web services using this architecture as "RESTful" web services. With RESTful web services, resource (data) is presented as Uniform Resource Identifiers (URIs), upon which HTTP protocol methods such as GET, POST, PUT, and DELETE operate to consume the data or perform an operation. A typical use case of a RESTful call is direct communication between a client browser and a web server.

On the other hand, SOAP XML web services are more complicated and suitable for enterprise environments in which multiple nodes exist between the message sender and the receiver. A SOAP message is an XML document with structured data to be placed in the HTTP body; therefore, it is generally larger than a RESTful HTTP message.

SOAP web services have existed as a major web service architecture for more than 10 years; however, RESTful web services are becoming more popular among major websites because of their resource-centric rationale and developer-friendly URI-based interface. Both require the client application to know the structure of the expected web response data before parsing it. For RESTful, this is usually documented at the web service website, as shown in Table 6-1. For SOAP XML web services, Web Service Description Language (WSDL) is the common language that describes the provided web services.

Social network applications are among the most popular applications on smart devices such as phones and tablets. Table 6-1 is a summary of major social websites that offer web services.

TABLE 6-1: A Summary of Major Social Websites

WEBSITE	WEB SERVICE	EXAMPLE
Twitter	http://dev.twitter.com/doc	Query latest tweets about WP7: http://search.twitter.com/search.json?q=wp7
Facebook	http://developers.facebook.com/docs/reference/api/	Query President Obama's profile: https://graph.facebook.com/barackobama
Flickr	www.flickr.com/services/api/	Get a list of photos with a tag "wp7": www.flickr.com/services/rest/?api_key=valid_api_key &auth_token=&method=flickr.photos.search&tags=wp7
Youtube	http://code.google.com/apis/youtube/getting_started.html#data_api	Get top 10 playlists matching the term "wp7": http://gdata.youtube.com/feeds/api/playlists/snippets?q=wp7&start-index=1&max-results=10&v=2
MySpace	http://wiki.developer.myspace.com/	Search people profile containing "android": http://api.myspace.com/opensearch/people?searchTerms=android

You might frequently search for information and news web services on a mobile device. Table 6-2 shows examples of these services.

TABLE 6-2: Examples of Information and News Web Services

WEBSITE	WEB SERVICE	EXAMPLE
Weather underground	http://wiki.wunderground.com/index.php/API_-_XML	Get weather forecast for Seattle, WA, in XML format: http://api.wunderground.com/auto/wui/geo/ForecastXML/index.xml?query=Seattle,WA
Google Places	http://code.google.com/apis/maps/documentation/places/	Get local information of a given location: https://maps.googleapis.com/maps/api/place/search/json?location=40.717859,-73.957790&radius=200&client=*clientId*&sensor=*true_or_false*&signature=*SIGNATURE*
Google Financial Service	http://code.google.com/apis/finance/docs/2.0/reference.html	Get portfolio feed: http://finance.google.com/finance/feeds/default/portfolios#id
Yelp	http://www.yelp.com/developers/documentation	http://api.yelp.com/business_review_search?term=pizza&location=Redmond%2A%20WA&ywsid=XXXXXXXXXXXXXXX

JSON and XML

RESTful web services use data payloads such as JavaScript Object Notation (JSON) and XML. Both are text-based data exchange formats. RFC 4627 defines JSON (`http://www.ietf.org/rfc/rfc4627.txt`), which is the dominating data format today because most web services support payload data in JSON format only, whereas some support both JSON and XML. This means that on mobile software platforms, developers need some helper classes to parse the JSON or XML payload returned from a web service.

JSON defines a simple representation of structured data with four primitive types (string, number, boolean, and null), and two structured types (object and array). An object element can contain primitive types or other objects and arrays. An example of a JSON message obtained from a Twitter web service follows. For simplicity, the message has been shortened with a reduced number of data items.

```
{
    Results: [
                    {
                     from_user_id: "1234567",
                     Profile_image_url:image_url,
                     From_user: "JohnDoe"
```

```
                }
                 {
                   ...
                 }
               ],
               query: "wp7",
               since_id: 0
}
```

In the preceding example, the message is written in an object named `Results` wrapped in curly brackets (`{}`). The `Results` object in turns contains an array, which is wrapped in square brackets (`[]`), and two primitive types: `query` and `since_id`. The array includes several objects that carry specific data fields.

CONSUMING WEB SERVICES ON IPHONE AND ANDROID

Fundamentally, consuming a web service encompasses two basic functions: making web-service calls and parsing response data. Both iOS and Android provide programming support for these basic scenarios so that developers can build sophisticated applications following published web service API specifications, such as those provided by Twitter and Facebook.

Consuming web services on the iPhone Operating System (iOS) is simple: the iOS SDK provides classes such as `NSURLConnection`, `NSMutableURLRequest` for HTTP-based web-request and response operations. The HTTP URL and associated headers are set in an `NSURL` object and an `NSMutableURLRequest` object. The delegate method for `connectionDidFinishLoading` processes the web response data found in the `NSMultableData` property.

If the response data received by the delegate is in JSON format, you can use some third-party JSON classes such as json-framework (`http://code.google.com/p/json-framework/`) or TouchJSON (`https://github.com/schwa/TouchJSON`) to parse it into a dictionary, because the iOS SDK does not provide a built-in JSON parser. On the other hand, if the response data is XML-based, you can use the iOS `NSXMLParser` class to parse it.

In some cases the message payload for either the request or response is large. You can add a compression HTTP header to provide better application performance.

Android offers more options for developers to implement web service calls:

- `HttpClient` classes in the `Apache` package (`Org.apache.http.client.HttpClient` interface and `org.apache.http.impl.client.DefaultHttpClient` class). A synchronous invocation looks like this:

```
HttpResponse org.apache.http.client.HttpClient.execute(HttpUriRequest
request)
throws IOExc.eption, ClientProtocolException
```

Here is an asynchronous invocation:

```
HttpResponse org.apache.http.client.HttpClient.execute (HttpUriRequest
request,ResponseHandler<? extends T> responseHandler)
```

➤ `HttpURLConnection` class in `java.net` package. Here is how you use this class:

```
URL url = new URL(request_url);
HttpURLConnection connection = (HttpURLConnection) url.
openConnection();
connection.setDoOutput(true);
connection.setRequestMethod("POST");
connection.setReadTimeout(20000);
connection.setConnectTimeout(20000);
OutputStream out = connection.getOutputStream();
```

For JSON messages, the Android SDK provides the `org.json` package for easy parsing. Here is an example of using `org.json` to parse a web response:

```
JSONObject json = new JSONObject(response_string);
JSONArray names = json.names();
JSONArray values = json.toJSONArray(names);
for(int i=0;i<valArray.length();i++)
{
     Log.i("JSON name: "+nameArray.getString(i)+" \n"
         +"JSON value:"+valArray.getString(i)+"\n");
}
```

Consuming Web Services on WP7

There are essentially two issues to consider when consuming web services from a mobile application:

- Performing HTTP web service calls synchronously and asynchronously
- Handling different data formats used by the web service

Some of the sample code snippets in this section are excerpted from the sample project SocialDemo, which is a simplified Twitter web service client. The SocialDemo sample project showcases a typical HTTP web client pattern, Open Authorization (OAuth) workflow pattern, as well as JSON data parsing and XML to Language-Integrated Query (LINQ) mapping. To try the sample project, you have to replace the application's consumer key and consumer secret key in the `TwitterConfig` class with the keys you registered at the Twitter developer site. It's helpful to have a basic understanding of OAuth 1.0a before you begin this example. For a good jumpstart guide, check out the Twitter OAuth sample at `http://dev.twitter.com/pages/auth`, and a getting-started guide at `http://oauth.net/documentation/getting-started`.

Making HTTP Calls

On WP7, you can use two classes to make an HTTP call:

- `System.Net.WebClient`
- `System.Net.HttpWebRequest`

The following sections describe how to use these two classes.

Using the System.Net.WebClient Class

The `WebClient` class provides two simple asynchronous methods to access web services with HTTP GET: `DownloadStringAsync()`, which places the response string in a parameter passed to the `DownloadStringCompleted()` event handler, and `OpenReadAsync()`, which uses a `Stream` object as a parameter of the `OpenReadComplete()` event handler. `WebClient` has limited HTTP control, making it mostly useful for simple web calls. The following example shows how to retrieve Twitter timeline tweets of a specific user in JSON format:

Available for download on Wrox.com

```
using System.Net;
...
try
{
    //Twitter streamline retrieval
    WebClient twitter = new WebClient();

    twitter.DownloadStringCompleted +=
        new DownloadStringCompletedEventHandler(
            twitter_DownloadStringCompleted);
    if(TwitterConfig.useJson == true)
        twitter.DownloadStringAsync(
            new Uri(TwitterConfig.StreamlineUri_json +
                TwitterConfig.screen_name));
    else
        twitter.DownloadStringAsync(
            new Uri(
                TwitterConfig.StreamlineUri_xml +
                TwitterConfig.screen_name));
}
catch (WebException we)
{
    Dispatcher.BeginInvoke(() =>
    {
        MessageBox.Show(we.Message);
    });
}
```

Code snippet SocialDemo\SocialDemo\Mainpage.xaml.cs

The example above specifies a `DownloadStringCompletedEventHandler` method named `twitter_DownloadStringCompleted()`. Note that it checks a `TwitterConfig` object used in the SocialDemo project to determine what URI (either JSON-based or XML-based Twitter streamline URI) is required for the `DownloadStringAsync()` call. You can also use the `WebClient` class for HTTP POST by using the `UploadStringAsync()` and `OpenWriteAsync()` methods.

> *Web service calls usually support JSON, XML, Really Simple Syndication (RSS), and Atom as the response data formats. The API URL should indicate which format a client expects to receive. For example, the URL of a Twitter user timeline API is defined as follows:*
>
> ```
> http://api.twitter.com/version/statuses/user_timeline.format
> ```
>
> *where the "format" string can be json, xml, rss, or atom.*

Using the System.Net.HttpWebRequest Class

Unlike `WebClient`, which adopts an event-based model for asynchronous communication, `HttpWebRequest` requires a `System.AsyncCallback` delegate. The requirement to use a delegate makes using `HttpWebRequest` more complex than using the event-based `WebClient`. `HttpWebRequest` is derived from `System.Net.WebRequest`. It provides fine-grained control over HTTP headers with class properties and the underlying asynchronous communication states. The following is an example of using `HttpWebRequest`. It uses two asynchronous patterns with corresponding `AsyncCallback` methods: one is `HttpWebRequest.BeginGetResponse()` and the other is `System.IO.Stream.BeginRead()`.

Available for download on Wrox.com

```
using System.IO; // Stream class
using System.Diagnostics; // Debug class
using System.Text; // Encoding class

// State class to pass data to async call backs
public class RequestState
{
    public const int BUFFER_SIZE = 2048;
    public byte[] buffer;
    public WebRequest request { get; set; }
    public WebResponse response { get; set; }
    public Stream responseData;
    public RequestState()
    {
        buffer = new byte[BUFFER_SIZE];
        request = null;
        responseData = null;
    }
}

public class HttpCall
{
    public void StartAsyncCall()
    {

        try
```

```
{
    System.Uri uri = new Uri("http://www.msdn.com");
    HttpWebRequest myHttpWebRequest =
        (HttpWebRequest)WebRequest.Create(uri);

    // Create a RequestState object to track state changes
    RequestState myRequestState = new RequestState();
    myRequestState.request = myHttpWebRequest;

    // Start an asynchronous Http request.
    IAsyncResult result =
        (IAsyncResult)myHttpWebRequest.BeginGetResponse(
        new AsyncCallback(ResponseCallback), myRequestState);

    // Now the main thread can wait
    // The MSDN online example has the "allDone.WaitOne()" to
    // keep the main thread waiting. Do NOT use it. It will block
    // your async threads from being called.

}
catch (WebException webe)
{
    Debug.WriteLine("Web exception status: " + webe.Status);
}
catch (Exception e)
{
    Debug.WriteLine(e.Message);
}
```

Code snippet SocialDemo\SocialDemo\HttpCall.cs

The corresponding call-back method for the HTTP request is shown below. This method ends the HTTP request and issues another asynchronous method, `Stream.BeginRead()`, to obtain the response data. The name of the call back in this case is `ReadDataCallBack`, which is also shown below.

Available for download on Wrox.com

```
private void ResponseCallback(IAsyncResult asynchronousResult)
{
    try
    {
        RequestState myRequestState =
            (RequestState)asynchronousResult.AsyncState;
        HttpWebRequest myHttpWebRequest2 =
            (HttpWebRequest)myRequestState.request;

        // Get the response stream
        myRequestState.response = myHttpWebRequest2.
            EndGetResponse(asynchronousResult);
        Stream responseStream =
            myRequestState.response.GetResponseStream();
        myRequestState.responseData = responseStream;

        // Start async call to read the response data
        // when the buffer is filled, the call back will be called
```

```
            IAsyncResult asynchronousInputRead =
                responseStream.BeginRead(
                myRequestState.buffer,
                0,
                RequestState.BUFFER_SIZE,
                new AsyncCallback(ReadDataCallBack), myRequestState);
        }
        catch (WebException webe)
        {
            Debug.WriteLine("Web exception status: " + webe.Status);
        }
        catch (Exception e)
        {
            Debug.WriteLine(e.Message);
        }
    }
```

Colde snippet SocialDemo\SocialDemo\HttpCall.cs

The following is the `ReadDataCallback()` method which eventually completes reading of the response data and closes the stream.

Available for download on Wrox.com

```
    private void ReadDataCallBack(IAsyncResult asynchronousResult)
    {
        try
        {
            RequestState myRequestState =
                (RequestState)asynchronousResult.AsyncState;
            Stream responseStream = myRequestState.responseData;
            int read = responseStream.EndRead(asynchronousResult);

            // Read the HTML stream
            if (read > 0)
            {
                // There is still data in the buffer, get it
                Debug.WriteLine(
                    Encoding.UTF8.GetString(myRequestState.buffer, 0, read));

                // Launch another read
                IAsyncResult result = responseStream.BeginRead(
                    myRequestState.buffer,
                    0, RequestState.BUFFER_SIZE,
                    new AsyncCallback(ReadDataCallBack),
                    myRequestState);
            }
            else
            {
                // Done. Close the stream and response.
                responseStream.Close();
                myRequestState.response.Close();
            }
        }
        catch (WebException webe)
```

```
        {
            Debug.WriteLine("Web exception status: " + webe.Status);
        }
        catch (Exception e)
        {
            Debug.WriteLine(e.Message);
        }
    }
```

Code snippet SocialDemo\SocialDemo\HttpCall.cs

Parsing JSON Data

Parsing JSON data is done using the `System.Runtime.Serialization.Json.DataContractJsonSerializer` class (`System.Servicemodel.Web.dll`) in Silverlight on WP7. `DataContractJsonSerializer` uses the underlying serialization engine in Microsoft Windows Communication Foundation (WCF), which has been discussed in Chapter 5. This class is able to deserialize a JSON string into a list of objects of a custom type that you define to match the data format in the string. It can also serialize these objects into a JSON string, although this feature isn't used very often for the examples in this book.

The key to using `DataContractJsonSerializer` is to parse JSON data returned from a web service by establishing a *data contract* for a published web service API between the client (the application) and the web service. To perform this task, you create a JSON data item class, which is the contract class that matches the defined JSON data format of the web call response. The `DataContractJsonSerializer` class can infer the data contract using the contract class, and thus can obtain a list of such class objects from the response data.

For example, as mentioned above, Twitter provides a web service call that returns the timeline — the latest tweets — of a given screen name. The following HTTP GET call will return the latest tweets from President Obama:

```
http://api.twitter.com/1/statuses/user_timeline.json?screen_name=barackobama
```

To see the response from Twitter in a web browser, install the JSONView add-on (`https://addons.mozilla.org/en-US/firefox/addon/jsonview/`) for Firefox, and then enter the URL. One example of the response for this call is shown below. To save space only some of the data items are shown.


```
[

        {
            o in_reply_to_status_id: null
            o truncated: false
            o created_at: "Sun Nov 28 17:03:28 +0000 2010"
            o geo: null
            o favorited: false
            o source: "<a href="http://www.hootsuite.com"
rel="nofollow">HootSuite</a>"
            o in_reply_to_status_id_str: null
            o id_str: "8928974595952640"
```

```
          o place: null
          o
            user: {
                + profile_background_color: "77b0dc"
                + description: "44th President of the United States"
                + geo_enabled: false
                + profile_use_background_image: true
                + favourites_count: 0
                + url: http://www.barackobama.com
                + follow_request_sent: false
                + lang: "en"
                + verified: true
                + profile_background_image_url:
http://a3.twimg.com/profile_background_images/174596417/newtwitter-OFA-v5.jpg
                + created_at: "Mon Mar 05 22:08:25 +0000 2007"
                + location: "Washington, DC"
                + profile_link_color: "2574ad"
                + notifications: false
                + profile_background_tile: false
                + id_str: "813286"
                + listed_count: 118778
                + following: true
                + profile_sidebar_border_color: "c2e0f6"
                + profile_image_url:
http://a3.twimg.com/profile_images/784227851/BarackObama_twitter_photo_normal.jpg
                + name: "Barack Obama"
                + time_zone: "Eastern Time (US & Canada)"
                + followers_count: 5989763
                + id: 813286
                + utc_offset: -18000
                + friends_count: 709157
                + screen_name: "BarackObama"
            }
          o retweet_count: null
          o in_reply_to_user_id: null
          o id: 8928974595952640
          o retweeted: false
          o text: "This West Wing Week: the traditional Thanksgiving turkey
pardon and a NATO summit in Portugal. http://OFA.BO/o3j5Wa"
      },
{...},
{...}
]
```


The response data for the Twitter streamline call is an array of tweet objects. Each tweet object contains some string fields and number fields, and a `user` object that includes several string and number fields to identify the user who posted the tweet.

The JSON data item class (our contract definition class) for a single tweet does not need to map all the fields in a response; only those that will be used by the application should be specified. For example, if the application will use the three fields `user.screen_name`, `user.profile_image_url`, and `text` (the tweet text), the data item class `TwitterStatusItem` should be as simple as the example class `TwitterStatusItem` shown in Listing 6-1. (Please note the embedded private class `TwitterUser`, which maps the `user` object in each tweet object. The `DataContract` and `DataMember` directives are critical for establishing the data contract.)

LISTING 6-1: Twitter status data contract class, SocialDemo\SocialDemo\TweetStatus.cs

Available for download on Wrox.com

```
using System.Runtime.Serialization; //DataContract

[DataContract]
public class TwitterStatusItem
{
    //"User" object
    [DataContract]
    public class TwitterUser
    {
        //"screen_name" item
        private string screen_name;
        [DataMember(Name = "screen_name")]
        public string Screen_name {
            get { return screen_name ?? "";}
            set { screen_name = value;}
        }

        //"profile_image_url" item
        private string profile_img;
        [DataMember(Name = "profile_image_url")]
        public string Profile_img {
            get { return profile_img ?? "";}
            set { profile_img = value;}
        }
    }

    private TwitterUser twitterUser;
    [DataMember(Name = "user")]
    public TwitterUser User
    {
        get {
            return twitterUser ??
                new TwitterUser { Screen_name = "", Profile_img = ""};
        }
        set { twitterUser =  value; }
    }

    //"text" item
    private string text;
    [DataMember(Name = "text")]
    public string Text
    {
        get { return text ?? ""; }
        set { text = value; }
    }
}
```

With the Twitter status data contract class in place, you can create a helper class to facilitate JSON string parsing. The helper class is agnostic in terms of the data contract class it uses, as shown in Listing 6-2.

LISTING 6-2: A JSON helper class, SocialDemo\SocialDemo\JsonHelper.cs

Available for download on Wrox.com

```
using System.Runtime.Serialization.Json; //DataContractJsonSerializer class
using System.IO; //MemoryStream class
using System.Text; //Encoding class

namespace SocialDemo
{
    //Helper methods for JSON string serialization and deserialization
    public class JsonHelper
    {
        public static string Serialize<T>(T obj)
        {
            string retVal = string.Empty;
            MemoryStream ms = null;
            try
            {
                DataContractJsonSerializer serializer =
                    new DataContractJsonSerializer(obj.GetType());
                ms = new MemoryStream();
                serializer.WriteObject(ms, obj);
                retVal = Encoding.UTF8.GetString(
                    ms.ToArray(), 0, (int)ms.Length);
            }
            catch
            {
                throw;
            }
            finally
            {
                if(ms != null)
                    ms.Dispose();
            }
            return retVal;
        }

        //Deserilize a JSON string
        public static T Deserialize<T>(string json)
        {
            MemoryStream ms = null;
            T obj = default(T);
            try
            {
                obj = Activator.CreateInstance<T>();
                ms = new MemoryStream(Encoding.Unicode.GetBytes(json));
                DataContractJsonSerializer serializer =
                    new DataContractJsonSerializer(obj.GetType());
                obj = (T)serializer.ReadObject(ms);
            }
```

continues

LISTING 6-2 *(continued)*

```
                catch
                {
                    throw;
                }
                finally
                {
                    if(ms != null)
                    {
                        ms.Close();
                        ms.Dispose();
                    }
                }
                return obj;
            }
        }
    }
```

To use the `JsonHelper` helper method, just pass the Twitter streamline response as a string object, and indicate the desired data type — a list of `TwitterStatusItem` objects — to the helper class, as shown below. A complete example of using both the JSON helper class and XML parsing is shown in the following section.

Available for download on Wrox.com

```
List<TwitterStatusItem> tweets =
   JsonHelper.Deserialize<List<TwitterStatusItem>>(e.Result);
listStatus.ItemsSource = tweets;
```

Code snippet SocialDemo\SocialDemo\MainPage.xaml.cs

The code above uses a `ListBox.ItemTemplate` data binding method for the listbox `listStatus`. Items of this listbox template are one-to-one mapped to a field of the `TwitterStatusItem` class in the XAML file (`User.Profile_img`, `User.Screen_name`, and `Text`). To show the data on the UI, you need to set the `ItemSource` property of the listbox object to a collection of `TwitterStatusItem` object as shown above. The following is an example of the listbox control in the XAML file.

```
<!--ContentPanel - place additional content here-->
<Grid x:Name="ContentPanel" Grid.Row="1" Margin="12,0,12,0">
    <ListBox Height="607" HorizontalAlignment="Left" Name="listStatus"
             VerticalAlignment="Top" Width="456"
             DataContext="{Binding}" ItemsSource="{Binding}">
        <ListBox.ItemTemplate>
            <DataTemplate>
                <StackPanel Orientation="Horizontal" Height="132">
                    <Image Source="{Binding User.Profile_img}"
                           Height="73" Width="73"
                           VerticalAlignment="Top" Margin="0,10,8,0"/>
                    <StackPanel Width="370">
                        <TextBlock Text="{Binding User.Screen_name}"
                                   Foreground="Gold" FontSize="28" />
                        <TextBlock Text="{Binding Text}"
```

```
                                        TextWrapping="Wrap" FontSize="24" />
                        </StackPanel>
                    </StackPanel>
                </DataTemplate>
            </ListBox.ItemTemplate>
        </ListBox>
    </Grid>
```

Parsing XML Data

Silverlight for WP7 provides two classes to parse XML data: `System.Xml.XmlReader` or `System.Xml.Linq`. Note that `XmlReader` provides forward-only, read-only access to a stream of XML data, meaning that you have to navigate through the entire XML stream to retrieve the nodes of interest.

The `System.Xml.Linq` option is easier to use and provides more structured output. With XML to Linq mapping, you can obtain node values of the XML string using a SQL-like syntax. For example, the Twitter streamline call for screen name BarackObama with XML as the data format is:

`http://api.twitter.com/1/statuses/user_timeline.xml?screen_name=barackobama`

The response in XML format is as follows:

```
<statuses type="array">

<status>
 <created_at>Thu Dec 02 20:04:24 +0000 2010</created_at>
 <id>10424062266310656</id>
 <text>
  Jews have lit Hanukkah candles as symbols of resilience in times of peace and
  in times of persecution. Their light inspires us to hope.
 </text>

 <source>
  <a href="http://www.hootsuite.com" rel="nofollow">HootSuite</a>
 </source>
 <truncated>false</truncated>
 <favorited>false</favorited>
 <in_reply_to_status_id/>
 <in_reply_to_user_id/>
 <in_reply_to_screen_name/>
 <retweet_count/>
 <retweeted>false</retweeted>
 <user>
  <id>813286</id>
  <name>Barack Obama</name>
  <screen_name>BarackObama</screen_name>
  <location>Washington, DC</location>
  <description>44th President of the United States</description>
  <profile_image_url>
    http://a3.twimg.com/profile_images/784227851/
    BarackObama_twitter_photo_normal.jpg
  </profile_image_url>
  <url>http://www.barackobama.com</url>
```

```
    <protected>false</protected>
    <followers_count>6019488</followers_count>
    <profile_background_color>77b0dc</profile_background_color>
    <profile_text_color>333333</profile_text_color>
    <profile_link_color>2574ad</profile_link_color>
    <profile_sidebar_fill_color>c2e0f6</profile_sidebar_fill_color>
    <profile_sidebar_border_color>c2e0f6</profile_sidebar_border_color>
    <friends_count>708982</friends_count>
    <created_at>Mon Mar 05 22:08:25 +0000 2007</created_at>
    <favourites_count>0</favourites_count>
    <utc_offset>-18000</utc_offset>
    <time_zone>Eastern Time (US & Canada)</time_zone>
    <profile_background_image_url>
      http://a3.twimg.com/profile_background_images/174596417/
      newtwitter-OFA-v5.jpg
    </profile_background_image_url>
    <profile_background_tile>false</profile_background_tile>
    <profile_use_background_image>true</profile_use_background_image>
    <notifications>false</notifications>
    <geo_enabled>false</geo_enabled>
    <verified>true</verified>
    <following>false</following>
    <statuses_count>1124</statuses_count>
    <lang>en</lang>
    <contributors_enabled>false</contributors_enabled>
    <follow_request_sent>false</follow_request_sent>
    <listed_count>119259</listed_count>
    <show_all_inline_media>false</show_all_inline_media>
  </user>
  <geo/>
  <coordinates/>
  <place/>
  <contributors/>
</status>
...
</statuses>
```


The root node is statuses, which contains an array of status nodes, each of which represents a tweet item. Each status node has a user sub-node carrying the user information, in addition to the tweet data sub-nodes such as text and created_at.

To model the data of interest in the above XML response, you define a data item class as follows. You'll use this class for the LINQ query.

Available for download on Wrox.com

```
public class TwitterStatusItem_forxml
{
    private string name;
    public string Name
    {
        get { return name ?? ""; }
        set { name = value; }
    }

    private string profile_img_url;
```

```
        public string Profile_img_url
        {
            get { return profile_img_url ?? ""; }
            set { profile_img_url = value; }
        }

        private string text;
        public string Text
        {
            get { return text ?? ""; }
            set { text = value; }
        }
    }
```

Code snippet SocialDemo\SocialDemo\TwitterStatus.cs

Then use LINQ to map the received XML data onto a listbox named `listStatus`, as shown in Listing 6-3. Note the example also shows how to use the JSON helper class created earlier.

LISTING 6-3: An example of using Linq with XML data, SocialDemo\SocialDemo\Mainpage.xaml.cs

Available for download on Wrox.com

```
using System.Xml.Linq; //Linq-to-xml

void twitter_DownloadStringCompleted(object sender,
    DownloadStringCompletedEventArgs e)
{
    if (e.Error != null) return;
    if (TwitterConfig.useJson == true) //The response data is JSON
    {
        try
        {
            var data = e.Result;
            List<TwitterStatusItem> tweets =
                JsonHelper.Deserialize<List<TwitterStatusItem>>(data);
            listStatus.ItemsSource = tweets;
        }
        catch (Exception ex)
        {
            Dispatcher.BeginInvoke(() =>
            {
                MessageBox.Show(ex.Message);
            });
        }
    }
    else //The response is XML
    {
        XElement xmlTweets = XElement.Parse(e.Result);
        //Binding the Linq data to the Listbox.
        //Note: you need to modify the listStatus data binding
properties
        //in the xaml to match the members of TwitterStatusItem_forxml
```

continues

LISTING 6-3 *(continued)*

```
            listStatus.ItemsSource =
                from tweet in xmlTweets.Descendants("status")
                select new TwitterStatusItem_forxml
                {
                    Name = tweet.Element("user").Element("screen_name").Value,
                    Text = tweet.Element("text").Value,
                    Profile_img_url =
                      tweet.Element("user").Element("profile_image_url").Value
                };
        }
    }
```

The `XElement` object `xmlTweets` represents the XML document in the web call response string. The `XElement.Descendants()` method with the parameter `status` returns an `IEnumerable<XElement>` interface instance that Linq uses to create a number of `TwitterStatusItem_forxml` objects for the received `status` nodes in the XML document. Those objects are then bound to a listbox control `listStatus`. Note that because the data item class has changed from `TwitterStatusItem` to `TwitterStatusItem_forxml`, in the XAML file the listbox template items must be modified to bind to the fields in the `TwitterStatusItem_forxml` class.

OAuth

OAuth is the standard authorization protocol used by most web services today. OAuth enables a web service and its client to engage in a well-defined workflow such that the client can access protected resources provided by the web service.

Before OAuth is widely used, a mobile application and a web service usually perform basic authentication that requires the application to pass a user's credential for each web service call. This also means the application must obtain and store a user's user name and password (encrypted) locally on the device, which could potentially be a security concern, as such sensitive information may be leaked. With OAuth, the application does not need to do that at all, since during OAuth the web service will push an authorization URL to the application with which the user's credential can be entered and posted directly back to the web service.

OAuth performs three basic steps as follows (assuming the developer has already received a consumer key and consumer secret from the web service provider for his or her application) from a client's perspective:

1. Pass the consumer key and the consumer secret to get an authorized request token and request token secret. The specification refers to the combination of both the request token and the request token secret as an oauth token at this stage of authorization process. Your code needs to compose a base string and sign it using the consumer secret. All required parameters can be either URL encoded or put in an HTTP header.
2. Launch the authorization URI with the request token attached to start the authorization flow so the user can authorize the application at the authorization website. The web service will reply with an `oauth_verifier` upon success.

3. Pass the request token and `oauth_verifier` to request an access token and an access token secret. Again the application code must compose a specific base string and sign it using the combined consumer secret key and request token secret.

Once the client receives the access token and access token secret (collectively called the OAuth token), it can launch web API calls that require an access token as one parameter. These calls must also have a signature attached. The signature is computed using a specific base string and the combined consumer secret key and access token secret.

There is no need to implement the OAuth workflow for your app from scratch. A third-party wrapper library would be a better choice. The sample project in this chapter, SocialDemo, relies on the Hammock OAuth library to facilitate OAuth with the Twitter web service (`http://hammock.codeplex.com/`).

WP7 PUSH NOTIFICATIONS

Many mobile applications enable a user to receive notifications from the Web even if the application is not running. This creates an always-on, always-connected user experience. The following section will discuss three types of WP7 push notifications, and how to use them in your applications.

Why Push Notifications?

The word "push" in push notifications tells you that the message is initiated on a web server and sent to an application running on a mobile device. This is in contrast to the other message-passing scheme in which it is the mobile application that periodically polls the server for updates. The reason the push scheme is widely used on mobile devices is that mobile devices have a limited battery life — if multiple applications on a mobile device wake up and poll servers, the device is constantly doing something, which reduces battery life.

Another important issue with getting notifications from the Web is that a mobile application most likely does not run all the time. Except for those that create a background service, applications will be closed when they are not in the foreground of the screen, in which case they will not be able to receive any notifications as they are not running at all.

The push notification approach addresses these issues by creating a single dedicated on-device service to communicate with a proxy service on the Internet. All the applications that require timely updates from their respective servers can share the same communication pipe between the on-device service and the proxy service. Using this approach, these applications do not need to actively poll their servers, as notifications will be pushed to them instead. In addition, they do not need to keep running in order to receive updates; the system can launch them when the on-device service receives messages for them. From the application's perspective, it looks as if the messages (notifications) are pushed from the Internet to the device and then dispatched by the device OS to the application, even if the application is not running at that moment.

Push Notification Architecture

Developers who are familiar with Windows Mobile 6.*x* should recall the push e-mail technology that Microsoft created for that platform. In essence, the device maintains a persistent network connection to the Exchange server (using heartbeats). This allows Microsoft Exchange Servers to deliver any new incoming e-mail notifications promptly to the e-mail application on the device.

WP7 (along with iOS and Android) has adopted a similar approach. The difference is that a systemwide dedicated communication pipe provides the persistent connection for all applications on the device. Any web service can post specific notifications to the proxy service, which in turn forwards the notifications to applications on the device. The overall architecture will look like Figure 6-1.

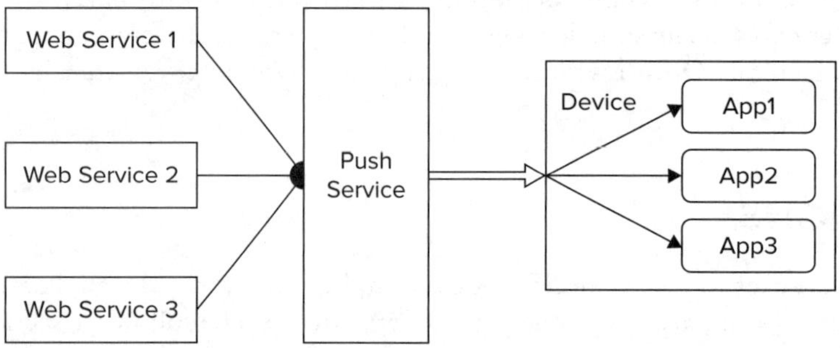

FIGURE 6-1: Push service architecture

Now let's look at the actual implementations of push notification services on iOS and Android. If you are quite familiar with these technologies, you can skip this section and move on to the discussion of WP7 notifications.

Apple APNs

Apple doesn't allow developers to write background services other than those it specifically supports. Such a limitation is designed to improve battery life and handle low-memory situations on the system. For an application that constantly communicates with a server on the Internet, even if it's not in the foreground, Apple provides a push notification system that leverages a single on-device software entity to communicate with a special-purpose server in a well-regulated manner. This limitation ensures that Apple has total control over what an application communicates and how communication occurs between an application and its server on the Internet when the application is in the background (i.e., not running).

Apple started to provide push notification services in iPhone OS 3.0. At the center of its architecture is the so-called Apple Push Notification service (APNs), which receives push notifications from providers, and then delivers them to specific applications on target devices using a best-effort approach. All messages originate from providers, which are specialized web servers used by mobile applications. Both providers and target devices establish secure persistent IP connections with APNs.

Suppose a provider receives a new message to deliver to an application on a particular device (for example, the provider might receive a notification that new tweets are available for a given topic).

It creates a notification that consists of a device token to identify the target device and specialized JSON data as a payload. This data includes a text message to display to the user, a sound to play, a badge on the application icon, and other tweet data. The notification is sent to APNs using a Secure Sockets Layer (SSL) channel. The APNs then routes the notification to the target device. If the target application isn't running when the notification arrives, the system plays or shows the alert text message, sound, or badge value. When the application is running, iOS delivers it to the application directly. The notification payload has a size limit of 256 bytes.

One of the big questions to answer is how the iOS knows which application is about to receive the notification from APNs.

The three parties use the device token as a unique identifier for an application running on a specific device. When the application is first installed on the device, the system registers it. During the registration process, APNs generates the device token and passes it back to the application, which in turns sends it to its provider. When the provider receives the token it can start to use the token as part of a notification. APNs maintains a mapping list between a device and its device tokens (a device may have several applications registered for different notifications and therefore have several device tokens). When APNs receives a new notification, it knows exactly where to forward that notification. If the device is not on, the system saves the push notifications and delivers them later when the device is on. In addition, APNs also provides feedback to providers, which may contain those device tokens that are no longer registered to receive specific notifications. Providers should use the feedback to remove the unregistered devices from the target device token list.

Developers don't need to develop providers from scratch, although, they can surely do so by programming on a web server and obtaining the required certificate from Apple. Instead, they can use third-party provider services to host their provider code. Boxcar (`http://boxcar.io`), Notifo (`http://notifo.com`) and MonoPush (`www.monopush.com`) are among the popular third-party provider services for iPhone.

Google C2DM

Google Android enables the traditional service for application development. Developers can write a background service that handles communication between an application and a server on the Internet. However, a poorly written background service may drain the battery quickly if it performs too many polling operations. In addition, it's fairly difficult for individual developers to build a push notification solution. For this reason Google started to provide Cloud To Device Messaging (C2DM) with Android 2.2 Froyo. Google applications such as Gmail and Calendar have used C2DM even before Froyo.

Three major parties are involved in C2DM: the application server, usually a web server where messages are generated; Google's C2DM service that forwards the message to the device; and applications running on a target device.

The identities in Table 6-3 are used in Android's C2DM architecture:

TABLE 6-3: Android C2DM Credentials

CREDENTIALS	DESCRIPTION	EXAMPLE
Sender ID	Used in the device application C2DM registration process as part of the intent data. A Google account developer chooses a sender identifier for the application he or she develops.	WileySampleApp@gmail.com
Application ID	Used in the device application C2DM registration process as part of the intent data.	Not exposed. The registration service will obtain the application ID.
Registration ID	Provided as an identifier C2DM that assigns to the application after a successful registration. The application should pass this information to its application server. C2DM may refresh registration IDs.	A 119-byte array: APA91bFT3_ LQS4w15wjSUcGyvLq- 0HbUhGyjvegKB7S92YWKIlfuvjlHq8Td WNXd2-UqmIx1AxFBNIHebhu9s OSa8IPfiUHBJ_uZnLcSDjflOva_zYFmjSk
Sender Auth Token	Assigned by Google as an authentication token for the application server in the ClientLogin response. The Sender Auth Token is also used in the POST header when a push notification is sent from the application server to C2DM.	In the ClientLogin response header: Auth=DQAAAGgA...dk3fA5N

To enable C2DM, an application on the device must first register with Google programmatically and get a registration ID. This is done in an `Intent` object call to `com.google.android.c2dm.intent.REGISTER`. The extra data items placed in the registration intent must have the strings `app` and `sender`, respectively, as shown in the following code:

```
Intent registrationIntent = new
    Intent("com.google.android.c2dm.intent.REGISTER");
registrationIntent.putExtra("app",PendingIntent.getBroadcast(context,
    0, new Intent(), 0));
registrationIntent.putExtra("sender", senderId);
context.startService(registrationIntent);
```

Upon a successful registration, the registration ID is assigned by C2DM and sent back to the device. The device broadcasts the `com.google.android.c2dm.intent.REGISTER` intent, and the application that registered the intent can retrieve it. The application can retrieve the intent using the following code and send it to the application server:

```
String registrationId = intent.getStringExtra("registration_id");
if (intent.getStringExtra("error") != null) {
        // Registration failed, backoff and retry later.
    } else if (intent.getStringExtra("unregistered") != null) {
        // unregistration done, new messages from the authorized sender
will be rejected
    } else if (registrationId != null) {
       // Send the registration ID to the 3rd party site that is sending
the messages.
}
```

When the application server is about to push a message to the application on the device, it posts an HTTP message to the C2DM service (`https://android.apis.google.com/c2dm/send`). The message contains the registration ID and message payload. The header of this HTTP message contains the auth token that was obtained from Google.

The C2DM service routes the received HTTP message to the device using the registration ID if the device is online. Otherwise, it will queue the message and send it once the device goes online. This portion of the communication is transparent to the application.

Once the device receives the HTTP message from C2DM, it sends an `Intent` broadcast to target applications. The application wakes up to process the message it has received in its `com.google.android.c2dm.intent.RECEIVE` intent receiver method, `onReceive()`. The intent received by the application encapsulates the payload in a number of extras. The app must define a `C2DMBaseReceiver` class that extends the `com.google.android.c2dm.C2DMBaseReceiver` class, instead of using the `com.google.android.c2dm.C2DMBaseReceiver` class directly. The following code shows a simple `onReceive()` implementation:

```
protected void onReceive(Context context, Intent intent) {
    if (intent.getAction().equals("com.google.android.c2dm.intent.RECEIVE")) {
       // Get the message from intent.extras
    }
}
```

The application server role is quite common — it just needs to send a specially formatted HTTPS POST message to the Google C2DM service. Developers can use Google App Engine or other web services to host their application servers. Of course, the application server must first authenticate itself against the C2DM server by passing its ClientAuth token to C2DM. The message has a size limit of 1,024 bytes.

When writing an Android application that uses C2DM, you need to set several permissions in the application's Manifest file, `AndroidManest.xml`, so the Android application run time can properly grant specific permissions.

Push Notifications on WP7

WP7's push notification architecture is similar to that on iOS and Android but much simpler in terms of the effort required to build the client on the device. The Microsoft Push Notification Service (MPNS) is actually a special Windows Live web service application running in the cloud. The application server can be a standard web application that communicates with the push server. Microsoft recommends using WCF, Microsoft's web service-oriented framework, for building the application server used in a push notification scenario. On the client side developers must understand three types of push notifications: *toast*, *tile*, and *raw*, as well as the steps used to register and receive these notifications from the application server.

The sample code referenced in this section is excerpted from the sample solution PushNotification, which includes two projects: PushNotificationClient and PushServerEmulator. The PushServerEmulator project implements a desktop program that emulates an application server to communicate with MPNS and send all three types of push notifications. The PushNotificationClient project is a device application that receives and processes all three types of supported push notifications. The contents of these notifications are event reminders, which are formatted and delivered to the device application using toast, tile, or raw notification.

To try the sample PushNotification solution, start to run the solution (which will run both projects) and make sure the desktop application runs on the PC, and the device application runs on the device emulator or a real device. Then you must let the PushServerEmulator application know the registered MPNS channel URI (for example, `http://sn1.notify.live.net/…`). You can copy this URI from the console output of the PushNotificationClient project and paste it into one of the three channel textboxes in the PushServerEmulator UI. Then from the PushServerEmulator application you can send notifications to the application on the device.

Basics

The key concepts in WP7 Push Notification architecture are explained as follows.

- **Notification channel:** A channel represents the application's registration with the MPNS. Before receiving push notifications, an application must first open a channel with MPNS so MPNS can deliver notifications to it. An app can only have one channel open per device. But a single channel can receive one, two, or all three types of push notifications. MPNS will keep the channel open for some time even if the application on the device is *tombstoned* (that is, deactivated because the user invokes a launcher, a chooser, or presses the Start button. For more details of the application execution model, please see Chapter 4). Therefore, the application should always use the channel name to check whether the channel is already open, and use it without reopening it. The `HttpNotificationChannel` class models a notification channel.
- **Toast notification:** A *toast* is a push notification that will appear on the top of the device screen when the receiving application is not in the foreground. If the receiving application is in the foreground, then the application should handle the toast, and no message will appear on the top of the screen. A toast is like the pop-up message on iOS and the floating message that appears in the notification area on Android. An app must first bind to toast in order to receive toast notifications. *Binding* indicates to the OS that the underlying app will receive

toast notifications; this is similar to the `com.google.android.c2dm.intent.RECEIVE` intent used by an Android app to receive push messages.

- **Tile notification:** A *tile* is a rectangular picture on the home screen that serves as a shortcut to an application. A tile is like the Home screen widget on Android. On WP7 the user can simply press and hold the application name in the program list, and then click Pin to Start to create a tile for it. A tile can have three types of data: a background image that can change dynamically or according to a schedule; a tile title string; and a number that appears in the top right corner of the tile. A tile notification is a push notification that is designated to change some or all the data associated with the tile for the underlying app. The app does not need to process received tile notifications.
- **Raw notification:** A *raw* notification is a user-defined HTTP message sent to the application (sometimes developers use the term raw http notification as well). In contrast to toast and tile notifications, where the WP7 shell takes control when it receives a notification, only the application can process a raw notification using a specific event handler. The format of a raw notification is completely up to the application developer. However, to make the processing easier, Microsoft recommends using either XML or JSON.

> *When an Android device receives a push notification from C2DM, its targeted application will wake up automatically and the system will call the corresponding intent receiver (even if the app doesn't come to the foreground). However, on iOS and WP7, if the targeting app is not in the foreground, the user can simply dismiss the toast notification, without waking up the application.*

Channel Establishment and Binding

As mentioned in the Basics section of this chapter, you should use the `HttpNotificationChannel` class to create and manage a notification channel. Once the application creates and opens a channel, it should set up event handlers to process various channel events. In addition, the application needs to tell the operating system what types of push notifications the channel should bind to. This will allow the system shell to handle received push notifications from this channel and display toast and tile updates on the device.

Your application should use the following steps to create the channel:

1. Call `HttpNotificationChannel.Find()` to determine if the named notification channel already exists.
2. If no such channel exists, the application creates the channel object, defines channel event handlers (so that the application can retrieve the assigned Channel URI), and then opens the channel by calling `HttpNotificationChannel.Open()`. At this point, the channel should bind to desired types of push notifications.
3. If the named channel already exists, use it as long as the channel URI isn't null. If it is, there may be a problem with the channel. Close the app and retry.

Listing 6-4 shows an example of the channel creation logic in the `MainPage` class in the PushNotificationClient project. Note the `SetupChannelEventHandlers()` routine and `BindChannelToTileToast()` routine. They are wrappers for setting up channel event handlers and binding the channel to toast and tile notifications.

Available for download on Wrox.com

LISTING 6-4: An example of channel establishment, PushNotification\PushNotificationClient\MainPage.xaml.cs

```
using Microsoft.Phone.Notification; //HttpNotificationChannel class
using System.IO; //Stream reader class
using System.Diagnostics; //Debug class

private void SetupChannel()
{
    Debug.WriteLine(
        "Checking if there is an existing notification" +
        "channel with the name available");
    httpChannel = HttpNotificationChannel.Find(channelName);
    if (null == httpChannel)
    {
        //We need to create the channel object first
        httpChannel = new HttpNotificationChannel(channelName);

        //Setup channel event handlers
        SetupChannelEventHandlers();

        //Open the notification channel
        try
        {
            httpChannel.Open();
        }
        catch (Exception e)
        {
            Debug.WriteLine("Exception: " + e.Message);
        }
    }
    else
    {
        if (null != httpChannel.ChannelUri)
        {
            //There is already a channel with the give name; reuse it
            ChannelUri = httpChannel.ChannelUri;
            Debug.WriteLine("There is already a channel");
            Debug.WriteLine("Channel Uri: " + ChannelUri.AbsoluteUri);
            SetupChannelEventHandlers();
            BindChannelToTileToast();
        }
        else
        {
            Debug.WriteLine("Error: channel exists but not opened");
            MessageBox.Show("Error: channel exists but not opened." +
                "Close the app and retry.");
```

```
                ChannelUri = null;
            }
        }
    }
```

The `SetupChannelEventHandlers()` should register four event handlers for the following events associated with the channel just created: channel URI change, raw (HTTP) notification received, shell and toast notification received, and error occurred.

Available for download on Wrox.com

LISTING 6-5: An example of setting up channel event handlers, PushNotification\PushNotificationClient\MainPage.xaml.cs

```
private void SetupChannelEventHandlers()
{
    //When there is a new URI for the channel
    httpChannel.ChannelUriUpdated +=
            httpChannel_ChannelUriUpdated);

    //When there is a raw notification
    httpChannel.HttpNotificationReceived +=
        new EventHandler<HttpNotificationEventArgs>(
            httpChannel_RawNotificationReceived);

    //When there is a toast notification
    //and if the app is in not running, system shell will handle it
    //Otherwise, the following handler will be called to handle it
    httpChannel.ShellToastNotificationReceived +=
        new EventHandler<NotificationEventArgs>(
            httpChannel_ShellToastNotificationReceived);

    //When something bad happens
    httpChannel.ErrorOccurred +=
        new EventHandler<NotificationChannelErrorEventArgs>(
            httpChannel_ErrorOccurred);
}
```

The `BindChannelToTileToast()` method basically calls `HttpChannelNotification.BindToShellToast()` and `HttpChannelNotification.BindtoShellTile()`. Here is an example of the `BindChannelToTileToast()` method:

Available for download on Wrox.com

LISTING 6-6: An example of toast channel binding, PushNotification\PushNotificationClient\MainPage.xaml.cs

```
private void BindChannelToTileToast()
{
   try
   {
      if (httpChannel.IsShellToastBound == false)
      {
         Debug.WriteLine("Binding to toast notifications");
         httpChannel.BindToShellToast();
```

continues

LISTING 6-6 *(continued)*

```
            }
        }
        catch (Exception)
        {
            //Already bound; do nothing
        }

        try
        {
            if (httpChannel.IsShellTileBound == false)
            {
                Debug.WriteLine(
                    "Binding to tile notifications with a remote picture");

                //Here we need to register all the tile background images that
                //may be used by this tile
                //Then later the push server can indicate which remote/local
                //image to use in the tile notification
                Collection<Uri> uris = new Collection<Uri>();
                Uri x = new Uri(
                    "http://media.wiley.com/assets/1135/98/event.gif");
                uris.Add(x);

                httpChannel.BindToShellTile(uris);
            }
        }
        catch (Exception)
        {
            //Already bound; do nothing
        }
    }
```

The caveat with `HttpChannelNotification.BindToShellTile()` is that there are two overloaded methods for this purpose: one that doesn't require any input arguments and the other that requires an input argument of `Collection` type. Both indicate to the system shell that the channel will be used to deliver some tile notifications. The difference between these two methods is that the latter provides a collection of local and remote URIs to the system. When a tile notification contains a tile image item that's in the specified collection, it will use the image item as the background image of the tile. Microsoft places some restrictions on remote pictures: the picture cannot be larger than 80 KB and the download time cannot exceed 15 seconds or the image will not be loaded.

The first time the channel is opened by the `open()` call from an application, MPNS will create a unique channel URI and assign it to the application. Then MPNS will pass that channel URI back to the application in the URI changed event handler. Both the application web server and MPNS use that unique channel URI to identify where to send a push notification (the particular app on a specific device).

Sending Push Notifications

To send push notifications to a WP7 application, the application server must generate specially defined HTTP messages and send them to the MPNS. Unlike iOS and Android, where a single push notification format is used, WP7 uses three types. The following section shows the steps required for sending notifications from an application server to MPNS using the sample code in the PushServerEmulator project.

A toast notification is defined as follows:

```
string toastMessage = "<?xml version=\"1.0\" encoding=\"utf-8\"?>" +
"<wp:Notification xmlns:wp=\"WPNotification\">" +
   "<wp:Toast>" +
      "<wp:Text1><string1></wp:Text1>" +
      "<wp:Text2><string2></wp:Text2>" +
   "</wp:Toast>" +
"</wp:Notification>";
```

A tile notification is defined as follows. The tile image path can be either a local or a remote path. If this is a remote path, then it should be listed in the `Collection` that passed as the only parameter to the `BindToTile`() call.

```
string tileMessage = "<?xml version=\"1.0\" encoding=\"utf-8\"?>" +
"<wp:Notification xmlns:wp=\"WPNotification\">" +
   "<wp:Tile>" +
      "<wp:BackgroundImage><background image path></wp:BackgroundImage>" +
      "<wp:Count><count></wp:Count>" +
      "<wp:Title><title></wp:Title>" +
   "</wp:Tile> " +
"</wp:Notification>";
```

A raw notification can be any format. But it is recommended that you should put your messages in XML or JSON so they can be easily parsed on the client.

The code that sends the push notification to MPNS simply launches an HTTP POST method to the channel URI that was passed from the application when it created the notification channel.

Note that the payload data for a toast notification contains one or more strings, whereas the payload of a tile notification includes a tile image URI, a number, and a text string combined. For raw notification, the payload can be any format, as long as the device app can recognize and process it. In the sample project PushServerEmulator, the raw notification payload is a well-formatted XML document consisting of some event data. The composition of the XML document is done in the `BuildRawNotification()` method. Listing 6-7 shows the `SendMessage()` method that accommodates all three notification types:

LISTING 6-7 An example of sending Push Notifications, PushNotification\PushServerEmulator\MainWindow.xaml.cs

Available for download on Wrox.com

```
// Send all three types of push messages.
// Parameter "type": message type. 1:toast, 2:tile, 3:raw.
// Parameter "channelUri": channel Uri.
// Parameter "message": text string as message body.
// Parameter "tileNum": number to be carried in the tile message.
private void SendMessage(
```

continues

LISTING 6-7 *(continued)*

```
int type,
string channelUri,
string message,
string tileNum)
    {
        if (null == channelUri || string.IsNullOrEmpty(message))
            return;

        HttpWebRequest sendNotificationRequest;
        try
        {
            sendNotificationRequest =
                (HttpWebRequest)WebRequest.Create(channelUri);
        }
        catch (Exception exc)
        {
            MessageBox.Show(exc.Message);
            return;
        }

        sendNotificationRequest.Method = "POST";
        //Indicate that you'll send toast notifications!
        sendNotificationRequest.ContentType = "text/xml;; charset=utf-8";
        sendNotificationRequest.Headers = new WebHeaderCollection();

        byte[] messagePayload = null; //The payload
        if (type == 1)
        {
            sendNotificationRequest.Headers.Add(
                "X-WindowsPhone-Target", "toast");
            //Possible batching interval values for toast
            //notifications:
            // 2: The message is delivered immediately.
            // 12: The message is delivered within 450 seconds.
            // 22: The message is delivered within 900 seconds.
            sendNotificationRequest.Headers.Add(
                "X-NotificationClass", "2");

            //Create xml envelope
            //The message will be formatted into the <wp:Text1> node
            string messageTemplate =
                "<?xml version='1.0' encoding='utf-8'?>" +
                "<wp:Notification xmlns:wp='WPNotification'>" +
                  "<wp:Toast>" +
                    "<wp:Text1>{0}</wp:Text1>" +
                  "</wp:Toast>" +
                "</wp:Notification>";
            //Wrap custom data into envelope
            messagePayload = UTF8Encoding.UTF8.GetBytes(
                string.Format(messageTemplate, message));
        }
        else if (type == 2)
        {
```

```
        //Indicate this is a tile notification
        sendNotificationRequest.Headers.Add(
            "X-WindowsPhone-Target", "token");
        sendNotificationRequest.Headers.Add(
            "X-NotificationClass", "1");

        //Create xml envelope
        //A Uri of a remote picture is sent.
        //Note this Uri must be first indicated to the device
        //system when the device app calls BindToShellTile().
        //The other two data items are an integer value as the
        //number attached to the tile and a string as tile title.
        string x = "http://media.wiley.com/assets/1135/98/event.gif";
        string messageTemplate =
            "<?xml version='1.0' encoding='utf-8'?>" +
            "<wp:Notification xmlns:wp='WPNotification'>" +
              "<wp:Tile>" +
                "<wp:BackgroundImage>" +
                  x +
                "</wp:BackgroundImage>" +
                "<wp:Count>{0}</wp:Count>" +
                "<wp:Title>{1}</wp:Title>" +
              "</wp:Tile>" +
            "</wp:Notification>";
        //Wrap custom data into envelope
        string messageText =
            string.Format(messageTemplate, tileNum, message);
        messagePayload = UTF8Encoding.UTF8.GetBytes(messageText);
    }
    else
    {
        //Raw notification does not need X-WindowsPhone-Target header
        sendNotificationRequest.Headers.Add(
            "X-NotificationClass", "3");

        //"message" is already formatted before being passed here
        messagePayload = UTF8Encoding.UTF8.GetBytes(message);
    }

    //Set Content Length
    sendNotificationRequest.ContentLength = messagePayload.Length;

    //Push data to stream
    using (Stream requestStream =
        sendNotificationRequest.GetRequestStream())
    {
        requestStream.Write(
            messagePayload, 0, messagePayload.Length);
    }

    //Start to get reponse
    HttpWebResponse response;
    try
```

continues

LISTING 6-7 *(continued)*

```
            {
                response =
                    (HttpWebResponse)sendNotificationRequest.GetResponse();
            }
            catch (Exception exc)
            {
                MessageBox.Show(exc.Message);
                return;
            }
            string notificationStatus =
                response.Headers["X-NotificationStatus"];
            string notificationChannelStatus =
                response.Headers["X-SubscriptionStatus"];
            string deviceConnectionStatus =
                response.Headers["X-DeviceConnectionStatus"];

            txtStatus.Text = notificationStatus;
            txtChannel.Text = notificationChannelStatus;
            txtConnection.Text = deviceConnectionStatus;
        }
```

Handling Push Notifications

On the device side, there is no need to process tile notifications in your application because the system shell handles all tile notifications. However, the application will most likely process toast messages and raw HTTP messages if any. These messages give the application an opportunity to update the UI and internal state with received notification data.

For a newly received toast notification, the handling occurs when the system raises the `HttpNotificationChannel.ShellToastNotificationReceived` channel event. Recall that when the application creates a new notification channel it also sets up some event handlers. One of the event handlers is for the toast notification event:

```
httpChannel.ShellToastNotificationReceived += new

EventHandler<NotificationEventArgs>(httpChannel_ShellToastNotificationReceived)
;
```

The `ShellToastNotificationReceived` event is raised only when the underlying application is in the foreground. When the application is in the background, the system shell displays the toast in the notification area and the application does not need to do anything.

A toast notification consists of a list of key-value pairs, making it handy to use `Dictionary<string, string>` to process it. The following Listing 6-8 shows how to retrieve all key-value pairs from the toast message and add them to a text box named `txtMessage`.

LISTING 6-8: An example of handing a toast notification, PushNotification\PushNotificationClient\MainPage.xaml,cs

Available for download on Wrox.com

```
void httpChannel_ShellToastNotificationReceived(
    object sender, NotificationEventArgs e)
{
    // A toast notification is a list of key-value pairs
    if (e.Collection != null)
    {
        Dictionary<string, string> collection =
            (Dictionary<string, string>)e.Collection;
        StringBuilder messageBuilder =
            new StringBuilder();

        Dispatcher.BeginInvoke(() =>
        {
            txtMessage.Text += "Received toast: ";
        });

        foreach (KeyValuePair<string, string> kvp in collection)
        {
            // Display the message
            Dispatcher.BeginInvoke(() =>
            {
                txtMessage.Text += kvp.Key + ":" + kvp.Value;
            });
        }
        Dispatcher.BeginInvoke(() =>
        {
            txtMessage.Text += "\n";
        });
    }
}
```

A raw notification is delivered to the application only when the application is in the foreground. Otherwise the raw notification is simply dropped. The handling of a raw HTTP message really depends on the way your server-side application builds the message. If the raw HTTP message is created as an XML document as suggested, then it's fairly simple to parse the receive message as an XML document on the device application side. Listing 6-9 in the following shows an example of this scenario. The code uses LINQ queries to parse a received XML document in the raw HTTP message. The XML document consists of an `EventReminder` element which has four string sub-elements: `Date`, `Time`, `Location`, and `Notes`. To see how the XML document is composed on the server side, please refer to `BuildRawNotification()` in the `MainWindow.xaml.cs` in the PushServerEmulator project.

LISTING 6-9: An example of handling raw notifictaions, PushNotification\PushNotificationClient\MainPage.xaml.cs

Available for download on Wrox.com

```
void httpChannel_RawNotificationReceived(
    object sender, HttpNotificationEventArgs e)
{
    XDocument document;
```

continues

LISTING 6-9 *(continued)*

```
        using (var reader = new StreamReader(e.Notification.Body))
        {
            string payload =
                reader.ReadToEnd().Replace('\0', ' ');
            document = XDocument.Parse(payload);
        }

        string eventDate =
            (from c in document.Descendants("EventReminder")
            select c.Element("Date").Value).FirstOrDefault();

        string eventTime =
            (from c in document.Descendants("EventReminder")
            select c.Element("Time").Value).FirstOrDefault();


        string eventLocation =
            (from c in document.Descendants("EventReminder")
            select c.Element("Location").Value).FirstOrDefault();

        string eventNotes =
            (from c in document.Descendants("EventReminder")
            select c.Element("Notes").Value).FirstOrDefault();

        StringBuilder eventDetails =
            new StringBuilder("Received Raw event: ");
        eventDetails.Append(eventDate + ", ");
        eventDetails.Append(eventTime + ", ");
        eventDetails.Append(eventLocation + ", ");
        eventDetails.Append(eventNotes);

        Debug.WriteLine(eventDetails);
        Dispatcher.BeginInvoke(() =>
        {
            txtMessage.Text += eventDetails + "\n";
        });
    }
```

MOBILE ADVERTISING

The core of the app store business model is that developers can earn money by selling their apps and share revenue with the platform provider. For those who are reluctant to pay for the apps, developers can make their application free in the application store and use advertising to recover the cost of development. This section discusses advertising in a mobile application.

Mobile Advertising Basics

Advertising is essential to online business. Google made a fortune by building an advertising business model upon its successful search engine technologies. Social network sites, news portals, and many other types of websites use text, pictures, video, or flash advertisements to create a revenue source based on the user's impressions (ads being viewed), clicks, and subsequent actions such as the user's purchase of the advertised item or decision to sign up for a service. The Yankee group has estimated that the U.S. online advertising market will grow to $50.3 billion in revenue in 2011 (`http://www.marketingcharts.com/interactive/us-online-advertising-market-to-reach-50b-in-2011-3128/`).

When it comes to mobile advertising, the Mobile Marketing Association (MMA) has published guidelines of media channels upon which ads are delivered:

- **Mobile web:** Websites that are optimized for mobile devices
- **Messaging:** Short Message Service (SMS) and Multimedia Messaging Service (MMS)
- **Mobile apps:** Applications (including games) that are downloaded from a variety of app stores (iPhone AppStore, Android Market, Windows MarketPlace, and Nokia Ovi) and installed on a device
- **Mobile video and TV:** Video clips that are optimized for mobile

The most commonly used advertising methods for mobile applications are banner ads and interstitial ads.

- **Banner ad:** A banner ad is an ad unit that is normally displayed at the top or bottom of the screen. It may consist of text, pictures, and flash images. Once the user clicks a banner ad, then a web page, an app store item, a phone call dial pad, or an application is displayed as the ad target.
- **Interstitial ad:** An interstitial ad is a full-screen image or video shown at a certain time when the app is running, such as at the beginning of the app run, when the app is about to exit, or when a game is paused by the user. Interstitial ads may include a menu to offer more choices than just being clickable. The developer normally provides a Skip menu item so the user can skip the video playback and jump to the app.

Both banner ads and interstitial ads are independent of the application's content and logic. There are other mobile advertising methods in which ad units are embedded in the application as part of the content. But the majority of today's mobile ads are either banner or interstitial ones, which are commonly supported by popular mobile advertising providers.

The mobile advertising industry consists of three parties: in the center are the ad providers such as Apple iAd and Google AdMob. Advertisers create ad campaigns with the ad provider, and pay for this. Ad publishers, such as application developers and website owners, get paid by displaying ads pulled from the provider's ad inventory. Figure 6-2 explains this model.

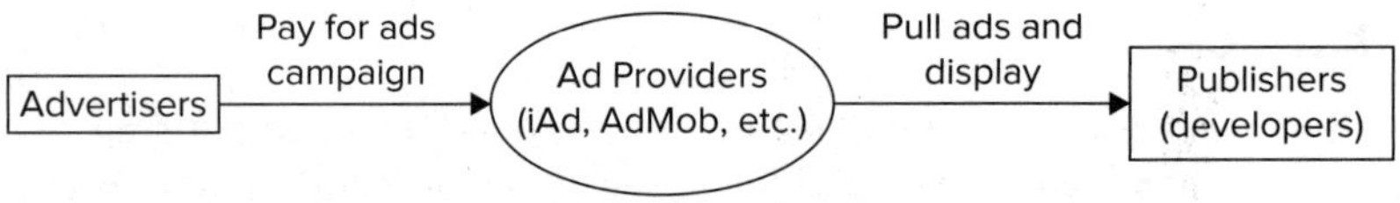

FIGURE 6-2: The model for purchasing and displaying ads

Some terms are frequently used in the mobile advertising domain. Table 6-4 shows a list of these terms.

TABLE 6-4: Terms Used in Mobile Advertising

TERM	DESCRIPTION
Impression	Number of impressions , i.e., the number of times the ad units are displayed and viewed by a user
CPM (Cost Per Mille)	Cost of 1,000 impressions that advertisers pay for. Also known as Cost Per Thousand (CPT)
CPC (Cost Per Click)	Cost per each user click that advertisers pay for. Also known as Pay Per Click (PPC)
CPA (Cost Per Action)	Cost per each user action (such as purchase, form submission, etc.) that advertisers pay for
eCPM (effective CPM)	Estimate of the amount of money a publisher can receive by switching to CPM from CPC or CPA
CTR (Click-Through Rate)	Ratio of number of clicks to number of impressions

Mobile Advertising Providers

For most mobile app developers, the easiest way to add mobile ads into their apps is to select a mobile advertising provider and follow its instructions to pull ads off the provider's ad inventory. Two of the most popular mobile advertising solutions are Apple iAd and Google AdMob. The Microsoft Ads Solution appears later in this chapter.

Apple iAd

Apple iAd came with iOS 4 in 2010. Ads delivered by iAd have a unique feature: the ad is *within* the app, meaning that once a user clicks a banner ad in an app, a full-screen advertisement will appear within the application. This is in contrast to AdMob's ad delivery method in which users are often directed to a web page in the browser, which is outside the application. The consequence of staying within the app is that users are more likely to click the banner ads because they know they aren't leaving the app. Figure 6-3 shows an iAd banner ad at the bottom of the screen in the KOMO news app, and the subsequent full screen ad once the banner ad is clicked.

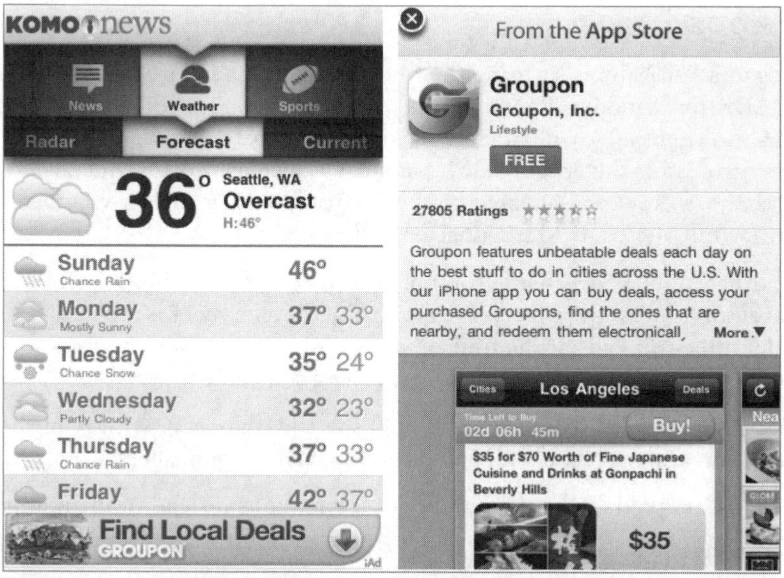

FIGURE 6-3: An iAd banner ad, in bottom-of-screen and full-screen mode (courtesy of KOMO News, Fisher Communications, Inc.)

With iAd, advertisers can build interactive, full-screen ads to improve CTR (Click-Through Rate) once the ads are shown.

The advertising revenue is split between Apple and mobile developers, with Apple taking 40 percent of the total amount generated.

There are concerns over the high cost of ad development by advertisers. In many cases the ad itself can be a full-featured application, which incurs additional cost compared to simply redirecting the user to a web URL.

Google AdMob

Google AdMob is the largest mobile advertising provider at the time of writing. In fact, when Google acquired AdMob for $750 million in 2009, the Federal Trade Commission put the deal on hold while conducting antitrust investigations on the acquisition. It was eventually cleared after Apple acquired the Quattro Wireless mobile ad network and launched its iAd platform. Unlike iAd, which can be used in iOS apps only, AdMob can be used on iOS, Android, and WP7. Figure 6-4 shows an example of an AdMob banner ad in a CNet news app.

FIGURE 6-4: AdMob banner ad in a CNet news app

Once registered at `www.admob.com`, developers can log in and submit information about the app in order to obtain the AdMob SDK. To place AdMob ads into the app, follow the instructions in the SDK to add the AdMob library into the project and use the appropriate SDK classes to request ads from the AdMob network.

Adding Ads to WP7 Apps

The official way of adding ads to a WP7 app is by using the Microsoft AdCenter AdControl provided by the Microsoft Advertising SDK for Windows Phone (`http://msdn.microsoft.com/en-us/library/ff973722(v=msads.10).aspx`). Download and install the SDK before adding ads to your WP7 applications. There are custom Silverlight AdMob controls (such as MoAds: `https://bitbucket.org/jacob4u2/moads/wiki/Home`) created by third parties, but it's unclear whether Microsoft will approve those ads for Microsoft MarketPlace certification.

Microsoft AdCenter is yet another mobile advertising solution which actually combines both Microsoft and Yahoo's ad inventory. Its support for WP7 application development is still preliminary as it does not offer interstitial ads at the time of writing — only banner ads are provided. Some unique features of the Microsoft AdCenter provider are listed below:

- Ad revenues are split 30-70 between Microsoft and publishers (developers), with publishers taking the 70 percent. Note that iAd and AdMob pay 60 percent to publishers.
- Ads in a WP7 application are paid for by the number of impressions instead of the number of clicks.
- Microsoft AdCenter boasts an "Ad Exchange" technology that is capable of choosing the most relevant and valuable ads among multiple ad networks (including Millennial Media, Where, InMobi, and MobClix). When an ad request with specific target information from a WP7 app is sent to AdCenter, several ad networks will bid, in real time, for the opportunity to show their ad for this request, and the highest bid will win.

The first step in using Microsoft AdCenter is to register at Microsoft PubCenter: `https://pubcenter.microsoft.com`. You need to obtain two piece of information before adding ads to your app:

- **Application ID:** Microsoft will assign a unique application ID to each app you want to use with AdCenter ads. The test app ID is test_client.
- **Ad Unit ID:** The ad unit ID is associated with the PubCenter account. At the time of writing, only banner ads are supported, which will result in either a click-to-web action or a click-to-call action. The test ad unit IDs are TestAd, Image480_80, and Image300_50.

Once you register your application at the PubCenter website, you need to create some ad units with the specified categories and unit format at the PubCenter website. The ad network will generate ad unit IDs and use the indicated categories to select ads for requests from the application.

In a WP7 Silverlight project, first add a reference to `Microsoft.Advertising.Mobile.dll` that should be in the directory of the Microsoft Advertising SDK for Windows Phone. On the application page where you want to add the AdCenter ad, you can either create an `AdControl` instance in the code or simply drag and drop the `AdControl` onto the page. `AdControl`'s application ID and ad unit ID must be specified in both cases.

ADD ADCONTROL TO IDE TOOLBOX

To add AdControl to the toolbox of the Visual Studio IDE, right-click on the Toolbox and select Choose Items. When promoted, select AdContro from the Windows Phone Components list.

The following code shows an example of generating an `AdControl` programmatically in the application. This code is excerpted from the AdControlDemo sample project. To try this code, replace the application ID parameter of the AdControl constructor with the one you received when you registered your application at PubCenter.

Available for download on Wrox.com

```
using Microsoft.Advertising.Mobile.UI; //AdControl class
...
AdControl adControl = new AdControl(
    "731820c8-xxxx-xxxx-xxxx-xxxxxxxxxxxx", //ApplicationID
    "26033", //ad unit ID registered at Pubcenter
    AdModel.Contextual, //Contextual is the only supported model
    true); //Enable auto refreshing of ads every 60s by default

adControl.AdSelectionKeywords = "iPad";

adControl.Width = 480;
adControl.Height = 80;
adControl.VerticalAlignment = VerticalAlignment.Bottom;

Grid grid = (Grid)this.LayoutRoot.Children[1];
grid.Children.Add(adControl);
```

Code snippet AdControlDemo\AdControlDemo\MainPage.xaml.cs

Alternatively, you can simply place an `AdControl` control onto your XAML page, and the IDE automatically adds the following code to MainPage.xaml. Note that it uses a different Ad Unit Id from the one that was created in the application code.

Available for download on Wrox.com

```
<!--ContentPanel - place additional content here-->
        <Grid x:Name="ContentPanel" Grid.Row="1" Margin="12,0,12,0">
            <my:AdControl
               Name="adControl1"
               Height="116"
               Width="480"
               HorizontalAlignment="Left"
               VerticalAlignment="Top"
               Margin="-16,43,0,0"
               ApplicationId="731820c8-xxxx-xxxx-xxxx-xxxxxxxxxxx"
               AdUnitId="26031"/>
        </Grid>
```

Code snippet AdControlDemo\AdControlDemo\MainPage.xaml

Figure 6-5 shows the output from this application, which includes two ad units.

The Microsoft PubCenter for Mobile online community (`http://community.microsoftadvertising.com/forums/32.aspx`) may provide some useful information on `AdControl` problems.

FIGURE 6-5: Adding ads to an application

SUMMARY

Consuming web services: Building web service clients on WP7 involves leveraging Silverlight .NET framework classes such as `WebClient` and `HttpWebRequest` that are included in Silverlight. JSON data serialization can be done using `DataContractJsonSerializer`, whereas XML data parsing can be done using LINQ. The sample project SocialDemo included in this chapter provides examples of these use cases.

In terms of supporting HTTP web methods, iOS, Android, and WP7 all provide APIs that can be used to implement frequently used web call scenarios. However, WP7 is unique in that its asynchronous communication pattern is easier to grasp and implement than that of Android and iOS. In addition, The Extensible Application Markup Language (XAML) data binding is also handy for UI update. On the other hand, a big limitation of WP7 compared to Android and iOS is that it does not provide APIs for network socket communication, which is in many cases the only way to build more sophisticated network applications.

Push notifications: A push notification service is essential to major mobile application frameworks because it creates the always-connected experience for the user. Battery life concerns for mobile devices have driven Apple, Google, and Microsoft to offer a dedicated push notification infrastructure, together with on-device components that facilitate push notification development. Developers have access to the Apple Push Notification service, the Google Cloud to Device Messaging, and the MPNS. Among the three, Microsoft's design is the most flexible in terms of the notification types it provides. Moreover, the push communication modeling using channels makes it easier to set up notification receivers in an application. In summary, here is a list of observations that might be useful to developers who want to consider WP7 push notifications in their applications:

- Consider using tile notifications for applications that are registered for push notification. Tile is a unique feature on WP7 (not seen on iOS and Android yet), and push notification will make it live.
- Remember that push notification delivery isn't guaranteed across the three platforms, meaning that the notifications may get lost. It may be necessary to build an acknowledgment mechanism between the application and the application server for critical notifications.

Mobile advertising: As demonstrated in this chapter, adding ads from major ads providers, such as iAd, AdMob, and AdCenter, is fairly straightforward. AdMob seems to have a larger mobile ad inventory and is supported on iOS, Android, and likely WP7 using a custom Silverlight control.

In addition, AdMob supports user-generated in-house ads with which developers can cross-promote their other apps by pointing to web URLs or app store URLs. Microsoft AdCenter for WP7 is still in its infancy with very limited ad types.

One of the issues you might wonder about is how you can leverage multiple providers in a single application. On Android, Google provides a solution called AdWhirl (`http://www.adwhirl.com`) that can retrieve ads from multiple ad networks, including AdMob, Google AdSense for Mobile Applications (AFMA), Millennial Media, Quattro, and ZestADZ. Microsoft AdCenter for WP7 provides similar capability with multiple built-in ads networks.

You might also wonder how to maximize your revenue. The rule of thumb is to choose and display ads that are most relevant to your application's target audience. For this the Microsoft AdCenter solution has built-in properties such as country, gender, location, postal code, and income range as demographic targeting parameters for the `AdControl` object. AdMob's AdView offers the `keywords` attribute in order to filter ads. Be sure to leverage the ad provider's online analytics tool to track the ad's display performance, and adjust the ad's filtering settings accordingly.

As more and more mobile applications embrace social capabilities over the Web, location-based services and applications have become popular. The next chapter will discuss creating WP7 applications that use location data and Bing maps.

7

Leveraging Location and Maps

WHAT'S IN THIS CHAPTER?

- Locating frameworks on smartphones
- Getting the current location
- Using GPS on the WP7 Emulator
- Using the Bing Maps control
- Showing locations on Bing Maps

Location-based applications are becoming one of the most popular categories of mobile application across major mobile OSs. GPS enables applications to obtain accurate geographic location information, which is essentially a built-in component on smartphones, together with the required cellular data and sometimes a Wi-Fi location database. Each mobile OS has to provide an easy-to-use location framework that allows developers to programmatically use location data, maps, and map overlays for a wide range of scenarios. Windows Phone 7 (WP7) is no exception.

This chapter starts with an overview of the location framework on three smartphone platforms: WP7, iOS, and Android. In addition, you'll see examples that show you how to get the location information on WP7 and demonstrate how to use the Bing Maps Control. In particular, you'll learn how to acquire current geographical location information using the `System.Device.Location` namespace and how to test your location-aware applications on WP7 emulators. In addition, you'll find out how to use Bing Maps web services to convert addresses (such as a street name) to geographic coordinates. You'll also discover how to use the `Microsoft.Phone.Controls.Maps` namespace to draw and navigate maps on WP7 devices. An example is also given to illustrate how to track your current location on a Bing map. At the end of the chapter, a brief introduction is given to help you leverage other Bing Maps web services to implement more advanced functions in your application, such as getting turn-by-turn directions.

LOCATION FRAMEWORKS ROUNDUP

Location-aware applications have gained momentum on the smartphone because they enable users to better connect with the surrounding world. To acquire location information, a smartphone can utilize three types of location services: Global Positioning System (GPS), cell-site triangulation, and Wi-Fi positioning.

Many high-end smartphones are now equipped with a GPS chip. This enables users to retrieve location information from satellites orbiting the earth, 24 hours a day, 7 days a week. Typical GPS units found in a smartphone belong to a type called assisted GPS (aGPS). Unlike a stand-alone GPS system that relies solely on radio signals from the satellites to retrieve information, an aGPS system also uses additional network resources, such as assistance servers, to speed up the locating process and to improve accuracy when the satellite signal strength is poor.

Compared to other positioning technologies, a GPS system can often retrieve location information more accurately. The precision of civilian GPS units is 10 to 20 meters (about 33 to 65 feet) in many cases. But these units also suffer from two major problems. One is the line-of-sight problem. If a GPS receiver is not able to see the satellites, it may not be able to receive any location data. Therefore, when a GPS-capable smartphone is inside a building or a tunnel, it simply cannot locate itself using just the GPS system. The other problem is power consumption. If your smartphone application requires the GPS sub-unit to remain running continuously, it will soon drain the battery.

Cell-site triangulation can track the position of a cell phone by measuring power levels and the antenna patterns a cell phone can receive from multiple cell base stations. In urban areas where there are more cell antenna towers, cell-site triangulation can achieve a precision of less than 50 meters (about 164 feet). Wi-Fi positioning is another technique that can be used to locate a cell phone. The idea is to use a public database that records the location of Wi-Fi hotspots. The accuracy of Wi-Fi positioning has not been fully studied, but it is considered to be one of the low-accuracy positioning technologies.

The current smartphone OS trend is to use a combination of all three positioning techniques to achieve better accuracy and reduce power consumption. On WP7, the `System.Device.Location` namespace provides a set of managed APIs to start and stop location services. As of this writing, the APIs allow developers to set two different levels of accuracy: `Default` and `High`. When the accuracy level is set to `High`, GPS positioning is generally used and it drains battery faster. When the accuracy level is set to `Default`, WP7 devices normally rely on a combination of cell-site positioning and Wi-Fi positioning to provide location information.

The geographical position information acquired from the positioning systems is in the format of geographical coordinates, in which the combination of a latitude value and a longitude value is used to pinpoint a unique location on the globe. For example, latitude 38.897687 and longitude -77.03654 represents the location of the White House. But geographical coordinates aren't easy for human beings to understand. People want to see the location on a map and require the location information in a format that's easier to comprehend, such as the mailing addresses we use in our daily lives.

WP7 provides the Microsoft Bing Maps services to developers. A number of APIs are available in the `Microsoft.Phone.Controls.Maps` namespace. Developers can use those APIs to show the location, add pushpins, and draw directions on a map. In addition, Microsoft Bing Maps services offer a set of web services using the Simple Object Access Protocol (SOAP). With the help of Bing Maps SOAP services, developers can easily translate the geographic coordinates to human-readable addresses, a process termed geocoding. Developers can also use the Bing Maps SOAP services to search for

Points of Interests (POI) and driving directions, and to retrieve imagery data. To use the service in your application, you'll need to acquire Bing Maps keys. The process of creating Bing Maps keys is explained later in this chapter.

Bing Maps web services are also offered in Representational State Transfer (REST) fashion, which allows the query to be sent as parameters in a URL. From a developer's point of view, using Bing Maps SOAP services and REST services are very similar except that the search service (such as POI) is not available through the REST interface, at least not as of this writing. For more information about the differences between SOAP and REST, please refer to Chapter 6. For illustration purposes, this chapter uses Bing Maps SOAP services in the sample code.

Because location is a piece of privacy-related information (you may not want everyone to know where you are), an application needs to obtain end-user permission to turn on location services. On WP7, this is handled by enabling the `ID_CAP_LOCATION` capability in the `WMAppManifest.xml` file.

Despite the platform and API differences, the development process for location-aware applications on WP7, iOS, and Android share a number of similarities. On iOS, you'll need to check if the location service is permitted by users and import the `CoreLocation.framework` into your Xcode project to acquire location data. Beginning with iOS 4.0, a low-power location-monitoring service is also supported that uses cell-site positioning to track user locations (`http://developer.apple.com/library/ios/#documentation/Miscellaneous/Conceptual/iPhoneOSTechOverview/CoreServicesLayer/CoreServicesLayer.html`). To display a map and support more advanced location features, such as routing services, iOS relies on Google Maps APIs. Developers need to register and acquire keys to use the Google Maps service.

Similar features and workflows can also be found on the Android platform. The application needs to claim either the `ACCESS_COARSE_LOCATION` or `ACCESS_FINE_LOCATION` permission in the Android manifest file to use this feature. After obtaining the user's permission, you can use the `android.location` package to access the location information, and use the Google maps external library to draw maps on Android-powered devices. Developers are also expected to register with and obtain keys to utilize Google Maps services.

Minimizing power consumption and maintaining reasonable accuracy and performance are always the major challenges for location-aware mobile applications. You can apply the following best practices to WP7, iOS, and Android:

- **Choose the right level of accuracy:** Choosing high accuracy or fine-location will turn on the smartphone's GPS units and drain the battery faster than under normal conditions. If your application doesn't need a high level of accuracy, consider using default or coarse accuracy.
- **Determine the frequency of updates:** Fewer updates will make your application consume less power. The downside is it will also reduce the precision of the results. You need to determine a reasonable trade-off to balance accuracy and power consumption.
- **Decide when to stop listening:** If your application isn't actively using the location services all the time, consider stopping listening to the status changes when they're not needed.
- **Prepare for a lack of data events:** In case the location data isn't available, you'll need to determine whether to wait for the data to become available or to use the cached data. In some situations, you may want to temporarily suspend your application. No matter what you decide, you should keep users informed about the situation and not staring at a frozen screen, wondering what's going on.

Testing location-aware applications on phone emulators is troublesome for developers. The Android platform seems to provide better support than the other two platforms. Android developers can use Eclipse, the Dalvik Debug Monitor Server (DDMS), or the geo command in the emulator console to simulate location-related events. However, Android is significantly better than the simulation environment on WP7 and iOS. In the next section, you'll learn how to use the Microsoft Developer Network (MSDN) GPS emulator to test your WP7 location-aware applications in the comfort of your office or home.

Table 7-1 summarizes the location and map services on three different platforms.

TABLE 7-1: Location and Maps Support on WP7, Android, and iOS

ITEMS	WP7	ANDORID	IOS
Requesting User Permission to turn on Location Service	`ID_CAP_LOCATION` capability in `WMAppManifest.xml`	`ACCESS_COARSE_LOCATION` or `ACCESS_FINE_LOCATION` in `AndroidManifest.xml`	`UIRequiredDevice` Capabilities in `Info` `.plist`
Location Service Providers	GPS, Cell, or Wi-Fi	GPS, Cell, or Wi-Fi	GPS, Cell, or Wi-Fi
Map	Bing Maps Control using Bing Maps services. Need API keys and free to register	Map kit framework using Google Maps/Google Earth API service. Need API keys and free to register	`com.google` `.android.maps` package using Google APIs add-on for the Android SDK. Need API keys and free to register
Geocoding Service	Both forward and reverse geocoding are available from Bing Maps web services. (Data is not comprehensive in Europe)	Both forward and reverse geocoding are supported using Google map services	Only reverse geocoding (converts a latitude and longitude into a placemark) are supported using Google map services
Turn-by-turn directions	Bing Maps routing web service	Google Directions API	Google Directions API
Point of Interest	Bing Maps search web service	Google Places API	Google Places API

ITEMS	WP7	ANDORID	IOS
Map annotations	Use a `Pushpin` object to set both the location information and visual representation, customizable	Add an image as the visual representation to a `MapOverlay` object. Set the location through this `MapOverlay` object and then add it to the `MapView` object	Use a `MKAnnotation` object to set the location information and use a `MKAnnotationView` object for visual representation
Emulator support	Third-party support	Using Eclipse, DDMS, or geo command in the emulator console	Third-party support

In short, from the developer's point of view, writing a location-aware application on WP7, iOS, or Android is very similar, in spite of the differences of each SDK.

GETTING CURRENT LOCATION

In this section, you'll learn how to acquire the current location on WP7. The first example shows how to get geographic coordinates and use the GPS emulator. The second example explains how to translate a human-readable address into geographic coordinates.

Geographical Data

The location information acquired using positioning technologies such as GPS is in the format of a geologic coordinate. In the first example you learn how to use the classes in the `System.Device.Location` namespace to retrieve information.

Many developers like to start testing location-aware applications in a simulated environment to save time and energy. It's necessary to address the simulation issue first. The WP7 phone emulator that comes with the SDK doesn't emulate location changes. However, it's possible for developers to write a mock application that can simulate the location changes. During the debugging process, you can let your phone application use the mock application as the location service provider. Alternatively, you can use one of the GPS emulator applications contributed by other developers.

The GPS emulator application you'll use in this chapter is from MSDN (`http://create.msdn.com/education/catalog/article/GPS-Emulator`). You can download the zip file directly from the MSDN site. However, the description provided by MSDN is a bit hard to follow and more information than necessary is presented to use the emulator. Therefore, simply download all the files from the GPS Emulator.

The two most important files in the folder are `GpsEmulator.exe` and `GpsEmulatorClient.dll`. You can use `GpsEmulator.exe` to simulate location movement and add `GpsEmulatorClient.dll` to your application to listen to the location events. Note that to launch `GpsEmulator.exe` in Windows Vista or Windows 7, you'll need to right-click on the program and choose Run as Administrator from the context menu. Click Yes when asked if you want to allow this program to make changes to your computer. A screenshot of the emulator program is shown in Figure 7-1.

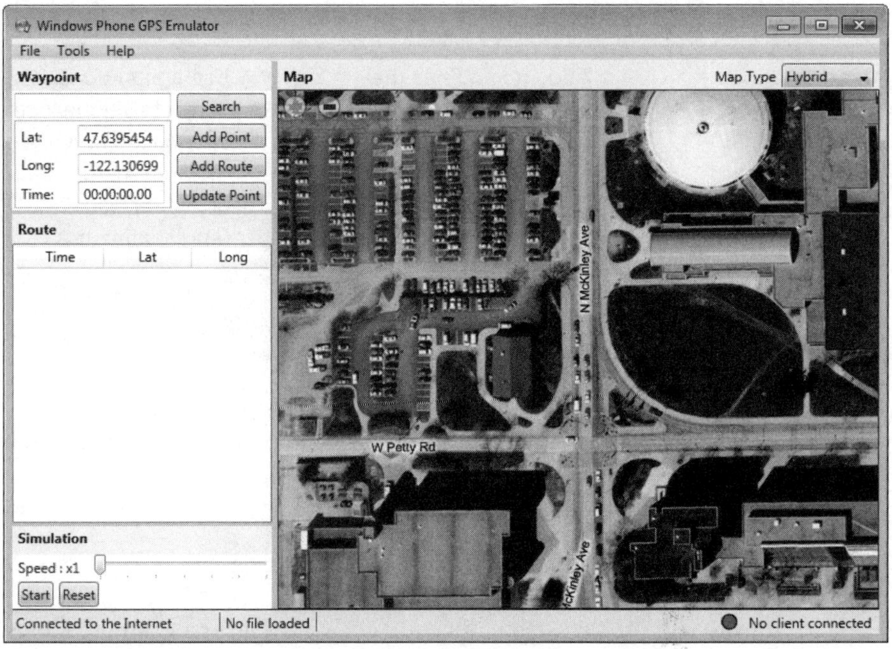

FIGURE 7-1: Windows Phone GPS Emulator

Using this emulator, you can create a simulated route by adding waypoints and clicking on the map. Alternatively, you can load the `BSU.route` file by clicking on File ➪ Open.

For now, you can put this emulator tool aside and start working on a simple location-aware application. Use the following steps to create the solution:

1. Start Microsoft Visual Studio Express 2010 for Windows Phones. Windows starts the Visual Studio Express 2010 IDE.
2. Choose File ➪ New Project. You'll see the New Project dialog box.
3. Choose the Visual C# Silverlight for Windows Phone templates from the left panel. Highlight the Windows Phone Application template. You'll see a description of this template in the right pane of the New Project dialog box.
4. Type **currentLocationTxt** in the Solution Name field and the Name field. Click OK. Visual Studio creates a blank application for you.

Change the page title, add 10 TextBlocks, and 3 Buttons as described in Listing 7-1.

Available for download on Wrox.com

LISTING 7-1: UI layout, currentLocationTxt\currentLocationTxt\MainPage.xaml

```
<!--TitlePanel contains the name of the application and page title-->
<StackPanel x:Name="TitlePanel" Grid.Row="0" Margin="12,17,0,28">
    <TextBlock x:Name="ApplicationTitle" Text="MY APPLICATION"
               Style="{StaticResource PhoneTextNormalStyle}"/>
```

```
        <TextBlock x:Name="PageTitle" Text="My Location" Margin="9,-7,0,0"
                   Style="{StaticResource PhoneTextTitle1Style}"/>
    </StackPanel>

    <!--ContentPanel - place additional content here-->
    <Grid x:Name="ContentPanel" Grid.Row="1" Margin="12,0,12,0">
        <TextBlock Height="40" Name="timeLabel" FontWeight="Bold" Text="Time"
                   Margin="20,20,0,0" VerticalAlignment="Top" Width="120"
                   FontSize="24" HorizontalAlignment="Left" />
        <TextBlock FontSize="20" FontWeight="Normal" Height="40"
                   Margin="140,20,0,0" Name="timeStr" Text="" Width="280"
                   HorizontalAlignment="Right" VerticalAlignment="Top" />
        <TextBlock FontSize="24" FontWeight="Bold" Height="40" Width="120"
                   HorizontalAlignment="Left" Margin="20,110,0,0"
                   Name="statusLabel" Text="Status" VerticalAlignment="Top" />
        <TextBlock FontSize="20" FontWeight="Normal" Height="40" Width="280"
                   HorizontalAlignment="Right" Margin="0,110,0,0"
                   Name="statusStr" Text="" VerticalAlignment="Top" />
        <TextBlock Height="40" Name="latLabel" FontWeight="Bold" Text="Latitude"
                   Margin="20,200,0,0" VerticalAlignment="Top" Width="120"
                   FontSize="24" HorizontalAlignment="Left" />
        <TextBlock FontSize="20" FontWeight="Normal" Height="40"
                   Margin="140,200,0,0" Name="latStr" Text="" Width="280"
                   HorizontalAlignment="Right" VerticalAlignment="Top" />
        <TextBlock Height="40" Name="longLabel" FontWeight="Bold" Text="Longitude"
                   Margin="20,290,0,0" VerticalAlignment="Top" Width="120"
                   FontSize="24" HorizontalAlignment="Left" />
        <TextBlock FontSize="20" FontWeight="Normal" Height="40"
                   Margin="140,290,0,0" Name="longStr" Text="" Width="280"
                   HorizontalAlignment="Right" VerticalAlignment="Top" />
        <TextBlock Height="40" Name="accuLabel" FontWeight="Bold" Text="Accuracy"
                   Margin="20,380,0,0" VerticalAlignment="Top" Width="120"
                   FontSize="24" HorizontalAlignment="Left" />
        <TextBlock FontSize="20" FontWeight="Normal" Height="40"
                   Margin="140,380,0,0" Name="accuStr" Text="" Width="280"
                   HorizontalAlignment="Right" VerticalAlignment="Top" />
        <Button Content="Stop" Height="72" HorizontalAlignment="Left"
                Margin="20,529,0,0" Name="stopBtn" VerticalAlignment="Top"
                Width="120" Click="stopBtn_Click" />
        <Button Content="High" Height="72" HorizontalAlignment="Left"
                Margin="165,529,0,0" Name="highBtn" VerticalAlignment="Top"
                Width="130" Click="highBtn_Click" />
        <Button Content="Low" Height="72" HorizontalAlignment="Left"
                Margin="319,529,0,0" Name="lowBtn" VerticalAlignment="Top"
                Width="131" Click="lowBtn_Click" IsEnabled="True" />
    </Grid>
```

The layout of the UI is shown in Figure 7-2.

Now from Solution Explorer, right-click on References, and choose Add Reference from the context menu, as shown in Figure 7-3.

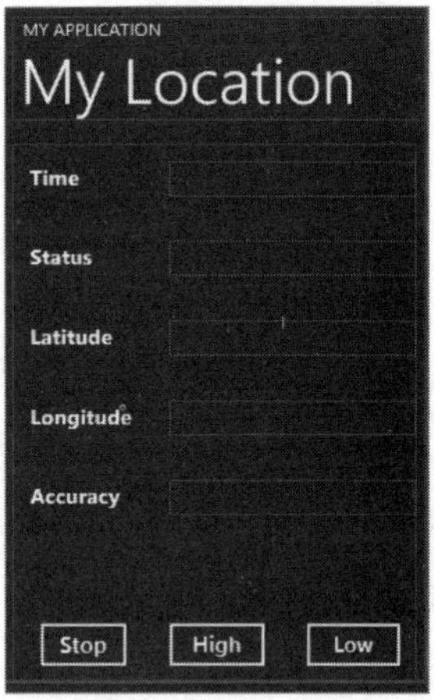

FIGURE 7-2: UI layout of currentLoctionTxt

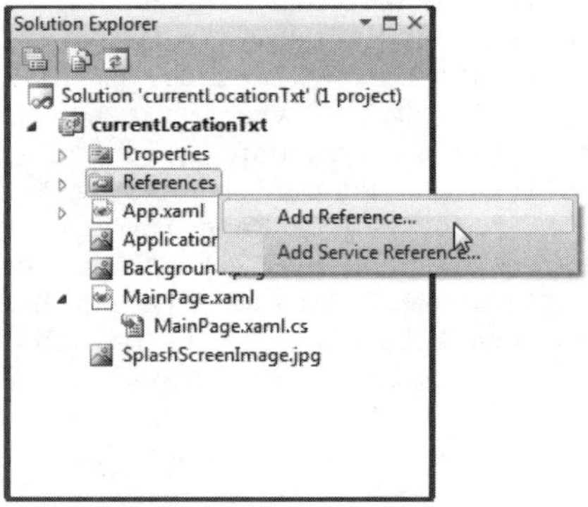

FIGURE 7-3: Add References

When you see the Add Reference dialog, choose the Browse tab and select the `GpsEmulatorClient.dll` you downloaded earlier. Click OK.

In the `System.Device.Location` namespace, the `GeoCoordinateWatcher` class exposes the location service. To use the GPS Emulator, however, you'll need to add the following three lines at the top of the `MainPage.xaml.cs` file.

Available for download on Wrox.com

```
#define GPS_EMULATOR          // defining a compiler GPS symbol.
using GpsEmulatorClient;      //Namespace for GpsEmulator
using System.Device.Location;   //Namespace for the actual location service
```

Code Snippet \currentLocationTxt\currentLocationTxt\MainPage.xaml.cs

The benefit of defining a `GPS_EMULATOR` macro is that it can make you quickly switch from the emulating environment to the actual testing environment. When the macro is defined, the new `GeoCoordinateWatcher` object uses the emulator namespace. If you comment out the first two lines of code, then the application instantiates a `GeoCoordinateWatcher` object (which is found in the `System.Device.Location` namespace). This is achieved by using a conditional compile statement.

The operation of this application is fairly simple. When a user clicks the High button, the location service starts in high-accuracy mode; otherwise, when the user clicks the Low button, the location service runs in low-accuracy mode. And the Stop button stops the location services. The returned geographic data is printed on the screen as text strings. The C# code of `currentLocationTxt` is shown in Listing 7-2.

Available for download on Wrox.com

LISTING 7-2: Displaying current geo-coordinates, currentLocationTxt\currentLocationTxt\MainPage.xaml.cs

```
public partial class MainPage : PhoneApplicationPage
{
    #if GPS_EMULATOR
            GpsEmulatorClient.GeoCoordinateWatcher watcher;
    #else
            System.Device.Location.GeoCoordinateWatcher watcher;
    #endif

    // Constructor
    public MainPage()
    {
        InitializeComponent();
    }

    private void stopBtn_Click(object sender, RoutedEventArgs e)
    {
        if (watcher != null)
        {
            watcher.Stop();
        }
        statusStr.Text = "Location service is turned off";
        timeStr.Text = System.DateTime.Now.ToString();
        latStr.Text = "";
        longStr.Text = "";
        accuStr.Text = "";
        highBtn.IsEnabled = true;
        lowBtn.IsEnabled = true;
    }

    //Start the location service with desired accuracy
    private void startLoation(GeoPositionAccuracy accu)
    {
        //change status info
        statusStr.Text = "Starting Location Service...";

        #if GPS_EMULATOR
            watcher = new GpsEmulatorClient.GeoCoordinateWatcher(accu);
        #else
            watcher = new System.Device.Location.GeoCoordinateWatcher(accu);
        #endif

        watcher.MovementThreshold = 20; //Set to 20 meters

        // Add event handlers for StatusChanged and PositionChanged events
        watcher.StatusChanged += new EventHandler
            <GeoPositionStatusChangedEventArgs>(statusChanged);
        watcher.PositionChanged += new EventHandler
            <GeoPositionChangedEventArgs<GeoCoordinate>>(positionChanged);

        // Start acquiring data
```

continues

LISTING 7-2 *(continued)*

```
        watcher.Start();

    }

    //event handler of position changes
    void positionChanged(object sender,
        GeoPositionChangedEventArgs<GeoCoordinate> e)
    {
        this.Dispatcher.BeginInvoke( () => changeUIposition(e) );

    }

    void changeUIposition(GeoPositionChangedEventArgs <GeoCoordinate> e)
    {
        if (watcher.Status == GeoPositionStatus.Disabled) return;

        latStr.Text = e.Position.Location.Latitude.ToString("0.00000");
        longStr.Text = e.Position.Location.Longitude.ToString("0.00000");
        timeStr.Text = System.DateTime.Now.ToString();
    }

    //event handler of status changes
    void statusChanged(object sender, GeoPositionStatusChangedEventArgs e)
    {
        this.Dispatcher.BeginInvoke( () => changeUIstatus(e) );
    }

    void changeUIstatus(GeoPositionStatusChangedEventArgs e)
    {
        switch (e.Status)
        {
            case GeoPositionStatus.Disabled:
                //service is disabled or unsupported.
                statusStr.Text = "location service is not enabled";
                break;

            case GeoPositionStatus.Initializing:
                //service is initializing.
                statusStr.Text = "initializing...";
                break;

            case GeoPositionStatus.NoData:
                //service is working, but it cannot get location data
                statusStr.Text = "Sorry, data unavailable";
                break;
            case GeoPositionStatus.Ready:
                //service is working and is receiving data
                statusStr.Text = "Ready";
                break;
```

```
            }

            timeStr.Text = System.DateTime.Now.ToString();
        }


        private void highBtn_Click(object sender, RoutedEventArgs e)
        {
            accuStr.Text = "High Accuracy";
            startLoation(GeoPositionAccuracy.High);

            //turn off high accuracy button and turn on low accuracy button
            highBtn.IsEnabled = false;
            lowBtn.IsEnabled = true;
        }

        private void lowBtn_Click(object sender, RoutedEventArgs e)
        {
            //turn on high accuracy button and turn off low accuracy button
            accuStr.Text = "Low Accuracy";
            startLoation(GeoPositionAccuracy.Default);

            //turn on high accuracy button and turn off low accuracy button
            highBtn.IsEnabled = true;
            lowBtn.IsEnabled = false;
        }
    }
```

In method `private void startLocation(GeoPositionAccuracy accu)`, the code begins by creating a `GeoCoordinateWatcher` object and `MovementThreshold` is set to 20 meters. This is a small threshold and should be accurate enough for many applications. Event handler `statusChanged()` is then added to the event queue when the `StatusChanged` event is triggered. Event handler `positionChanged()` is added to the event queue when the `PositionChanged` event occurs. Because getting geographic data is a background process, it can't update the contents of a control that is created on the UI thread. To solve this problem, you call the `Dispatcher.BeginInvoke()` method and declare the actions needed to update the UI content. In this example, when the current position is changed, `changeUIposition()` is invoked to update the latitude and longitude information, and `changeUIstatus()` is invoked to update the status information.

Four different statuses are defined in the `GeoPositionStatus` enumeration to reflect the status of a location provider:

- `Ready`: A location provider is available and can supply data.
- `Initializing`: A location provider is initiating. For a GPS receiver, it may take a few minutes to obtain its signal.
- `NoData`: A location provider is enabled but has no data.
- `Disabled`: A location provider is disabled.

While `GpsEmulator.exe` is still running, you can start debugging the application. The result is shown in Figure 7-4.

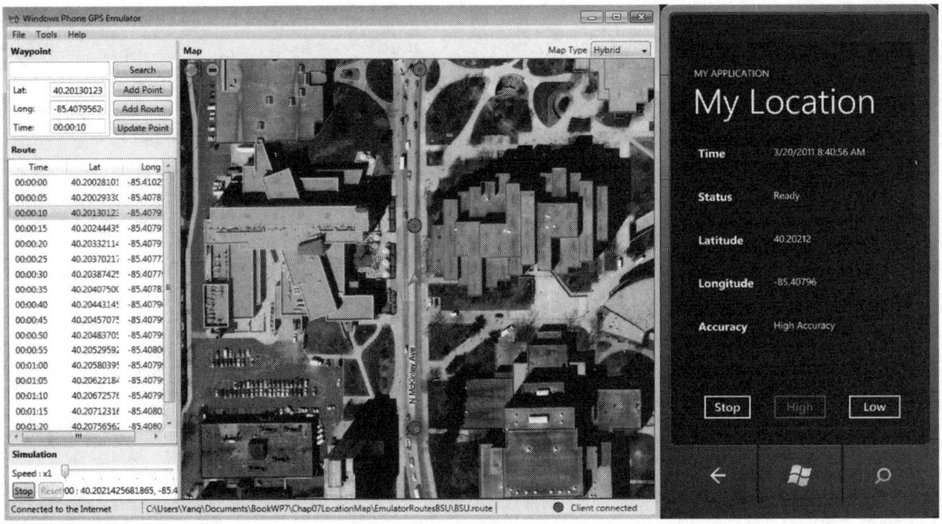

FIGURE 7-4: Running currentLocationTxt

Civilian Data

Imagine someone asks you, "Hey, what's your home address in geographic coordinates?" Your response is probably, "What?!" Well, we human beings remember and find locations based on street addresses, or what could be termed as civilian data. So if your application requires that users input location information, expect users to enter a street address. Since most of the location-related APIs speak only geographical data, it's necessary to create a process that translates the information between geographical data and civilian data. This translation process is often called *geocoding*. The forward geocode service translates street names or entity names into geographical data, and reverse geocoding locates possible street names from geographic coordinates.

On WP7, geocoding is offered as a web service. You can add the Bing Maps geocode service to your application as a service reference. Before you do so, be sure to create a Bing Maps account and create an application ID for your application. Otherwise, your application won't be able to get responses from the Bing Maps web services.

Open your web browser and type the following URL: `https://www.bingmapsportal.com/`. Follow the online instructions and create a free Bing Maps account with your Windows Live ID. Once you sign in, you can then create up to five application IDs for your map applications. Figure 7-5 shows a screenshot of the web page.

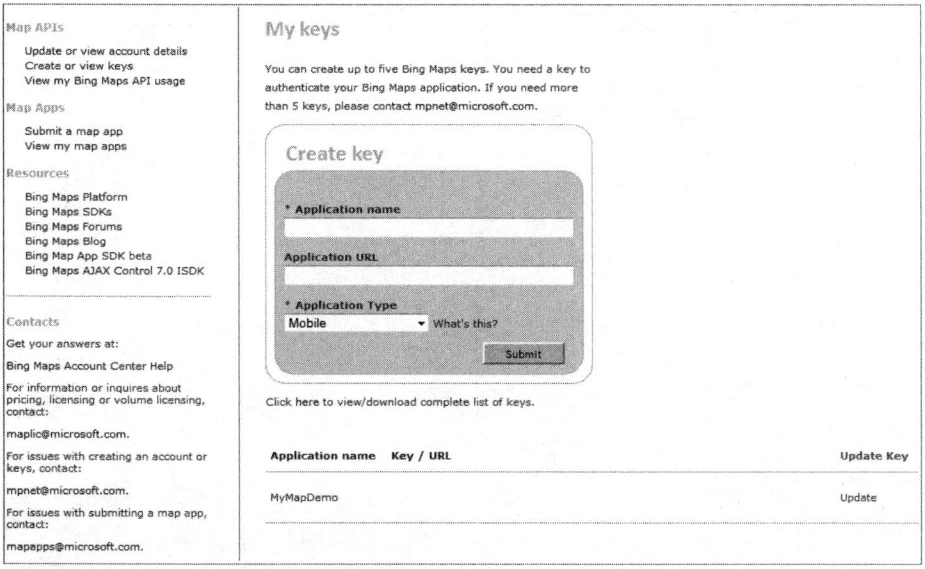

FIGURE 7-5: Create or view Bing Maps key

After obtaining a Bing Maps key, use the following steps to create a new project:

1. Start Microsoft Visual Studio Express 2010 for Windows Phones. Windows starts the Visual Studio Express 2010 IDE.
2. Choose File ⇨ New Project. You'll see the New Project dialog box.
3. Choose the Visual C# Silverlight for Windows Phone templates from the left panel. Highlight the Windows Phone Application template. You'll see a description of this template in the right pane of the New Project dialog box.
4. Type **street2Geo** in the Solution Name field and the Name field. Click OK. Visual Studio creates a blank application for you.
5. In Solution Explorer, right-click on Service References. You'll see the dialog box shown in Figure 7-6.
6. Choose Add Service Reference. You will see the Add Service Reference dialog box.
7. Type **`http://dev.virtualearth.net/webservices/v1/geocodeservice/geocodeservice.svc`** in the Address field, and type **BingGeoCodeService** in the Namespace field, as seen in Figure 7-7.

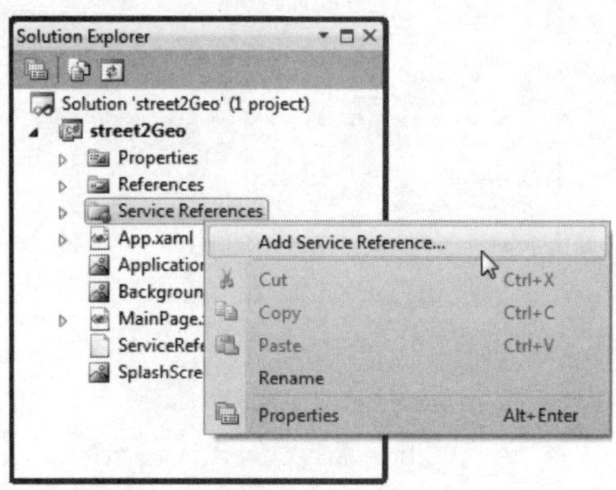

FIGURE 7-6: Add service references.

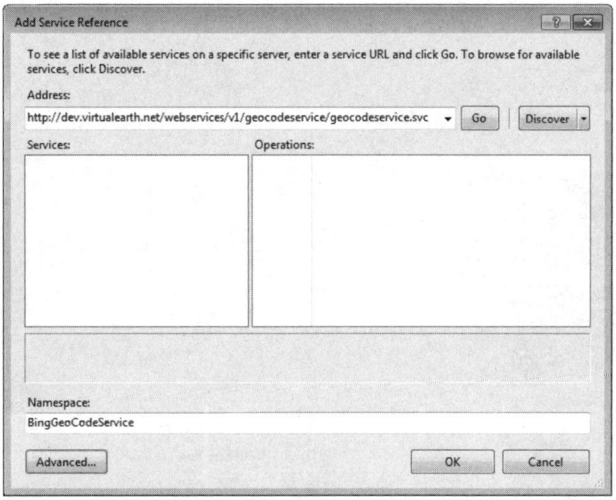

FIGURE 7-7: Enter web service URL.

8. Click Go and you'll see that the IDE discovers a new service. This confirms the geocode web service is available to your application, as seen in Figure 7-8.

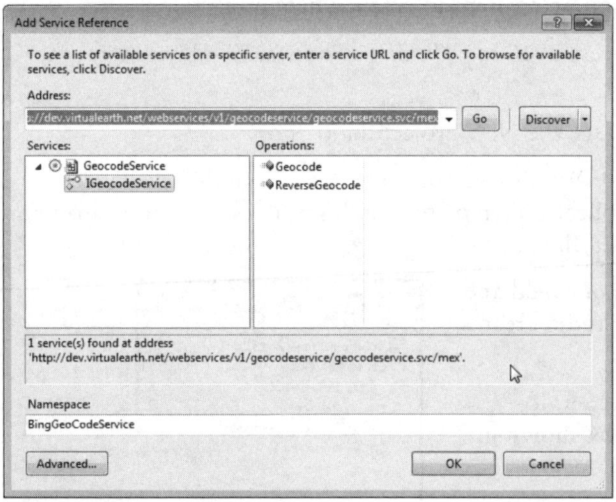

FIGURE 7-8: Geocode service found.

9. Click Advanced, and you'll see a Service Reference Settings dialog box.
10. Uncheck the Reuse types in the referenced assemblies checkbox, as shown in Figure 7-9. If you don't uncheck this option, you'll get a warning message in your IDE: "Custom tool warning: Unable to load one or more of the requested types..."
11. Click OK twice. You'll see that the web service is now added to the solution, as shown in Figure 7-10.

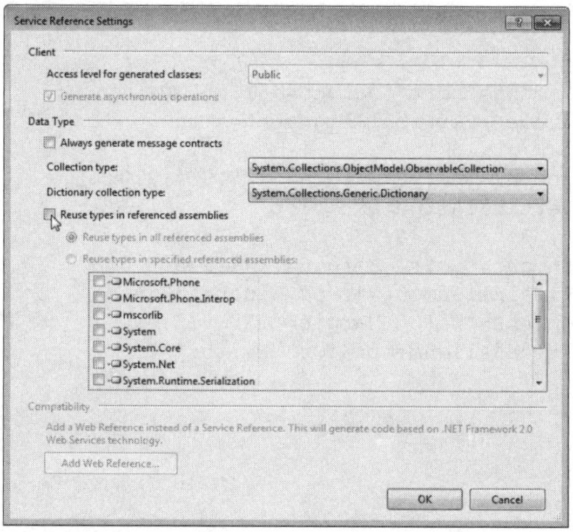

FIGURE 7-9: Clear type reuse.

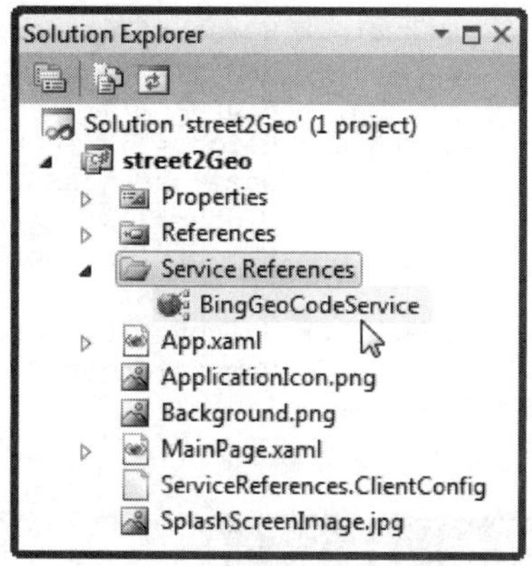

FIGURE 7-10: GeoCodeService is added to your project.

At this point, you can add a few UI controls to the project, as shown in Listing 7-3. Figure 7-11 shows the UI layout of this project.

Available for download on Wrox.com

LISTING 7-3: UI Layout of street2Geo project, currentLocationTxt\currentLocationTxt\MainPage.xaml

```
<!--TitlePanel contains the name of the application and page title-->
<StackPanel x:Name="TitlePanel" Grid.Row="0" Margin="12,17,0,28">
    <TextBlock x:Name="ApplicationTitle" Text="MY APPLICATION"
               Style="{StaticResource PhoneTextNormalStyle}"/>
    <TextBlock x:Name="PageTitle" Text="Street Address to Coordiantes"
               Margin="9,-7,0,0" FontSize="32" FontWeight="Bold"
               Style="{StaticResource PhoneTextTitle1Style}" />
</StackPanel>

<!--ContentPanel - place additional content here-->
<Grid x:Name="ContentPanel" Grid.Row="1" Margin="12,0,12,0">
    <TextBox Height="72" HorizontalAlignment="Left" Margin="-4,42,0,0"
             Name="streetInput" Text="2000 W. University Ave, Muncie, IN"
             VerticalAlignment="Top" Width="460" />
    <TextBlock Height="30" HorizontalAlignment="Left" Margin="6,6,0,0"
               Name="inputLbl1" Text="Enter address below"
               VerticalAlignment="Top" Width="423" />
    <Button Content="Find Geo Data" Height="72" HorizontalAlignment="Left"
            Margin="127,103,0,0" Name="findBtn" VerticalAlignment="Top"
            Width="323" Click="findBtn_Click" />
    <TextBlock Height="36" HorizontalAlignment="Left" Margin="6,248,0,0"
               Name="statusLbl" Text="Status" VerticalAlignment="Top"
               Width="115" />
    <TextBlock Height="36" HorizontalAlignment="Left" Margin="9,345,0,0"
               Name="latLbl" Text="Latitude" VerticalAlignment="Top"
               Width="115" />
```

continues

LISTING 7-3 *(continued)*

```
    <TextBlock Height="36" HorizontalAlignment="Left" Margin="12,437,0,0"
               Name="longLbl" Text="Longitude" VerticalAlignment="Top"
               Width="115" />
    <TextBlock Height="30" HorizontalAlignment="Left" Margin="127,254,0,0"
               Name="statusStr" Text="" VerticalAlignment="Top"
               Width="323" />
    <TextBlock Height="30" HorizontalAlignment="Left" Margin="127,345,0,0"
               Name="latStr" Text="" VerticalAlignment="Top" Width="323" />
    <TextBlock Height="30" HorizontalAlignment="Left" Margin="127,437,0,0"
               Name="longStr" Text="" VerticalAlignment="Top" Width="323" />
</Grid>
```

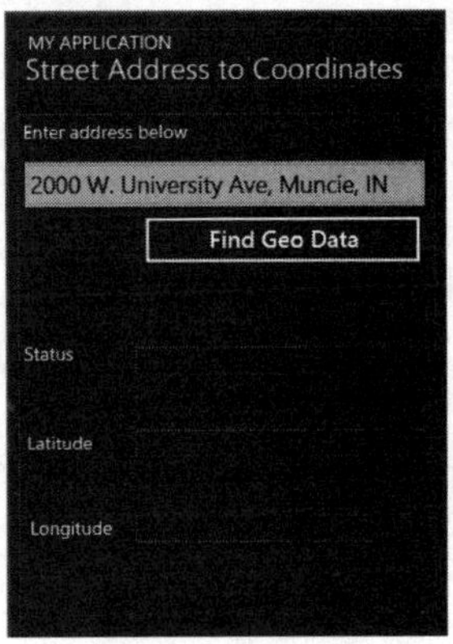

FIGURE 7-11: UI Layout of street2Geo

Now implement the constructor method and button-click event handler. You will also need to create a method (`showResult()` in this example) as the event handler when the results of the query are returned. The C# code of `MainPage.xaml.cs` is shown in Listing 7-4.

LISTING 7-4: Using Bing Maps geocode web service, street2Geo\street2Geo\MainPage.xaml.cs

Available for download on Wrox.com

```
using System;
using System.Collections.Generic;
using System.Linq;
using System.Net;
using System.Windows;
using System.Windows.Controls;
using System.Windows.Documents;
using System.Windows.Input;
using System.Windows.Media;
using System.Windows.Media.Animation;
using System.Windows.Shapes;
```

```
using Microsoft.Phone.Controls;
using street2Geo.BingGeoCodeService;  //namespace for the web services
using System.Collections.ObjectModel; //namespace for filters collection

namespace street2Geo
{
    public partial class MainPage : PhoneApplicationPage
    {
        GeocodeServiceClient geocodesvc;

        // Constructor
        public MainPage()
        {
            InitializeComponent();

            //create a new service client object.
            geocodesvc =
                new GeocodeServiceClient("BasicHttpBinding_IGeocodeService");

            //Add an event handler when query is completed
            geocodesvc.GeocodeCompleted += new EventHandler
                <GeocodeCompletedEventArgs>(showResult);

        }

        private void showResult(object sender, GeocodeCompletedEventArgs e)
        {
            //Confidence.High=0, Medium=1; Low=2;
            //Select the most confident result
            var geoResult = (from r in e.Result.Results
                             orderby (int)r.Confidence ascending
                             select r).FirstOrDefault();

            if (geoResult == null)
            {
                statusStr.Text = "No results found. Try again.";
                latStr.Text = "";
                longStr.Text = "";
                return;
            }
            statusStr.Text = "Yes, found it!";
            latStr.Text = geoResult.Locations[0].Latitude.ToString();
            longStr.Text = geoResult.Locations[0].Longitude.ToString();
        }

        private void findBtn_Click(object sender, RoutedEventArgs e)
        {
            GeocodeRequest req =  new GeocodeRequest();

            req.Credentials = new BingGeoCodeService.Credentials();
            req.Credentials.ApplicationId = "Enter your Bing Maps Key here";
            req.Query = streetInput.Text;

            req.Options = new GeocodeOptions();
```

continues

LISTING 7-4 *(continued)*

```
                req.Options.Filters= new ObservableCollection<FilterBase>();

                ConfidenceFilter filter = new ConfidenceFilter();
                filter.MinimumConfidence = Confidence.High;

                req.Options.Filters.Add (filter);

                try
                {
                    geocodesvc.GeocodeAsync(req);
                }
                catch //Get error
                {
                    statusStr.Text = "Bad address. Try again.";
                    latStr.Text = "";
                    longStr.Text = "";
                }

            }
        }
    }
```

> *To get this application running, you'll need to pass your Bing Maps key to the* `req.Credentials.ApplicationId` *property.*

The basic idea behind this code is to create a new `GeocodeServiceClient` object named `geocodesvc` in the constructor and add an event handler `showResult()` to update the UI content when the results are returned. In the button-click event handler `findBtn_Click()`, you create a new `GeocodeRequest` object `req`; set the `Query` field based on users' input; set the confidence level to high; then use the `GeocodeAsync()` method to call the Bing Maps SOAP web services.

The `GeocodeServiceClient` class has an overloaded constructor. Don't use the most basic form of the constructor that requires no parameters even though the application will compile. This is because without knowing the type of services offered, a `GeocodeServiceClient` object may not be able to establish a connection to the web service provider. In this example, the connection type is set to `BasicHttpBinding_IGeocodeService`.

Web services may return more than one result for a query. If you only care about the result with the most confidence, whose numeric value is 0, you can order the results by the confidence level in ascending order and print the first result from the array.

When you compile and run the program, you'll be able to get the latitude and longitude information from a street name or entity name, as shown in Figure 7-12.

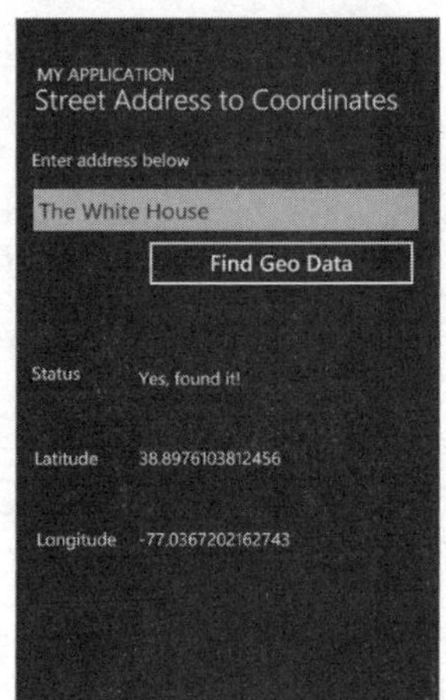

FIGURE 7-12: Running street2Geo

USING MAPS

So far you've learned how to acquire location information and how to translate civilian data to geographical data. To end users, however, nothing is more straightforward than seeing the location on a map. In this section, you'll learn how to use the Bing Maps control to present location information.

Using the Bing Map Control

To use the Bing Map control in your application, you'll need to have a Bing Maps account and create a key (`ApplicationID`) for your application. In the following example, you'll learn how to use the Bing Map control to easily center a map, zoom in and out, and add a pushpin to the map. Use the following steps to create a new project:

1. Start Microsoft Visual Studio Express 2010 for Windows Phones. Windows starts the Visual Studio Express 2010 IDE.
2. Choose File ➪ New Project. You'll see the New Project dialog box.
3. Choose the Visual C# Silverlight for Windows Phone templates from the left panel. Highlight the Windows Phone Application template. You'll see a description of this template in the right pane of the New Project dialog box.
4. Type **simpleMap** in the Solution Name field and the Name field. Click OK. Visual Studio creates a blank application for you.
5. From Toolbox, select the map control and drop it to the center of the Designer. Performing this step automatically loads a few libraries needed for the Bing Map control, such as `Microsoft.Phone.Controls.Map`, `System.Device`, `System.Runtime.Serialization`, and `System.ServiceModel`.
6. Add a button and a slider to the Designer and follow Listing 7-5 to change the name, size, and other attributes of the controls. The UI layout is shown in Figure 7-13.

LISTING 7-5: UI Layout of simpleMap project, simpleMap\simpleMap\MainPage.xaml

Available for download on Wrox.com

```
<!--TitlePanel contains the name of the application and page title-->
<StackPanel x:Name="TitlePanel" Grid.Row="0" Margin="12,17,0,28">
    <TextBlock x:Name="ApplicationTitle" Text="MY APPLICATION"
               Style="{StaticResource PhoneTextNormalStyle}"/>
    <TextBlock x:Name="PageTitle" Text="Bing Map" Margin="9,-7,0,0"
               Style="{StaticResource PhoneTextTitle1Style}"/>
</StackPanel>

<!--ContentPanel - place additional content here-->
<Grid x:Name="ContentPanel" Grid.Row="1" Margin="12,0,12,0">
    <my:Map Height="526" HorizontalAlignment="Left" Margin="0,6,0,0"
            Name="map1" VerticalAlignment="Top" Width="456"
            DataContext="{Binding}" LogoVisibility="Collapsed">
        <my:Map.CredentialsProvider>
            <my:ApplicationIdCredentialsProvider
                ApplicationId="Enter your key here"/>
        </my:Map.CredentialsProvider>
```

continues

LISTING 7-5 *(continued)*

```
        </my:Map>
        <Button Content="Satellite" Height="72" HorizontalAlignment="Left"
                Margin="0,529,0,0" Name="modeBtn" VerticalAlignment="Top"
                Width="160" Click="modeBtn_Click" />
        <Slider Height="95" HorizontalAlignment="Left" Margin="172,529,0,0"
                Name="zoomSlider" VerticalAlignment="Top" Width="284"
                Maximum="20" Value="15" BorderThickness="1"
                Background="Black" Foreground="Blue"
                ValueChanged="zoomLevel_ValueChanged" Minimum="1" />
    </Grid>
```

Note that you'll need to enter your own Bing Map key in the `ApplicationID` line. In this simple map application, you can click the button to toggle between the road mode and aerial mode (satellite mode), and use the slider to zoom in and out. The Bing Map zoom level ranges from 0 to 21, with 0 showing the entire world and 21 showing close-ups of a particular location. The Bing Map Control has a `ZoomBarVisibility` property and is set to `Collapsed` on WP7 by default. This is because the zoom bar is too small to operate on a phone. In addition, if the viewable area of the map is too small, only a portion of the zoom bar will be visible on the map, thus making the built-in zoom control useless. In this example, we use a larger slider control to change the zoom level of the map.

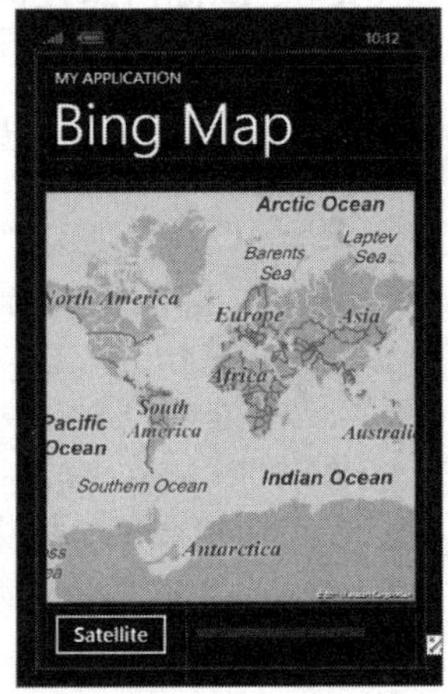

FIGURE 7-13: UI Layout of simpleMap project

Listing 7-6 shows how you can use just a few lines of code to easily manipulate a Bing Map control.

LISTING 7-6: Using Bing Maps control, simpleMap\simpleMap\MainPage.xaml.cs

Available for download on Wrox.com

```
using System;
using System.Collections.Generic;
using System.Linq;
using System.Net;
using System.Windows;
using System.Windows.Controls;
using System.Windows.Documents;
using System.Windows.Input;
using System.Windows.Media;
using System.Windows.Media.Animation;
using System.Windows.Shapes;
using Microsoft.Phone.Controls;
//add Maps namespace
using Microsoft.Phone.Controls.Maps;
using System.Device.Location;

namespace simpleMap
{
    public partial class MainPage : PhoneApplicationPage
```

```
    {
        bool isRoadmode = false;
        Pushpin pinBSU; //

            //Add the pushpin to the map
        // Constructor
        public MainPage()
        {
            InitializeComponent();

            //Set the location of the pushpin

            pinBSU = new Pushpin();
            pinBSU.Location = new GeoCoordinate
                    (40.2002810100341, -85.4102575778961);
            pinBSU.Content = "BSU";

            //Initial Location
            map1.ZoomLevel = zoomSlider.Value;
            map1.Center = pinBSU.Location;

            map1.Children.Add(pinBSU);
        }

        private void modeBtn_Click(object sender, RoutedEventArgs e)
        {
            if (isRoadmode) {
                map1.Mode = new RoadMode();
                modeBtn.Content = "Satellite";
                isRoadmode = false;
            }
            else
            {
                map1.Mode = new AerialMode();
                modeBtn.Content = "Road";
                isRoadmode = true;
            }

        }

        private void zoomLevel_ValueChanged(object sender,
            RoutedPropertyChangedEventArgs<double> e)
        {
            if (zoomSlider != null)
                map1.ZoomLevel = zoomSlider.Value;
        }
    }
}
```

The constructor method creates a `Pushpin` object and sets the `Location` property to the geographic coordinates of Ball State University. You can then add this `Pushpin` object to the list of `map1.Children`, which is of type `UIElementCollection`.

When you compile and run the project, you'll see that the location of Ball State University is shown on a map with a "BSU" pushpin at the center of the map. The screenshot of this project is shown in Figure 7-14.

Now that you've gained a better understanding of the Bing Map control, you're ready to explore how to use both the location service and the map control in your application.

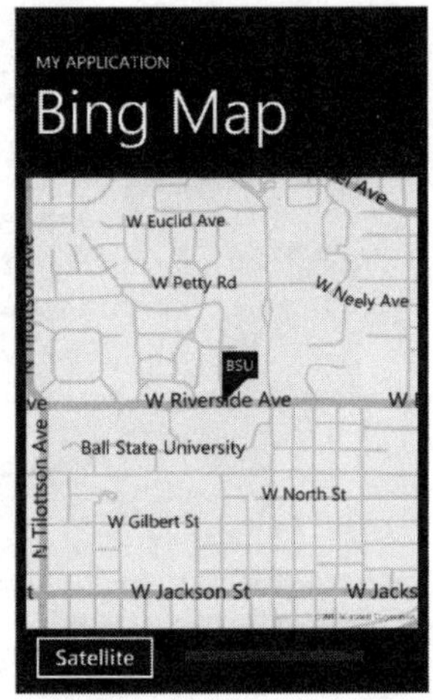

FIGURE 7-14: Running simpleMap

Combining the Location Service and Bing Map

In this example, you'll write an application that can track the movement of your phone on the map. Most of the knowledge required to build this application has already appeared previously, so this section doesn't require detailed instructions.

Create a new Silverlight Windows Phone project titled TrackMe. Add a Bing Map control and two buttons to the Designer, and add `GpsEmulatorClient.dll` to the References. Follow Listing 7-7 and change the locations and other properties of the controls.

LISTING 7-7: UI Layout of TrackMe project, TrackMe\TrackMe\MainPage.xaml

Available for download on Wrox.com

```
<!--TitlePanel contains the name of the application and page title-->
<StackPanel x:Name="TitlePanel" Grid.Row="0" Margin="12,17,0,28">
    <TextBlock x:Name="ApplicationTitle" Text="MY APPLICATION"
           Style="{StaticResource PhoneTextNormalStyle}"/>
    <TextBlock x:Name="PageTitle" Text="Track me on map" Margin="9,-7,0,0"
           Style="{StaticResource PhoneTextTitle1Style}" FontSize="48" />
</StackPanel>

<!--ContentPanel - place additional content here-->
<Grid x:Name="ContentPanel" Grid.Row="1" Margin="12,0,12,0">
    <my:Map Height="547" HorizontalAlignment="Left" Margin="0,6,0,0"
            Name="map1" VerticalAlignment="Top" Width="456"
            CopyrightVisibility="Collapsed" LogoVisibility="Collapsed">
        <my:Map.CredentialsProvider>
            <my:ApplicationIdCredentialsProvider
                ApplicationId="Enter your Key Here"
        </my:Map.CredentialsProvider>
    </my:Map>
    <Button Content="Satellite" Height="72" HorizontalAlignment="Left"
           Margin="0,561,0,0" Name="modeBtn" VerticalAlignment="Top"
           Width="223" Click="modeBtn_Click" />
    <Button Content="Start" Height="72" HorizontalAlignment="Left"
           Margin="227,561,0,0" Name="onOffBtn" VerticalAlignment="Top"
           Width="223" Click="onOffBtn_Click" />
</Grid>
```

Note that you'll need to enter Bing Map key in the `ApplicationID` line. And the UI layout is shown in Figure 7-15.

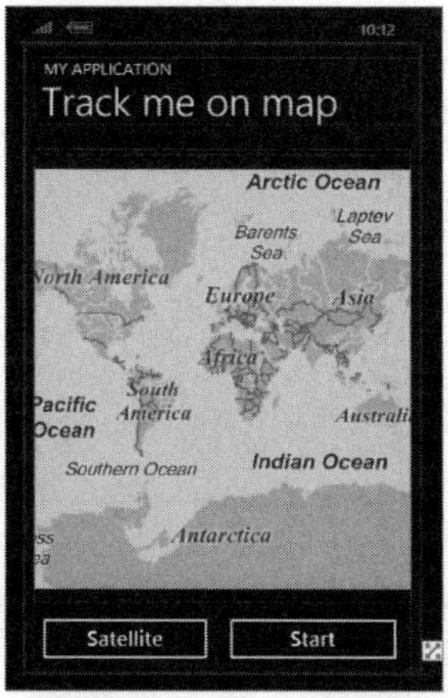

FIGURE 7-15: UI Layout of TrackMe

The C# code for this project is shown in Listing 7-8.

LISTING 7-8: Display current location on map, TrackMe\TrackMe\Mainpage.xaml.cs

Available for download on Wrox.com

```
#define GPS_EMULATOR // defining a compiler GPS symbol.

using System;
using System.Collections.Generic;
using System.Linq;
using System.Net;
using System.Windows;
using System.Windows.Controls;
using System.Windows.Documents;
using System.Windows.Input;
using System.Windows.Media;
using System.Windows.Media.Animation;
using System.Windows.Shapes;
using Microsoft.Phone.Controls;
using Microsoft.Phone.Controls.Maps; //for Maps
using System.Device.Location;   //Namespace for the actual location service
using GpsEmulatorClient;        //Namespace for GpsEmulator

namespace TrackMe
{
    public partial class MainPage : PhoneApplicationPage
    {
```

continues

LISTING 7-8 *(continued)*

```
#if GPS_EMULATOR
    GpsEmulatorClient.GeoCoordinateWatcher watcher;
#else
    System.Device.Location.GeoCoordinateWatcher watcher;
#endif

bool isRoadmode;
bool isStarted;

Pushpin curLoc;

// Constructor
public MainPage()
{
    InitializeComponent();

    isRoadmode = true;
    isStarted = false;

    curLoc = new Pushpin();
    curLoc.Content ="Me";
    curLoc.Location = new GeoCoordinate
        (40.2002810100341, -85.4102575778961);

    map1.Center = curLoc.Location;
    map1.ZoomLevel = 18;
    map1.Children.Add(curLoc);
}

private void onOffBtn_Click(object sender, RoutedEventArgs e)
{
    if (isStarted)  //Turn off GPS
    {
        if (watcher != null)
        {
            watcher.Stop();
        }
        isStarted = false;
        onOffBtn.Content = "Start";
    }
    else          // Turn on GPS
    {
        #if GPS_EMULATOR
            watcher = new GpsEmulatorClient.
                GeoCoordinateWatcher(GeoPositionAccuracy.High);
        #else
            watcher = new System.Device.Location.
                GeoCoordinateWatcher(GeoPositionAccuracy.High);
        #endif

        watcher.MovementThreshold = 20;
```

```
                // Add event handlers for PositionChanged events
                watcher.PositionChanged += new EventHandler
                        <GeoPositionChangedEventArgs<GeoCoordinate>>
                        (positionChanged);
                watcher.Start();

                onOffBtn.Content = "Stop";
                isStarted = true;
            }
        }

        //event handler of position changes
        void positionChanged(object sender,
            GeoPositionChangedEventArgs<GeoCoordinate> e)
        {
            this.Dispatcher.BeginInvoke( () => changeUIposition(e) );
        }

        void changeUIposition(GeoPositionChangedEventArgs <GeoCoordinate> e)
        {
            curLoc.Location = e.Position.Location;
            map1.Center = curLoc.Location;
        }

        private void modeBtn_Click(object sender, RoutedEventArgs e)
        {
            if (isRoadmode) {
                map1.Mode = new AerialMode();
                modeBtn.Content = "Road";
                isRoadmode = false;
            }
            else{
                map1.Mode = new RoadMode();
                modeBtn.Content ="Satellite";
                isRoadmode = true;
            }
        }


    }
}
```

In the constructor method, a Pushpin object, `curLoc`, is created and its `Content` and `Location` properties are initialized; `curLoc` is also added to the map control `map1`. The map is re-centered at the location of `curLoc` with a zoom level of 18, which will provide enough details about the surroundings.

When the Start button is clicked, a `GeoCoordinateWatcher` object, `watcher`, is instantiated with the accuracy set to high. The `MovementThreshold` is set to 20 meters so that a relatively small movement will raise a `PositionChanged` event, which in turn will trigger the event handler `positionChanged()`. To update the foreground UI control, `Dispatcher.BeginInvoke()` is called to explicitly invoke the `ChangeUIPosition()` method, and the map is re-centered at the current location.

Now start the `GpsEmulator.exe` program and then execute this project; you'll see that the position of the phone is now shown at the center of a map. A sample screenshot of the WP7 emulator running with GpsEmulator is shown in Figure 7-16.

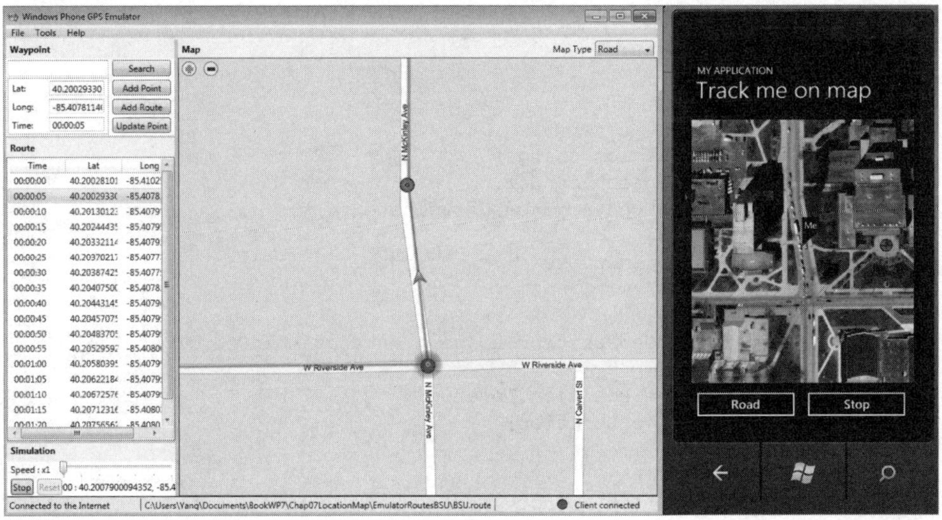

FIGURE 7-16: Running TrackMe

If you need to display a route on a map, such as driving directions or traffic information, you can add the waypoints to a `LocationCollection` object and then use a `MapPolyline` object to display it on the map. If shapes are necessary in your application, you can add additional `MapLayer` objects to the map and use `MapPolygon` objects to draw the shapes over the map.

In addition to displaying location information on a map, you are also encouraged to leverage location and maps to build cool new applications. For example, you can develop applications that combine social networking with location information to track friends and families (e.g., `http://www.locimobile.com/android/`); applications that use GPS data to track a hike (e.g., My Tracks `http://www.geardiary.com/2010/04/07/review-my-tracks-for-android/`); and applications that combine location with mobile commerce (e.g., shopkick, `http://www.shopkick.com/`). With the advent of NFC in mobile landscape, you may also be interested in developing applications that can combine location, social networks, and mobile commerce to offer ubiquitous mobile experiences to end users.

SUMMARY

This chapter introduces the fundamentals of location services and the Bing Map control on WP7. You have learned how to acquire the current location and how to test your application with a GpsEmulator program. You've also discovered how to use the Bing Maps SOAP web services to translate civilian data to geographical data. This chapter also explains how to use the Bing Map control and how to integrate the location service with the map control in your application.

From the developer perspective, developing location-aware applications on the WP7 platform is on a par with developing similar applications on iOS and Android. And strategies to develop a good location-aware application are almost identical on all three platforms; you need to find the trade-off point between accuracy, performance, and power consumption. In addition to the geocode web service, Bing Maps also offers Imagery service, Route service, and Search service. In addition, you can also use Microsoft Expression Blend 4 to quickly develop a map application. Readers are encouraged to try more map services and alternative development tools to gain deeper understanding of this topic.

In the next chapter, you'll learn about an exciting topic: Graphics. And you will also find out about a very powerful programming framework: XNA.

8

Graphics

WHAT'S IN THIS CHAPTER?

- Defining basic 2D and 3D graphics terms
- Understanding when to use XNA
- Using the XNA game loop
- Drawing 2D and 3D graphics with XNA

Developers have greatly improved the user experience on mobile devices by leveraging high-performance graphics and animations. And not surprisingly, many best-selling mobile applications have also relied on great graphics. It's important for developers to understand how to draw graphics in the Windows Phone 7 (WP7) platform.

Graphics is an area that even many seasoned software developers are unfamiliar with. That's why this chapter begins with a basic glossary of graphics terms. However, if you've already created graphics on the iOS or Android platform, you can certainly skip the "Graphics Fundamentals" section. Next, you'll learn how graphics are handled in WP7 at a conceptual level, and how techniques differ when working with the iOS and Android.

The most important concept you'll learn in this chapter is how to work with the graphics API in the XNA (New generation Architecture) framework. XNA is a set of programming tools that's built on top of the .NET framework and provides both powerful graphics and a formidable gaming engine. Unlike Silverlight, XNA is not an event-driven framework. Instead, it utilizes the game loop concept. In this chapter, you'll learn how to draw 2D or 3D graphics using XNA and how to implement animations and enable special effects in WP7.

> **RUNNING XNA GRAPHICS ON WP7 EMULATOR**
>
> To run XNA graphics on the WP7 emulator, the graphics card of your computer needs to support Direct X version 10 or above and the Driver mode needs to be WDDM 1.1 or above. Otherwise, you will still be able to compile the code, but when you debug your program in the emulator you will get the following error message: "The current display adapter does not meet the emulator requirements to run XNA Framework applications."
>
> Note that all the sample code in this chapter will NOT run on an emulator that does not meet the driver requirements. However, you will be able to run the sample applications on a physical Windows Phone 7 device.
>
> To find out the display adapter driver of a Windows computer, you can use the DirectX Caps Viewer tool (`http://msdn.microsoft.com/en-us/library/ee417852(VS.85).aspx`). Or use the DxDiag command and then choose the Display tab. If the driver of your computer does not meet the requirements, a possible work-around is to update the display driver. For example, you can go to the Microsoft Download Center (`http://www.microsoft.com/downloads/en/default.aspx`) and type "DirectX End-User Runtime Web Installer" in the search box and then choose the right installer for your computer.

GRAPHICS FUNDAMENTALS

In this section, you'll learn about the basics of drawing graphics on mobile platforms. The discussion includes everything from graphics-related terms to the fundamental techniques used to manipulate 2D and 3D graphics. This section also prepares you to use the APIs in the XNA framework to display graphics on WP7.

Basic 2D and 3D Graphics Glossary

Let's begin by studying a little of the jargon used with computer graphics. The most basic display element of a computer monitor or a smartphone screen is the *pixel*, which is the smallest unit in the digital imaging world. Each pixel is used to present a tiny portion of an image. In color image systems, each pixel is typically represented by three basic color elements: Red (R), Green (G) and Blue (B). The total number of bits used to represent the information of a color is called *color depth*. For example, if 8 bits are used to present R, G, and B, respectively, then the color depth is 24 bits. And the total number of colors it can represent is 2^{24}, which is about 16 million colors. On current WP7 devices, the screen size is 480 pixels × 800 pixels, but there are plans to support screens with fewer pixels (480 × 320) in the future.

In the 2D graphics world, images are rendered using screen *coordinates* to define pixel position. Figure 8-1 shows the 2D coordinate system of a WP7 screen in landscape mode. The x axis represents the width. The numbers start from 0 at the left corner and stop at 799 at the right corner.

The y axis represents the height, with 0 as the top and 479 as the bottom. In such 2D coordinate systems, the top left pixel is represented by (0,0) and bottom right pixel is represented by (799,479).

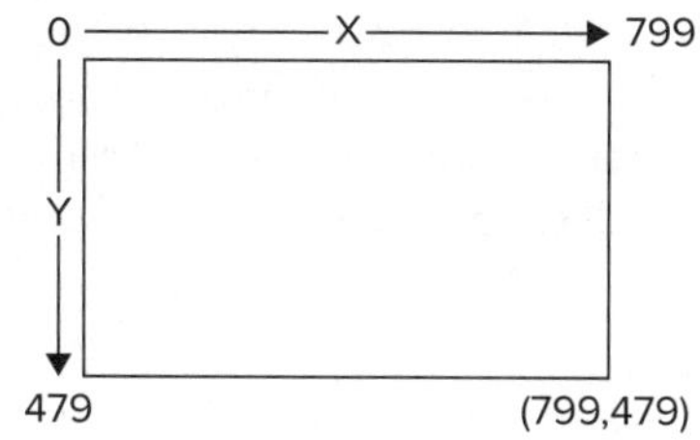

FIGURE 8-1: 2D coordinates on Landscape mode WP7

The term *sprite* refers to a kind of 2D image that you render on the screen. It differs from pictures that you display on the screen, even though you can use a single sprite to render the entire screen. (*Rendering* is the process of drawing an image on the screen in a way that emulates a 3D environment or provides a realistic presentation.) For example, if a simple gaming application is composed of 20 small pictures (or scenes), you can construct 20 sprites to represent those pictures in your program. Each sprite represents one of those 20 small pictures. In a different scenario, you can build a very large sprite, but display only a portion of it on the screen at any given time. This is actually a popular technique used to generate 2D animations.

The concept of 3D graphics discussed in this chapter refers to a 2D picture that also shows the depth of an image using shadows and variations in color and shade to make it appear as a real object in the real world. It is not the type of 3D that you've experienced when watching a movie such as *Avatar* with a pair of giant goggles. To better understand 3D techniques in the digital imaging world, it is necessary to introduce the 3D coordinate system (Figure 8-2).

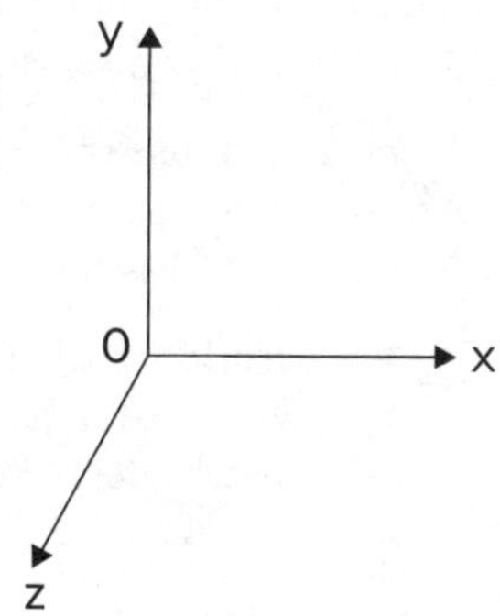

FIGURE 8-2: Right-handed 3D coordinates system

Figure 8-2 illustrates the right-handed three-dimensional coordinate system. The value of x increases from left to right; the value of y increases from bottom to right; and negative z values refer to the forward direction. Thus the position of a point in a 3D coordinate system can be determined by the values of its 3D coordinates (x, y, and z).

In mathematical terms, a *vector* is a group of values. A set of 3D coordinates is a vector of 3 that identifies the position of a point in 3D space. You can also use vectors to specify the magnitude and direction of object movement. People can then use various vector calculations, such as addition, subtraction, and multiplication, to describe the movement of a point object. To facilitate calculation, a homogeneous component called the *w* property is also introduced in 3D graphics.

An object may move or rotate in a 3D space. To calculate the new coordinates of an object, computer graphics rely on a *matrix*. A typical 4 × 4 matrix (16 float values) is defined as follows to assist in computing any 3D transformation:

$$\begin{bmatrix} R_x & R_y & R_z & R_w \\ U_x & U_y & U_z & U_w \\ -F_x & -F_y & -F_z & -F_w \\ T_x & T_y & T_z & T_w \end{bmatrix}$$

The first row represents the right vector of the coordinate space, the second row represents the up vector of the coordinate space, the third row represents the backward direction of the coordinate space, and the last row represents the translation vector. For example, the following matrix describes an object that rotates Θ degrees (in radians) on the X axis:

$$\begin{bmatrix} 1 & 0 & 0 & 0 \\ 0 & cos & -sin & 0 \\ 0 & sin & cos & 0 \\ 0 & 0 & 0 & 1 \end{bmatrix}$$

You might get intimidated by all the equations behind 3D graphics. However, the good news for developers is that APIs provide all the different transformation matrices. As a result, you don't have to memorize different equations and specify all 16 float values used in a transformation matrix in your program. The more important consideration is how to use the built-in methods to manipulate a 3D object shown on the screen.

To be able to use the 3D libraries in the WP7 developer tools and the developer tools for iOS and Android OS, you need to understand the process of projecting a moving 3D object on a 2D screen. A typical process involves three fundamental concepts:

- **World space:** The world space defines the geometry of an object using 3D coordinates. When an object moves, rotates, or scales in 3D space, you can use the transformation matrix described earlier to compute the geometry of the new location.
- **View:** In a 3D space, an object may appear differently depending on where the viewer is located. For example, when you look at your car from the side, it looks different than when you look at it from the front. In 3D graphics, the term *camera* is borrowed from the real world as the viewer. To correctly render 3D graphics, you need to specify the location of the camera.
- **Projection:** As the name implies, the projection process simply projects a 3D scene onto a 2D space. This is also the last step of 3D rendering in your program.

You can understand this process better by using a photographer analogy. Let's say you want to take a picture of your WP7 phone. The first thing you need to do is to put your phone in a place with good lighting conditions. The 3D location and physical layout of your phone can be described in world space. You can describe the act of moving the phone, such as rotating it, by calculating the movement using the transformation matrix in the world space. Next you need to set up your camera. You need to ask yourself questions such as, "Do you need a side view, a front view or a top view? Do you need a close-up of a button or a simple portrait?" By positioning the camera differently, you can get the picture you want. Adjusting the camera location defines the process of specifying the view location. The last action is pressing the shutter, which is equivalent to projecting a 3D object onto a 2D space. And the final picture you get from your camera is the 2D graphics rendered on the screen.

Now you know the basics of 2D graphics and 3D graphics, let's take a quick overview of how WP7, iOS and Android handle 2D and 3D graphics in application development.

Drawing Graphics on Smartphones

When working with WP7, both the Silverlight framework and the XNA framework support drawing graphics. The focus of this chapter is drawing graphics using the XNA framework because XNA utilizes the Microsoft low-level graphic engine DirectX and is better suited to rapidly creating 2D and 3D graphics with animations.

The Silverlight framework uses the Extensible Application Markup Language (XAML) to create graphics. With just a few lines of code, you can easily create different lines or shapes. The following example uses the `System.Windows.Shapes.Ellipse` class to create a 100-pixel by 100-pixel circle with 2 pixels of black outline and is filled with blue color on the screen (light gray in this illustration):

```
<Ellipse Fill="Blue"
   Height="100"
   Width="100"
   StrokeThickness="2"
   Stroke="Black"
/>
```

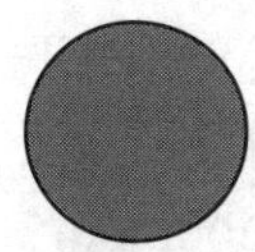

FIGURE 8-3: Create a circle with Silverlight

The result of the image is illustrated in Figure 8-3.

Although the Silverlight framework offers a rich collection of basic shapes and lines and allows you to draw and animate 2D graphics with ease, it doesn't provide full 3D graphics support. The 3D transformation applies only to Silverlight elements, which are currently 2D elements. So if your application needs to animate a 3D model, such as the human body, you'll need to use the XNA framework to fully unleash the graphics power and perform 3D transformation on 3D objects. The "When Do You Use XNA?" section of this chapter provides more details about the differences between the two frameworks and helps you decide what framework to use to better fit the needs of your application.

To work with 2D and 3D graphics in XNA, there are basically two approaches. One option is to build the application graphics using XNA primitives. The other option is to rely on a third-party application to generate 2D pictures or 3D models, and then add them in your program as references. The second approach is better because drawing a graphic or building a 3D model with XNA is not a straightforward process — you don't get to see the results until you compile and execute the application. In addition, if you want to create the same mobile applications on WP7, iOS, and Android platforms, you'll have to spend a significant amount of time rewriting the code used to re-generate the 2D picture or 3D model on different platforms. You'll learn how to work with 2D and 3D graphics with XNA in the "Drawing Graphics with XNA" section of this chapter.

You also have several choices when rendering graphics on the iOS platform. When working with 2D graphics you can use a slightly different markup language called Extensible Hypertext Markup Language (XHTML). Together with JavaScript, XHTML allows a developer to handle basic shapes, videos, and animations with ease. You can also use the iOS SDK to render graphics in your iOS application. The Cocoa layer uses the Objective-C language to manipulate high-level GUI objects such as `UIImage`, `UIColor`, and `UIFont`. And the Quartz Core Graphics layer uses the C language to draw basic vector graphics, such as lines and shading.

When creating high-performance 3D graphics, the iOS resorts to using Open Graphics Library for Embedded Systems (OpenGL ES). OpenGL offers a collection of cross-language and cross-platform APIs that applications can use to produce high-performance 2D and 3D graphics. OpenGL ES is a lightweight version of OpenGL developed for embedded systems such as smartphones. Two widely used OpenGL ES versions are 1.X and 2.X. OpenGL ES 2.X introduces a new shading language and offers more controls for developers. The benefit of using OpenGL ES 2.X is that it can minimize the cost and the power consumption of graphics subsystems. But it isn't backward compatible with OpenGL ES 1.X and using this functionality also requires that the developer write more code to achieve the same functionality that OpenGL ES 1.X provides (see `http://www.khronos.org/opengles/2_X/` for details). Devices such as the iPhone 3GS and 4G support openGL ES 2.0, but devices such as the first generation iPod touch, iPhone, and iPhone 3G support only OpenGL ES 1.1. Developers will need to determine which version of the OpenGL ES APIs to use in their programs, based on platform requirements.

Similarly the Android platform provides several techniques to render graphics. For basic shapes and images, you can use a custom 2D graphics library. The `android.graphics.drawable` and `android.view.animation` packages provide you sufficient functions to manipulate 2D graphics. For example, the following code snippets show how to draw a circle with a radius of 100 pixels centered at (200,300) and filled with blue color:

```
Paint paint = new Paint();
paint.setColor (Color.BLUE);
paint.setStyle (Paint.Style.FILL);
canvas.drawCircle (200,300,100,paint);
```

To support high-performance graphics, Android also relies on APIs from OpenGL ES. At a very high level, the following steps show how to use the OpenGL ES APIs in Android applications (see `http://developer.android.com/guide/topics/graphics/opengl.html` for additional details):

1. Set up a custom view, such as a `GLSurfaceView`.
2. Obtain a handle to an `OpenGLContext`.
3. In your view's `onDraw()` method, get a handle to a `GL` object, and use its methods to perform `GL` operations.

As of this writing, Android only fully supported OpenGL version 1.0. So features that appear in version 1.1, such as a vertex buffer, aren't supported on all Android devices. For developers who want their applications to reach more Android users, you'd better stick to OpenGL ES 1.0 for now.

In summary, all three platforms offer two major techniques to render graphics. One is for simple 2D shapes and animations, such as Silverlight XAML and XHTML. The other is for high-performance graphics, such as OpenGL ES and DirectX.

There are a lot of debates among developers as to which 3D APIs are better, OpenGL or DirectX. The major benefits of OpenGL are obvious: it's royalty-free, it's cross-platform, and it arguably supports more features. When developing high-performance graphics applications on WP7, your only viable option is DirectX with XNA. And the good news is that DirectX is a lot easier to learn and to use in your WP7 program.

DRAWING GRAPHICS WITH XNA

You've learned the basics of handling graphics on smartphones. It's time to work through a few examples to understand how to render 2D and 3D graphics on WP7. But before jumping into the code, you need to make a decision about using the Silverlight framework or the XNA framework.

When Do You Use XNA?

When you create a WP7 application using Visual Studio 2010, you need to choose between the Silverlight Framework and the XNA framework. And no, you can't use both frameworks in a single application, even though you can use a number of Silverlight APIs in your XNA applications and vice versa.

As stated earlier in this chapter, Silverlight is perfect for creating 2D shapes and animations. But this fact doesn't mean you should always use Silverlight if your application only needs 2D graphics. As a matter of fact, many 2D games use the XNA framework because the XNA framework graphics engine provides a 3D graphics pipeline. It leverages the graphic card's hardware acceleration, particularly the GPU (Graphic Processing Units), to render both 2D graphics and 3D graphics on screen. So if there are a lot of 2D animations in your applications, using Silverlight will overwhelm the WP7 CPU with graphics calculations and slow your application down.

Table 8-1 is a good guideline provided by Microsoft to help you decide which framework to choose in your application (see `http://msdn.microsoft.com/library/ff402528.aspx` for details).

TABLE 8-1: Silverlight or XNA

USE SILVERLIGHT	USE XNA
For an event-driven application.	For a high performance game framework.
To rapidly create Rich Internet Application-style user interface.	To rapidly create multi-screen 2D and 3D games.
To use Windows Phone controls.	To manage models, meshes, sprites, textures, and effects in the XNA Content Pipeline.
To embed video content inside your application.	
To use a web browser control.	

If you decide to use the Silverlight framework, you can still call many XNA framework APIs except those in the `Microsoft.Xna.Framework.Game` class and the `Microsoft.Xna.Framework.Graphics` namespace.

If, on the other hand, you pick the XNA framework, you can also use many classes in the Silverlight framework except many (but not all) classes in the `System.Windows` namespace.

Game Loop

Enough said about graphics and XNA; let's start a WP7 application and learn the game loop concept in the XNA framework. Use the following steps to create the solution:

1. Start Microsoft Visual Studio Express 2010 for Windows Phones. Windows displays the Visual Studio Express 2010 IDE.
2. Choose File ➪ New Project. You'll see the New Project dialog box shown in Figure 8-4.
3. Choose the Visual C#/XNA Game Studio 4.0 folder from the left panel. You can see a list of templates in the middle pane.
4. Highlight the Windows Phone Game (4.0) template. You'll see a description of this template in the right pane of the New Project dialog box.
5. Type **rolling2D** in the Solution Name field and the Name field. Click OK. Visual Studio creates a blank application for you.

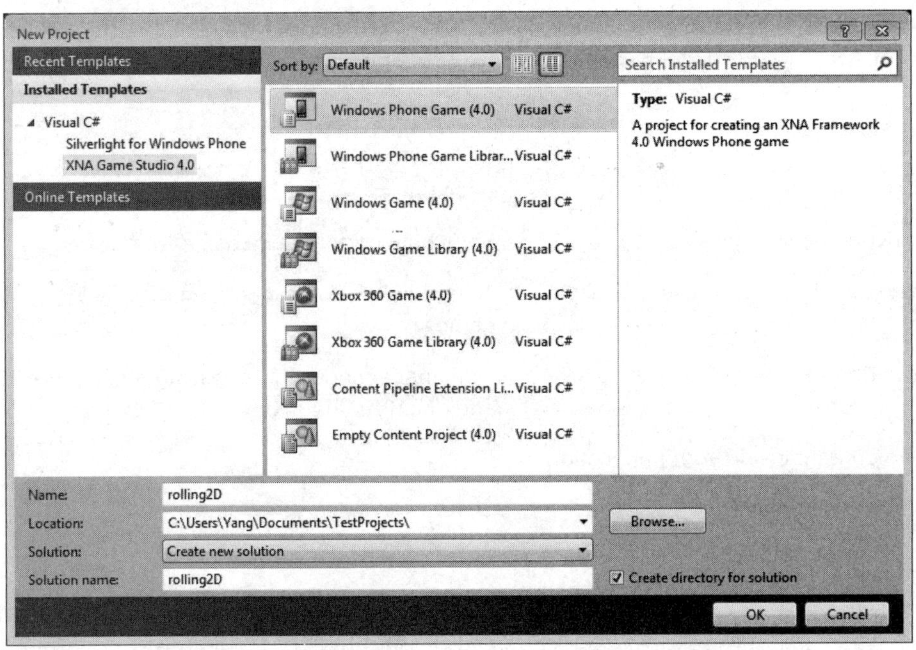

FIGURE 8-4: Creating a new XNA Project

Double-click the `Game1.cs` file in Solution Explorer. The IDE opens the `Game1.cs` file for editing. You'll notice XNA has already generated some lines of code for you. Listing 8-1 shows the code used to create the `rolling2D` namespace.

LISTING 8-1: Code Automatically Generated XNA Code, Game1.cs

```
public class Game1 : Microsoft.Xna.Framework.Game
{
    GraphicsDeviceManager graphics;
    SpriteBatch spriteBatch;

    public Game1()
    {
        graphics = new GraphicsDeviceManager(this);
        Content.RootDirectory = "Content";

        // Frame rate is 30 fps by default for Windows Phone.
        TargetElapsedTime = TimeSpan.FromTicks(333333);
    }

...

    protected override void Initialize()
    {
        // TODO: Add your initialization logic here
        base.Initialize();
    }

    ...
    protected override void LoadContent()
    {
        // Create a new SpriteBatch, which can be used to draw textures.
        spriteBatch = new SpriteBatch(GraphicsDevice);

        // TODO: use this.Content to load your game content here
    }

...

    protected override void UnloadContent()
    {
        // TODO: Unload any non ContentManager content here
    }

...

    protected override void Update(GameTime gameTime)
    {
        // Allows the game to exit
```

continues

LISTING 8-1 *(continued)*

```
            if (GamePad.GetState(PlayerIndex.One).Buttons.Back ==
                ButtonState.Pressed)
            this.Exit();

            // TODO: Add your update logic here

            base.Update(gameTime);
        }

...

        protected override void Draw(GameTime gameTime)
        {
            GraphicsDevice.Clear(Color.CornflowerBlue);

            // TODO: Add your drawing code here

            base.Draw(gameTime);
        }
    }
```

The layout of an XNA application looks very different from a Silverlight application. The automatically generated code contains the constructor for your `Game1` class and five methods: `Initialize()`, `LoadContent()`, `UnloadContent()`, `Update()`, and `Draw()`. There are no UI controls, no event-driven methods, and no main function. In fact, an XNA application is simply a loop, termed a *game loop*, that repeats itself until the application meets certain exit criteria.

The XNA Framework supports two types of game loops: a fixed time-step loop and variable time-step loop. In the fixed time-step mode, each loop always consumes the same amount of time. The default loop time on a WP7 system is one thirtieth of a second, or 30 loops per second, which is fast enough to vividly produce action-oriented video games. If a loop has performed all the requested actions, and it isn't time for the next loop, the system simply stays idle until the time expires for the current loop.

In the variable time-step mode, WP7 systems simply execute all the actions in your program. When it reaches the end of the loop, it starts from scratch without checking game time. By default, WP7 uses a fixed time-step game loop. If you want to write an XNA application in the variable time-step mode, you need to set the `IsFixedTimeStep` property of the game class to `false`. Benchmarking software is a good example that uses the variable time-step game loop.

A majority of the XNA applications are written using the fixed time-step game loop. This is also the focus of this chapter. Figure 8-5 illustrates how a fixed time-step game loop is processed in XNA.

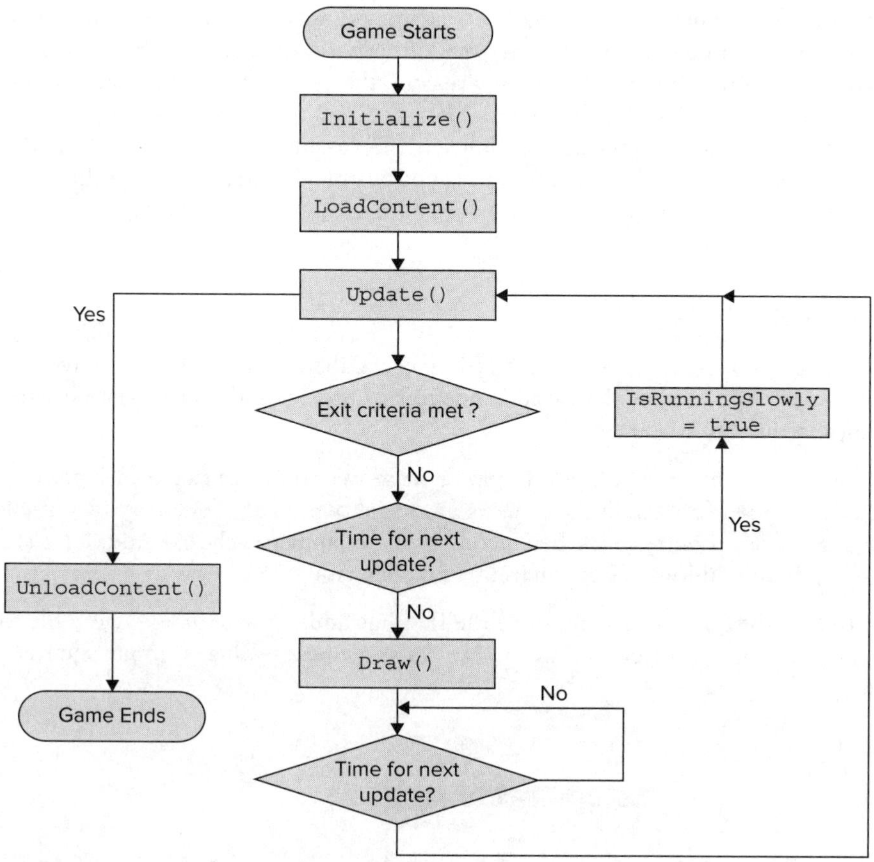

FIGURE 8-5: Fixed time-step game loop in XNA

An XNA application starts with an `Initialize()` method call that occurs just once during application execution. For many applications, you don't have to add any code to the default `Initialize()` method. The application then executes the `LoadContent()` method (called once during application execution). In this method, you typically need to load resources into your application, such as 2D pictures, or 3D models.

At this point, the application enters the game loop by executing the `Update()` method, which is where you implement your game logic. The program checks to determine whether it meets the exit criteria in the `Update()` method. The default behavior occurs when the user clicks the Back button, and the program executes the `UnloadContent()` method and terminates itself.

After the `Update()` method executes, the program checks to determine whether it's time for the next update. If there is still time available for the current game loop, the application executes the `Draw()` method, in which the application updates the screens if necessary.

Another check on the time for the next game loop occurs at this point. If it isn't time for the next game loop, the application stays idle until the time arrives. Otherwise, the application starts the loop again by running the `Update()` method. If for any reason, the `Update()` method takes more than a loop time to run, the application skips the `Draw()` method and the loop restarts by calling the `Update()` method again. In addition, the application sets the `IsRunningSlowly` property to `true`. If you're concerned that your XNA application may not be able to run as fast as it should on a slower WP7 phone, you'll need to check the status of `IsRunningSlowly` property in your `Update()` method.

Textures and Sprites

In the game loop subsection, you created an empty XNA project called "rolling2D" and have learned how the game loop works. You will now add code to this project and understand how to manipulate 2D graphics in the XNA framework.

In the Visual Studio 2010 IDE, open the Solution Explorer window. You'll see two projects, as shown in Figure 8-6. One is `rolling2D` and the other is `rolling2DContent(Content)`. Right-click the `rolling2DContent(content)` entry and when a context menu appears, choose Add ➪ Existing Item. In the Add Existing Item window, select square60 and click Add.

After you add the picture to the content, you'll notice the IDE has added the `square60.png` file to the `rolling2DContent(Content)` project, and the `Asset Name` property value is simply square60, as shown in Figure 8-7.

FIGURE 8-6: Adding resources to an XNA Project

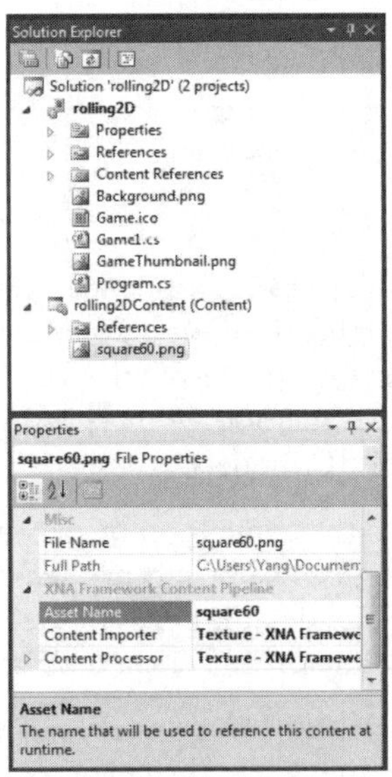

FIGURE 8-7: Asset Name of newly added picture

Now it's time to edit the `Game1.cs` file as shown in Listing 8-2.

LISTING 8-2: WP7 2D graphics demo, rolling2D\rolling2D\Game1.cs

Available for download on Wrox.com

```
using System;
using System.Collections.Generic;
using System.Linq;
using Microsoft.Xna.Framework;
using Microsoft.Xna.Framework.Audio;
using Microsoft.Xna.Framework.Content;
using Microsoft.Xna.Framework.GamerServices;
using Microsoft.Xna.Framework.Graphics;
using Microsoft.Xna.Framework.Input;
using Microsoft.Xna.Framework.Input.Touch;
using Microsoft.Xna.Framework.Media;

namespace rolling2D
{

    public class Game1 : Microsoft.Xna.Framework.Game
    {
        GraphicsDeviceManager graphics;
        SpriteBatch spriteBatch;

        public Game1()
        {
            graphics = new GraphicsDeviceManager(this);

            //Make the graphic occupy the entire screen
            graphics.PreferredBackBufferHeight = 480;
            graphics.PreferredBackBufferWidth = 800;

            //Enable only Landscape Mode
            graphics.SupportedOrientations = DisplayOrientation.LandscapeLeft;
            Content.RootDirectory = "Content";
            TargetElapsedTime = TimeSpan.FromTicks(333333);
        }

        protected override void Initialize()
        {
            base.Initialize();
        }

        Texture2D rollingPic;
        Vector2 position = new Vector2(20, 100);
        Random rnd = new Random();

        protected override void LoadContent()
        {
            // Create a new SpriteBatch, which can be used to draw textures.
```

continues

LISTING 8-2 *(continued)*

```
            spriteBatch = new SpriteBatch(GraphicsDevice);

            // TODO: use this.Content to load your game content here
            rollingPic = this.Content.Load<Texture2D>("square60");
        }

        protected override void UnloadContent()
        {
            // TODO: Unload any non ContentManager content here
        }

        protected override void Update(GameTime gameTime)
        {
            // Allows the game to exit
            if (GamePad.GetState(PlayerIndex.One).Buttons.Back
                ==  ButtonState.Pressed)
                this.Exit();

            //Increase X value by 3 in each iteration
            position.X += 3;
            if (position.X >= 800)
            {
                position.X = 0;
                //set the new Y value to a random number
                position.Y = rnd.Next(1, 420);
            }

            base.Update(gameTime);
        }

        protected override void Draw(GameTime gameTime)
        {
            GraphicsDevice.Clear(Color.CornflowerBlue);

            // TODO: Add your drawing code here
            this.spriteBatch.Begin();

            this.spriteBatch.Draw(this.rollingPic,
                new Rectangle((int) position.X,(int) position.Y,60,60),
                Color.White);

            this.spriteBatch.End();

            base.Draw(gameTime);
        }
    }
}
```

In the `Game1()` constructor, the `PreferredBackBufferHeight` property is set to `800` and the `PreferredBackBufferWidth` property is set to `480`. This makes the graphic occupy the entire screen. The next property, `SupportedOrientations`, is set to `DisplayOrientation.LandscapeLeft`.

This makes the application work in only one orientation: the display is rotated counterclockwise 90 degrees into a landscape mode. If more orientations are needed in your application, you can use the bitwise or operation to set the corresponding bit flags of supported orientations. For instance, to support both `LandscapeLeft` and `LandscapeRight`, you can use the following code snippet:

```
graphics.SupportedOrientations =
    DisplayOrientation.LandscapeLeft | DisplayOrientation.LandscapeRight;
```

Three objects are declared in the body of the `Game1` class:

- `rollingPic` is a `Texture2D` object used to hold the 2D image that you want to render on the screen.
- `position` is a `Vector2` object and has an initial value of (20,100). `Vector2` is a class in the XNA framework that contains two values. It's typically used to represent the x and y value on a 2D plane.
- `rnd` is a `Random` object that you use to generate a random y value.

The application uses the `Content.Load()` method to load the asset to `rollingPic` in the `LoadContent()` method.

In the `update()` method, the value of `X` is increased by 3 three times. When the value of `X` is greater than or equals to 800, the picture is outside the display area. At this point, the value of `X` is reset to 0, and the value of `Y` is set to a random number within the height of the screen.

In the `draw()` method, the program begins by setting the color of the background to `CornflowerBlue` by calling `GraphicsDevice.Clear(Color.CornflowerBlue)`. It then uses the `SpriteBatch` class methods to draw the pictures on screen. `SpriteBatch` is the class in the XNA framework that can draw one or many sprites directly on the screen. The sequence is simple:

1. Call the `SpriteBatch.Begin()` method to start the drawing process.
2. Call the `SpriteBatch.Draw()` method to display content on screen.
3. Call the `SpriteBatch.End()` method to end the drawing process.

The `SpriteBatch.Draw()` method is a heavily overloaded method that can take seven different parameters. This program uses the most basic form: `SpriteBatch.Draw(Texture2D, Rectangle, Color)`. This call adds a sprite to a batch of sprites for rendering using the specified texture, destination rectangle, and color. Note that the last parameter, `Color`, is not the color you use to paint a picture; instead it is a color to tint a picture. What it does is multiply the color value of the Texture2D object with the value of the `Color` parameter and render the new color on the screen. When `Color.White` is used in the program, the call retains the original color. If you build and run the application, you'll see a square box moving from the top to the bottom (in portrait mode) of your WP7 screen. Figure 8-8 shows typical output from this example.

FIGURE 8-8: Rolling 2D square box in action

This simple example demonstrates the basics of rendering 2D graphics using the XNA framework. It's a fairly straightforward process. You add the picture (or asset) to the project and read it to a `Texture2D` object. You then define how you'd like to treat the picture as time elapses. Finally, the code draws the picture in the timeline you define, using the `SpriteBatchDraw()` method.

Animation

The basic idea behind animation is to display multiple pictures in sequence during a short timeframe. This is similar to the process used to make motion pictures.

In the next example, you'll see how to create animation using the XNA framework. To better illustrate the concept, I took six pictures of a toy in different positions. Then I resampled each picture to 480 pixels × 320 pixels and tiled the pictures together in a single .jpg file, as shown in Figure 8-9.

FIGURE 8-9: Six frames of a toy

This single picture, `toyAnimation.jpg`, contains the six frames needed to create the sample animation application. Use the following steps to create a new project:

1. Start Microsoft Visual Studio Express 2010 for Windows Phones. Windows displays the Visual Studio Express 2010 IDE.
2. Choose File ➪ New Project. You'll see the New Project dialog box.
3. Choose the Visual C#/XNA Game Studio4.0 folder from the left panel. You'll see a list of templates in the middle pane.
4. Highlight the Windows Phone Game (4.0) template. You'll see a description of this template in the right pane of the New Project dialog box.
5. Type **ToyAnimation** in the Solution Name field and the Name field. Click OK. Visual Studio creates a blank application for you.
6. Right-click the ToyAnimationContent (Content) project in Solution Explorer and choose Add ➪ Existing Items from the context menu. You'll see the Add Existing Item dialog box.
7. Choose toyAnimation and click Add. You'll see the `toyAnimation.jpg` file added to the ToyAnimationContent (Content) project.

In Game1.cs, add a Texture2D object, and edit the LoadContent() and Draw() methods as shown in Listing 8-3.

Available for download on Wrox.com

LISTING 8-3: Displaying a portion of a sprite to get the animation effect, ToyAnimation\ToyAnimation\Game1.cs

```
GraphicsDeviceManager graphics;
SpriteBatch spriteBatch;

Texture2D texture;

...

protected override void LoadContent()
{
    // Create a new SpriteBatch, which can be used to draw textures.
    spriteBatch = new SpriteBatch(GraphicsDevice);

    // TODO: use this.Content to load your game content here
    texture = Content.Load<Texture2D>("toyAnimation");
}

protected override void Draw(GameTime gameTime)
{
    GraphicsDevice.Clear(Color.CornflowerBlue);

    // TODO: Add your drawing code here

    int num = (int)(gameTime.TotalGameTime.TotalMilliseconds/500) % 6;

    spriteBatch.Begin();

    spriteBatch.Draw(texture, new Vector2(180,80),
        new Rectangle(0, num * 320, 480, 320), Color.White);
    spriteBatch.End();

    base.Draw(gameTime);
}
```

The texture object is used to hold the 2D picture file, toyAnimation.jpg. You load the image by calling the Content.Load<>() method. In this case num is an integer that ranges from 0 to 5 and increases by one every half second. When it reaches 5, it returns to 0 during the next loop. The key part of the program that generates the animation effect is this line of code:

```
spriteBatch.Draw(texture, new Vector2(180,80),
    new Rectangle(0, num * 320, 480, 320), Color.White);
```

This call draws a picture on screen with the top left of its corner located at (180,180). The source picture (or sprite) is `texture`, but code doesn't render the entire picture on screen. Instead, it only renders a rectangular area expressed by the coordinates (0, num * 320, 480, 320).

If you build the application and execute it, you'll see that the wheel of the toy is rotating slowly.

3D Graphics

In this example, you learn how to manipulate 3D graphics using the XNA framework.

As stated earlier in this chapter, you should use third-party software to create a 3D model and then import this model into the XNA application. Otherwise, you'll spend more time working with the graphics primitives the XNA Framework provides to create the image, and the 3D model you build is unlikely to export to other programs.

There are many 3D model formats. XNA supports the .x and .fbx format, which are owned and developed by Autodesk. There are also many 3D modeling applications that you can choose from. For example, Blender (`http://www.blender.org/`) is a free application you can use to create a model in .fbx format. In this example, I used Softimage Mod Tools (`http://autodesk.com/softimage`) and created something very similar to a Rubik's cube but without any colors. The text rendering of the model is shown in Figure 8-10.

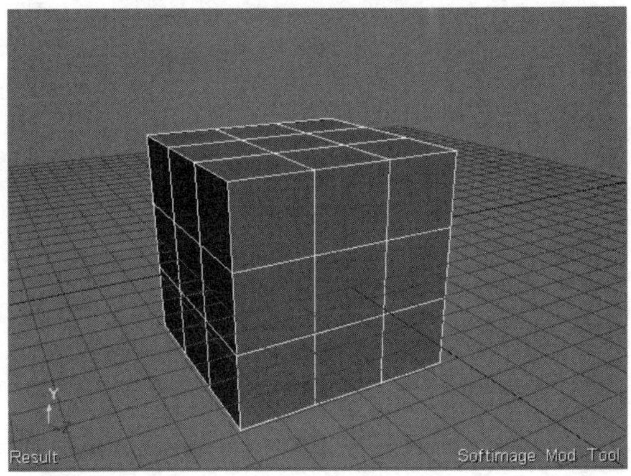

FIGURE 8-10: 3D model of a cube

Now follow the steps below to create a new project to render 3D graphics on your WP7 phones.

1. Start Microsoft Visual Studio Express 2010 for Windows Phones. Windows displays the Visual Studio Express 2010 IDE.
2. Choose File ➪ New Project. You'll see the New Project dialog box.
3. Choose the Visual C#/XNA Game Studio 4.0 folder from the left panel. You see a list of templates in the middle pane.
4. Highlight the Windows Phone Game (4.0) template. You'll see a description of this template in the right pane of the New Project dialog box.
5. Type **cube3D** in the Solution Name field and the Name field. Click OK. Visual Studio creates a blank application for you.
6. Right-click the "cube3DContent(Content)" project in Solution Explorer and choose Add ➪ Existing Items from the context menu. You'll see the Add Existing Item dialog box.
7. Choose `cube1.fbx` (which you can download from the Wrox.com site for this chapter) and click Add. You'll see the `cube1.fbx` file added to the Cube3DContent(Content) project.

In `Game1.cs`, add the variables and edit `LoadContent()`, `Update()`, and `Draw()` methods as shown in Listing 8-4.

LISTING 8-4: XNA 3D graphics demo, cube3D\cube3D\Game1.cs

Available for download on Wrox.com

```
public class Game1 : Microsoft.Xna.Framework.Game
{
    GraphicsDeviceManager graphics;
    SpriteBatch spriteBatch;

    Model cube3d;

    float aspectratio;
    float rotation = 0.0f;
    Vector3 cubepostion = new Vector3(0, 2, 5);
    Vector3 cameraposition = new Vector3(0.0f, 15.0f, 30.0f);

...

    protected override void LoadContent()
    {
        // Create a new SpriteBatch, which can be used to draw textures.
        spriteBatch = new SpriteBatch(GraphicsDevice);

        // TODO: use this.Content to load your game content here
        cube3d = Content.Load<Model>("cube1");
        aspectratio = graphics.GraphicsDevice.Viewport.AspectRatio;
    }
    ...

    protected override void Update(GameTime gameTime)
    {
        // Allows the game to exit
        if (GamePad.GetState(PlayerIndex.One).Buttons.Back
            == ButtonState.Pressed)
            this.Exit();

        // TODO: Add your update logic here
        rotation += (float)gameTime.ElapsedGameTime.TotalMilliseconds *
            MathHelper.ToRadians(0.05f);

        base.Update(gameTime);
    }

    protected override void Draw(GameTime gameTime)
```

continues

LISTING 8-4 *(continued)*

```
        {
            GraphicsDevice.Clear(Color.CornflowerBlue);

            // TODO: Add your drawing code here
            Matrix[] trans = new Matrix[cube3d.Bones.Count];
            cube3d.CopyBoneTransformsTo(trans);

            foreach (ModelMesh mmesh in cube3d.Meshes)
            {
                foreach (BasicEffect ef in mmesh.Effects)
                {
                    ef.LightingEnabled = true;

                    ef.DirectionalLight0.Enabled = true;
                    ef.DirectionalLight0.DiffuseColor = Color.Violet.ToVector3();

                    ef.World = trans[mmesh.ParentBone.Index]*
                        Matrix.CreateScale(1.0f) *
                        Matrix.CreateRotationX(rotation) *
                        Matrix.CreateRotationY(rotation*0.3f)*
                        Matrix.CreateRotationZ(rotation*0.7f) *
                        Matrix.CreateTranslation(cubepostion);

                    ef.View = Matrix.CreateLookAt
                        (cameraposition, Vector3.Zero, Vector3.Up);

                    ef.Projection = Matrix.CreatePerspectiveFieldOfView
                        (MathHelper.ToRadians(45.0f), aspectratio, 1.0f, 200.0f);
                }
                mmesh.Draw();

            }

            base.Draw(gameTime);
        }
    }
```

Two float values, `aspectratio` and `rotation`, describe the aspect ratio of the display and how many degrees in radians you want the model to rotate after each animation frame. The code instantiates two `Vector3` objects to define the positions of the cube and the camera. A `Model` object, `cube3d`, is declared in the body of `Game1` class. It is later used in the `LoadContent()` method to read the 3D model from `cube1.fbx`. The `update()` method doesn't do much other than update the rotational value in degrees. The equation implies the rotation speed is about 50 degrees every second.

In XNA, a 3D model consists of multiple parts called `ModelMesh`. For example, a 3D model of a four-door passenger car will have a `ModelMesh` for the body of the vehicle, four `ModelMesh` instances of wheels, and four `ModelMesh` instances of doors. The `ModelBone` class in XNA describes how the `ModelMesh` objects are positioned. Each `ModelMesh` has a `ParentBone` property, which contains the transform matrix of this `ModelMesh` to its parents. To render a 3D model in the XNA framework, you will need to call the `Draw()` method of each `ModelMesh`.

The key functions of this application are all in the `Draw()` method. First, the code creates an array of matrices based on the number of bones (a collection of `ModelBone`) in the 3D model. Next, the code calls the `cube3d.CopyBoneTransformsTo(trans)` method to store the transformation information to the `trans` matrices. Finally, the program iterates every `ModelMesh` of `cube3d`.

At this point, you may want to place a spotlight on the rotating cube. The following three lines show how to project a violet directional light onto the cube.

Available for download on Wrox.com

```
ef.LightingEnabled = true;
ef.DirectionalLight0.Enabled = true;
ef.DirectionalLight0.DiffuseColor = Color.Violet.ToVector3();
```

code snippet cube3D\cube3D\Game1.cs

The task of rotating the cube comes next. The following code shows how to define the rotation action and then perform the actual rotation:

Available for download on Wrox.com

```
ef.World = trans[mmesh.ParentBone.Index]*
    Matrix.CreateScale(1.0f) *
    Matrix.CreateRotationX(rotation) *
    Matrix.CreateRotationY(rotation*0.3f)*
    Matrix.CreateRotationZ(rotation*0.7f) *
    Matrix.CreateTranslation(cubepostion);
```

Code snippet cube3D\cube3D\Game1.cs

Note that `ef.world` uses transformation matrix to calculate the new 3D coordinates of the transformation. `Matrix.CreateScale(1.0f)` keeps the size of the model unchanged. `Matrix.CreateRotationX()` rotates the model on the x axis. `Matrix.CreateRotationY()` rotates the model on the y axis. And `Matrix.CreateRotationZ()` rotates the model on the z axis. `Matrix.CreateTranslation()` moves the model to the specified location (3D coordinates).

In order to display the graphic on screen, you must define a camera location and then tell the application to render the 3D object onto a 2D projection. The following code shows how to perform this task:

Available for download on Wrox.com

```
ef.View = Matrix.CreateLookAt
    (cameraposition, Vector3.Zero, Vector3.Up);

ef.Projection = Matrix.CreatePerspectiveFieldOfView
    (MathHelper.ToRadians(45.0f), aspectratio, 1.0f, 200.0f);
```

Code snippet cube3D\Game1.cs

The `Matrix.CreateLookAt()` method defines the location of the camera, and the `Matrix.CreatePerspectiveFieldOfView()` method finally renders the 2D projection of the 3D model.

If you build and run this project, you'll see a violet cube rotating on all three axes. A sample screenshot is shown in Figure 8-11.

FIGURE 8-11: 3D model of a cube

SUMMARY

This chapter explains the basics of rendering 2D and 3D graphics on WP7. Even though the math behind the 3D graphics pipeline is complex, using the APIs is not that difficult. You've discovered how to create an XNA application and how the XNA game loop works. You've also learned how to render a 2D image on WP7, and how to make it animate. The last example of this chapter also demonstrates the power of 3D graphics and how easy it is to manipulate 3D graphics using the XNA Framework.

You're probably amazed by how simple it is to render complex graphics using the XNA Framework. Don't forget that you can easily create a copy of your XNA program for either the Windows platform or the Xbox360 platform. If you need to build a high-performance graphics application from scratch, WP7 has a clear advantage because it's easy to learn and simple to use. Graphics are typically associated with gaming applications. You'll want to read other books or search online for discussions on how to detect collisions, and react to user inputs in the WP7 environment.

In the next chapter, you'll learn about writing applications that provide a multimedia experience to the user.

9

Multimedia

WHAT'S IN THIS CHAPTER?

- Understanding multimedia concepts
- Working with audio
- Working with video

Today's smartphone original equipment manufacturers (OEMs) and mobile operators (MOs) use multimedia functionality as their main selling point. Most smartphones have a good camera with 5 or 8 megapixel capabilities, an FM radio, a variety of sounds like different ringtones, and video capture functions. Some smartphones even have a surround sound system and can be regarded as mobile theaters.

With users relying on mobile devices increasingly as entertainment devices, Windows Phone 7 (WP7) has a stronger emphasis on multimedia technology including music, pictures, and video.

This chapter begins with an architectural overview of the common multimedia components offered by three OSs: iPhone OS (iOS), Google Android, and WP7. The chapter then compares the common features and differences among the three OSs. The latter part of the chapter focuses mainly on WP7. You'll discover what WP7 provides for multimedia features in details. The accompanying code illustrates how to use WP7's multimedia APIs in your applications.

MULTIMEDIA OVERVIEW

This section discusses the fundamentals and similarity of multimedia technologies on the three platforms. Some examples are shown to illustrate the differences in architecture, APIs, and some implementation details.

Multimedia Architectural Overview

A modern multimedia framework across major mobile platforms adopts a paradigm similar to that illustrated in Figure 9-1. This is a high-level architectural abstraction of the multimedia framework used on WP7, Android, and iOS. The concrete implementations of each component shown in this figure may have different names on the specific platform, but no matter what they are called, you can find the conceptual mappings here. The components inside the dash-line circle are usually encapsulated as one class object to handle the multimedia functions.

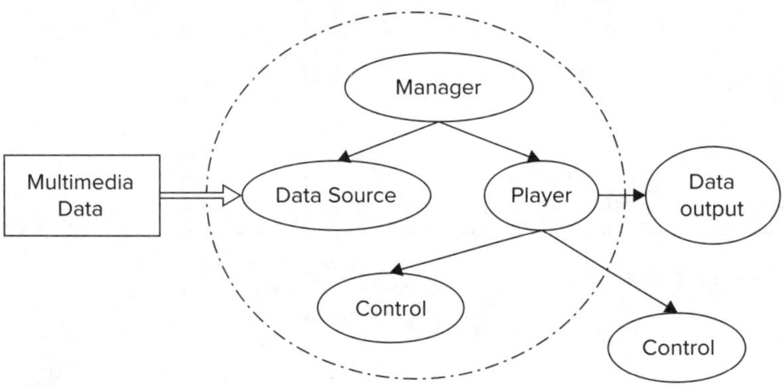

FIGURE 9-1: Conceptual multimedia framework architecture

There are basically four software components:

- **`Manager:`** The main coordinator for multimedia processing, it provides the factory used to create `DataSource` and `Player` objects.
- **`Data Source:`** This is the main input for receiving multimedia content, which could be a data file, an input stream, or another source.
- **`Player:`** It manages the multimedia content play actions like start, play, stop, pause, and resume, and controls the life cycle through its methods.
- **`Control:`** It enables seeking, skipping, forwarding, and rewinding functionalities, and controls the flow of multimedia contents.

The conceptual architecture just discussed in Figure 9-1 mainly applies to audio and video media types. Image is another important media type which has a relatively simpler manipulation mechanism. In general, all three platforms provide a view-based control or container for displaying images. On WP7 it is `Image` control, while it is `ImageView` widget on Android, and the `UIImageView` class on iOS.

WP7 Multimedia

Let's look at how the components shown in Figure 9-1 are implemented on WP7.

To implement the `Player` functions and add the media playback function into a WP7 application, you have two options. One is to use the Silverlight `MediaElement` class defined in the `System.Windows.Controls` namespace (`http://msdn.microsoft.com/library/ms611595.aspx`). The other option is to use the `MediaPlayer` class in the `Microsoft.Xna.Framework.Media` namespace (`http://msdn.microsoft.com/library/dd254868.aspx`). You'll find an explanation of the WP7 sound and video playing details later in this chapter. Figure 9-2 shows the WP7 version of the `Player` object states.

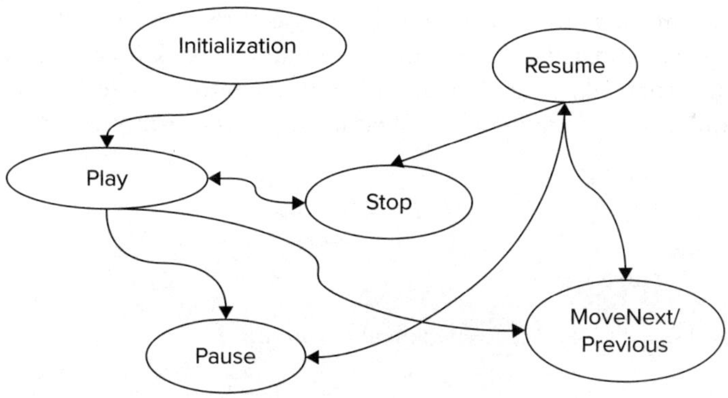

FIGURE 9-2: Player object state model on WP7

The `DataSource` on WP7 is an abstract layer for handling different types of codecs. It's responsible for generating data and content for the `MediaPlayer` object to use. `DataSource` hides the details of data payload formats. The WP7 `DataSource` refers to the `Source` property of the `MediaElement` class or the `MediaLibrary` class defined in the `Microsoft.Xna.Framework.Media` namespace (`http://msdn.microsoft.com/library/microsoft.xna.framework.media.medialibrary.aspx`)

On WP7, the `Control` functions shown in Figure 9-1 are actually included in the `MediaElement` and `MediaPlayer` classes, which not only expose methods and properties to play, pause, resume, and stop media, but also provide shuffle, repeat, play position, and visualization capabilities. For example, to control song playback, the `MoveNext()` and `MovePrevious()` methods move to the next or previous song in the queue and the `IsMuted`, `IsRepeating`, `IsShuffled`, and `Volume` properties are used to get and set playback options.

The `Image` control is used on WP7 to display an image in JPEG or PNG formats. Because the JPEG decoder in WP7 is faster than the PNG decoder, you may want to use JPEG format if you need to display opaque images. However, for transparent images, PNG is the only choice. When you include images in your WP7 application, as in some sample applications demonstrated previously, you can set their Build Type as either `Resource` or `Content`. When the `Resource` type is used, the images are built in the assemblies (.dll files). The larger the application's dll, the slower the application will launch. However, once the application launches, the images will show up more quickly than using `Content` as the build type. Thus you should take tradeoffs to select the proper build type based on the requirements.

iOS Multimedia

On iOS, you can use its *Media Player Framework* to select and play audio and video. For example, AVAudioPlayer is a sound-playing object. Additionally, a commonly used media player on iOS is the MPMusicPlayerController class, which is responsible for playing audio media items. The MPMusicPlayerController object methods are play(), beginSeekingForward(), beginSeekingBackward(), endSeeking(), skipToNextItem(), skipToBegining(), skipToPreviousItem(), and stop().

If MPMusicPlayerController is instantiated as an iPodMusicPlayer object, the system will launch the system default iPod music player application to handle the audio playback functions. The user then can use the playback controls provided by the iPod application to control its iPod music player states such as repeat, shuffle, and now-playing item. For iPod music player instantiation, the code is as follows:

```
(void) setMusicPlayState {
MPMusicPlayerController *iPodController =
    [MPMusicPlayerController iPodMusicPlayer];
    playPauseButton.selected =
    (iPodController.playbackState == MPMusicPlaybackStatePlaying);
}
```

If MPMusicPlayerController is instantiated as an applicationMusicPlayer object, a new player control is created as part of the application. It plays music locally within the user's application, but doesn't act as the iPod music player. Furthermore, the user interactions with this application player will not affect the system iPod player's state. When the application moves to the background, the player stops if it was playing. The following code shows the instantiation of the normal application music player.

```
(void) createMyMusicPlayer {
MPMusicPlayerController *myMusicPlayer =
    [MPMusicPlayerController applicationMusicPlayer];
}
```

iOS uses the UIImageView class to provide a display of one or more images. This object can also set the duration and frequency of animation. To view a series of images, you set this object's animationImages property to hold one or more UIImage objects. Then you set the contentMode property (display mode) to the desired display type (such as UIViewContentModeScaleAspectFit), and the animationDuration property to the number of seconds to display the series of images. Here is an example of a typical setup:

```
NSArray *myimagecollection = [NSArray arrayWithObjects:
   [UIImage imageNamed:@"IOS.png"],[UIImage imageNamed:@"Android.png"],
   [UIImage imageNamed:@"WP.png"],[UIImage imageNamed:@"WP7.png"],nil];
CGRect animatedframe = CGRectMake(0,0,600,800);
UIImageView *myanimationView = [[UIImageView alloc]
   initWithFrame:animatedframe];
```

```
myanimationView.animationImages = myimagecollection;
myanimationView.contentMode = UIViewContentModeScaleAspectFit;
myanimationView.animationDuration = 6;
```

This code creates `myimagecollection` as `INSArray`'s instance and creates `myanimationView` as the `UIImageView`'s instance. After you create the `UIImageView` object, you can set the repeat count and start the animation using the following code:

```
myanimationView.animationRepeatCount = 6;
[myanimationView startAnimating];
```

When the `UIImageView` control displays the PNG or JPEG image array, you can touch the screen to launch the program that starts the animation slideshow.

Before iOS 4, you could play video only in full-screen mode. When working with iOS 4 you can embed videos in the view window as shown here:

```
(void)playMovie:(NSURL*) myaddress {
   MPMoviePlayerController* media = [[MPMoviePlayerController alloc]
     initWithContentURL:myaddress];
   media.scalingMode = MPMovieScalingModeNone;
   media.movieControlMode = MPMovieControlModeDefault;
   [[NSNotificationCenter defaultCenter] addObserver:self
   selector:@selector(mediacallback:)
   name:MPMoviePlayerPlaybackDidFinishNotification
   object:media];
   [media play];
}

(void) mediacallback:(NSNotification*)mynotice {
   MPMoviePlayerController* media = [mynotice object];
   [[NSNotificationCenter defaultCenter] removeObserver:self
   name:MPMoviePlayerPlaybackDidFinishNotification
   object:media];
   [media release];
}
```

This example shows how to use the `MPMoviePlayerController` class to manage the video player. The code uses `NSNotificationCenter` to register a notification so the system can tell the application when the movie is done playing. The `mediacallback()` method is designed to unregister and release the movie object.

Android Multimedia

Android uses the `MediaPlayer` class (described at `http://developer.android.com/reference/android/media/MediaPlayer.html`) to control playback of audio and video files and streams. Figure 9-3 shows the media playback state diagram.

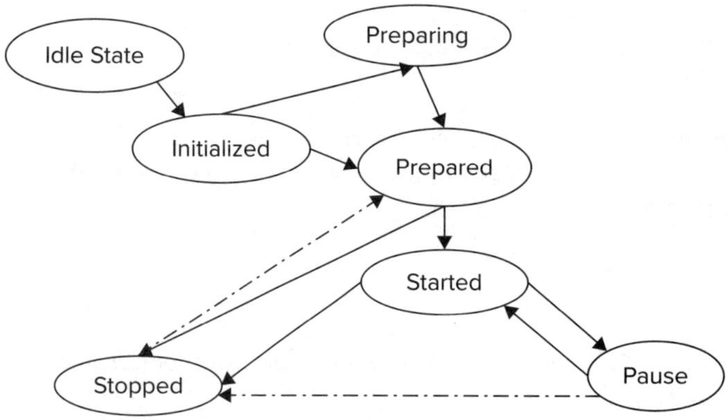

FIGURE 9-3: Android media player state diagram

When working with Android, the `MediaPlayer` object has the following states:

- **Idle:** Using the `reset()` method, calling the code creates a `MediaPlayer` object in the idle state.
- **Initialized:** When the code calls `setDateSource()`, it changes the `MediaPlayer` object from the idle state to the initialized state.
- **Preparing:** This state is a temporary state, where the `MediaPlayer` object behavior is not deterministic. Only when the code calls `prepareAsync()` does the system change the `MediaPlayer` object state to preparing. The `prepareAsync()` method does not take any parameter, and this method immediately returns the programming control and will not block code execution in this running thread. The working thread can continue to finish other tasks, and wait for its registered callback function to return and handle the deterministic state's information.
- **Prepared:** After the `prepareAsync()` method returns and the internal player engine's preparation work is completed, the `MediaPlayer` object enters the prepared state. As an alternative, a call to the `prepare()` method can also place the `MediaPlayer` object in the prepared state. The `prepare()` method will block the current thread execution until the method returns; the next line of code can continue to run, with the benefit that the code logic will be protected in sequential order.
- **Started:** At this point, the code calls the `start()` method and waits for the call to return successfully. When the call is successful, the `MediaPlayer` object is in the started state. You can call `isPlay()` to verify if the `MediaPlayer` object is in the started state.
- **Pause:** Use the `pause()` method to place the `MediaPlayer` object in pause state. While `MediaPlayer` is in the pause state, you can reposition the current playback. When the code calls `pause()`, the `MediaPlayer` object state transition occurs asynchronously in the player engine. Transitioning from the pause to started state also occurs asynchronously. Calling the `start()` method places a paused `MediaPlayer` object in the started state.
- **Stopped:** Calling the `stop()` method stops playback. It also places a `MediaPlayer` object that is in the started, paused, or prepared states into the stopped state.

On Android, the `MediaPlayer` class assumed the role of the `manager` and `player` functions in Figure 9-1. It has the `setDataSource()` method to enable the object to access the multimedia data sources.

The Android SDK provides a set of audio and video APIs for developers to build rich multimedia functionality into an application. The `android.media.MediaPlayer` class can be used to play audio files, video files, and media streams. Your application can use any of the formats listed in the "Supported Media Codecs" section of this chapter. These media files can appear in an Android Package (.apk) file, Secure Digital (SD) card, or a smartphone's NAND flash and Embedded MultiMediaCard (EMMC) storage. The following code shows how to play an MP3 stored in an .apk file. By using the `MediaPlayer` object's `create()` method, you can obtain the handle to the `MediaPlayer` object:

```
MediaPlayer mySounder = MediaPlayer.create(this, r.raw.sound);
If (mySounder != null)
    mySounder.stop();
```

Before starting the sound playback, you must call the `prepare()` method, and then you can call the `start()` method to play the music as shown here:

```
mySounder.prepare();
mySounder.start();
```

When the music files are stored on an SD card, in NAND flash, or in EMMC storage, you use the following code to obtain a handle to the `MediaPlayer` object:

```
MediaPlayer mySounder = new MediaPlayer();
mySounder.setDataSource("/AndroidSD/yourfun.jpeg");
```

Once you have access to the file, `mySoundPlayer` can call `prepare()` and `start()` to set up and play the sound. Similarly, you can call `pause()` and `stop()` to pause and stop the playback. The `MediaPlayer` object also supports event-driven notifications during playback. For instance, once the `MediaPlayer` object completes the playback, it will fire an `onCompletion` event. You can release the `MediaPlayer` object and other related resources as part of the event handler as shown here:

```
Public void onCompletion(MediaPlayer mySounder) {
    mySounder.release();
}
mySounder.setOnCompletionListener(this);
```

You can also use the `android.widget.VideoView` class, which is a wrapper for `MediaPlayer`, to perform video playback on the Android. This class can load videos from resources or content providers. It offers various display options such as scaling and tinting and is managed by a layout manager. The `VideoView` object can also calculate video measurements and use a number of formats as described in the "Supported Media Codecs" section of this chapter. To use this object, you can add a `VideoView` element to the layout XML file:

```
<VideoView android:id="@+id/myViewer" android:layout_width="600px"
android:layout_height="300px"/>
```

You can implement the video playback by calling the `setVideoURI()` method with a file path and filename. You must also set the controller by calling `setMediaController()`, and then call the `start()` method to begin playback as shown here:

```
myViewer.setVideoURI(Uri.parse("file///androidSD/fun.wav"));
myViewer.setMediaController(new MediaController(this));
myViewer.start();
```

If you want to pause or stop the video player, you can use the following code:

```
myViewer.pause();
myViewer.stopPlayback();
```

In addition, you can use the `android.view.SurfaccView` class to manage a drawing surface in a view hierarchy. It's possible to adjust the format and size as well. This class can place the surface at the proper position on the screen, which is something you can't achieve using the `VideoView` component. To make the video player more flexible, you can integrate `MediaPlayer` with `SurfaceView`. Before using the `SurfaceView` component, you need to create a `SurfaceHolder` object as shown here:

```
SurfaceView myViewer = (SurfaceView) findViewById(R.id.surfaceView);
SurfaceHolder myHolder = myViewer.getHolder();
myHolder.setFixedSize(200, 200);
myHolder.setType(SurfaceHolder.SURFACE_TYPE_PUSH_BUFFERS);
```

If you don't call `setFixedSize()`, the rest of the `SurfaceView` is black except for the default video area. When you call the `SetFixedSize()` method, it sets the video player area to the size you specify. Whether you view this functionality as a bug or a feature, the `SetFixedSize()` method always fills the whole `SurfaceView` area once it's called.

After the code completes this setup, you can use the `MediaPlayer.setDisplay()` method to put the video player in the `SurfaceView` area, as shown here:

```
MediaPlayer myVideoPlayer = new MediaPlayer();
myVideoPlayer.setAudioStreamType(AudioManager.STREAM_MUSIC);
myVideoPlayer.setDisplay(myHolder);
myVideoPlayer.setDataSource("AndroidSDCard/fun.3gp");
myVideoPlayer.prepare();
myVideoPlayer.start();
```

You can call `setAudioStreamType()` to get the sound stream. By calling `setDataSource()`, the `myVideoPlayer` object can load a video file stored at the location you specify. For pause and stop, `myVideoPlayer` can use `pause()` and `stop()` methods individually.

Supported Media Codecs

Smartphone users care a lot about the sound, voice, and video quality when they play multimedia like video clips. Thus developers need to pay attention to choosing the media codecs the OS supports, because it can affect the user experience in ways like latency, well-encoding color, and surface texture. For example, the sports media needs to encode motion in high quality. Moreover, different types of codecs can have different effects on the device itself, affecting power consumption, talk time, and other things.

There are audio codecs and video codecs. Some media streams contain audio data, video data, and metadata that will be used for synchronization of audio and video.

The iOS supports the following audio formats:

- Advanced Audio Coding (AAC)
- Protected AAC (From iTunes Store)
- High-Efficiency Advanced Audio Coding (HE-AAC)
- Motion Picture Experts Group (MPEG) Layer-3 Audio (MP3)
- MP3 Variable Bit Rate (VBR)
- Audible (formats 2, 3, and 4, Audible Enhanced Audio, Audible Enhanced Audiobook File (AAX), and AAX+)
- Apple Lossless
- Audio Interchange File Format (AIFF)
- WAV (wave)

The iOS also supports the following video formats:

- H.264 video, up to 720p (or 1280×720 pixels), 30 frames per second, Main Profile level 3.1 with Advanced Audio Coding Low-Complexity (AAC-LC) audio up to 160 Kbps, 48kHz, stereo audio in MPEG-4 Visual (.m4v), MPEG Layer-4 Audio (.mp4), and .mov (movie) file formats
- MPEG-4 video, up to 2.5 Mbps, 640 by 480 pixels, 30 frames per second, Simple Profile with AAC-LC audio up to 160 Kbps per channel, 48kHz, stereo audio in .m4v, .mp4, and .mov file formats
- Motion Joint Photographic Experts Group (M-JPEG) up to 35 Mbps, 1280 by 720 pixels, 30 frames per second, audio in μ law, Pulse-Code Modulation (PCM) stereo audio in .avi file format

For additional details about the supported media format on iOS, please refer to `http://www.apple.com/iphone/specs.html`.

Android provides built-in supports for the following audio formats:

- MP3
- AAC
- AMR
- AAC +
- PCM/WAV

- Musical Instrument Digital Interface (MIDI)
- Ogg Vorbis

The Android system also supports the following video formats:

- H.264 AVC
- H.263
- MPEG-4 SP
- VP8

For additional details about the supported media format on Android, please refer to `http://developer.android.com/guide/appendix/media-formats.html`.

WP7 offers broad support on audio and video codecs as described in `http://msdn.microsoft.com/library/ff462087.aspx`

WP7 supports the following audio formats:

- Low Complexity AAC (AAC-LC)
- Linear Pulse Code Modulation (LPCM)
- Microsoft Custom Technology Differential Pulse Code Modulation (MS ADPCM)
- Interactive Multimedia Association ADPCM (IMA ADPCM)
- Version 1 of HE-AAC using spectral band replication (HE-AAC v1)
- Version 1 of HE-AAC using both spectral band replication and parametric stereo to enhance the compression efficiency of stereo signals (HE-AAC v2)
- MP3
- MP3 Variable Bit Rate (VBR)
- Adaptive Multi-Rate Audio Codec (AMR-NB)
- Windows Media Audio which was developed by Microsoft for audio data compression technology (WMA 10 Professional)
- Version 9 of Windows Media Audio (WMA Standard v9)

WP7 supports the following video formats:

- MPEG-4
- Windows Media Video's Video Codec Simple Profile (WMV VC-1 SP)
- Windows Media Video's Video Codec Main Profile (WMV VC-1 MP)
- Windows Media Video's Video Codec Advanced Profile (WMV VC-1 AP)
- MPEG-4 Part 2 Simple Profile
- MPEG-4 Part 2 Advanced Simple Profile

- MPEG-4 Part 10 (MPEG-4 AVC, H.264) Level 3.0 - Baseline Profile
- MPEG-4 Part 10 (MPEG-4 AVC, H.264) Level 3.0 - Main Profile
- MPEG-4 Part 10 (MPEG-4 AVC, H.264) Level 3.0 - High Profile
- H.263, which is the video compression standard as a low bit rate compressed format for videoconferencing

All three of these platforms also support the JPEG, Portable Network Graphics (PNG), Graphic Interchange Format (GIF), Bitmap (BMP), and Tagged Image File (TIF) picture codecs.

PLAYING AUDIO ON WP7

This section discusses how to play audio on WP7. You'll discover how to build applications that offer integrated sound, photo, and graphics experience.

WP7 audio development is based on two assembly models: the Silverlight object model (`System.Windows.Controls.MediaElement`) and the Game SDK, XNA model (`Microsoft.Xna.Framework.Audio.SoundEffect`).

The difference between the two models is that when you try to design a good web, business, or personal .NET application, it's easier to use Silverlight's `MediaElement` object. But when you design for game effects, you might want to consider using XNA's `SoundEffect` object.

Let's start with `MediaElement` class. It should be noted that the `MediaElement` object is also designed to play videos.

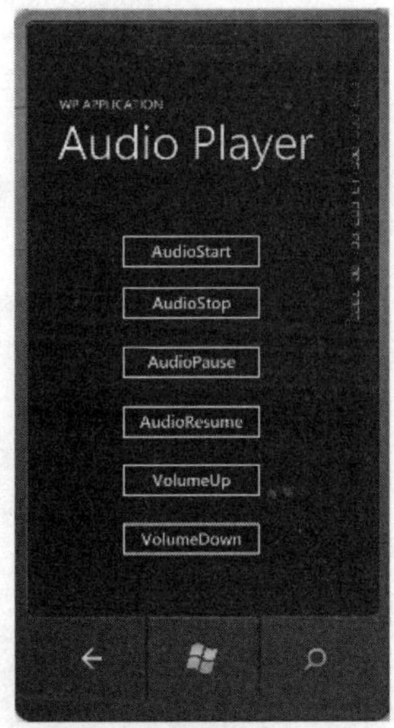

FIGURE 9-4: Sample application using MediaElement to play audio

Playing Sounds Using MediaElement

The simplest way to play Audio in WP7 is using the `MediaElement` object, which is a .NET Framework object that appears as part of the Silverlight object model. This Windows control appears as a rectangle in the Visual Studio and Expression Blend designer window. Now let's use a sample application to demonstrate how to use it to play sounds on WP7. The UI of this sample application is shown in Figure 9-4.

The sample application `WP7AudioPlayerDemo` has a simple UI with six buttons to demonstrate the basic audio playback controls (start, stop, pause, and resume) and volume control (volume up and volume down). You can use the following steps to build this sample application.

1. Create a Windows Phone project and type **`WP7AudioPlayerDemo`** as the project name. For details of how to build a WP7 application, please refer to Chapter 2.
2. Drag UI components for the application to the UI designer canvas. The main UI components used in this sample application are `Button` controls and a `MediaElement` control. As shown in Figure 9-4, the button label clearly presents the button function.

3. Create event handlers for action commands or events associated with each button. Listing 9-1 shows the core UI layout for the content panel and the event handler for each button.

Available for download on Wrox.com

LISTING 9-1: Core UI components, media element, and event handlers for the sample audio player, WP7AudioPlayerDemo\WP7AudioPlayerDemo \MainPage.xaml

```
<Grid x:Name="ContentPanel" Grid.Row="1" Margin="12,0,12,0">
   <Button Content="AudioStart" Name="button1" Click="AudioStart_Click"
      IsEnabled="True" />
   <Button Content="AudioStop"  Name="button2" Click="AudioStop_Click"
      IsEnabled="True"/>
   <Button Content="AudioPause" Name="button3" Click="AudioPause_Click"
      IsEnabled="True"/>
   <Button Content="AudioResume" Name="button6" Click="AudioResume_Click"
      IsEnabled="True"/>
   <Button Content="VolumeUp" Name="button4" Click="VolumeUp_Click"
      IsEnabled="True"/>
   <Button Content="VolumeDown" Name="button5" Click="VolumeDown_Click"
      IsEnabled="True"/>
</Grid>
<MediaElement x:Name="Audio1" AutoPlay="False"
   MediaOpened="Audio1_MediaOpened"/>
```

The `MediaElement` tag shown in Listing 9-1 has the `AutoPlay` attribute set to `False`, which means the application won't play the media until the user takes some action. The `MediaOpened` attribute defines a connection to the `Audio1_MediaOpened()` event handler. Here's the event handler code for the `MediaOpened` event.

Available for download on Wrox.com

```
private void Audio1_MediaOpened(object sender, RoutedEventArgs e)
{
    Audio1.Play();
}
```

Code snippet WP7AudioPlayerDemo\WP7AudioPlayerDemo\MainPage.xaml.cs

Table 9-1 contains a summary of the main events for the `MediaElement` class.

TABLE 9-1: MediaElement Main Events

EVENT	DEFINITION
BufferingProgressChanged	This event fires whenever the `BufferingProgress` property changes. The code can take the event and implement its business logic
CurrentStateChanged	This event fires whenever the `CurrentState` property changes.
DownloadProgressChanged	This event fires whenever the `DownloadProgress` property changes.

EVENT	DEFINITION
GotFocus	The application fires this event whenever a `UIElement` object receives focus.
KeyDown/UP	The system fires this event whenever the user presses a keyboard key and a `UIElement` object is in focus.
LayoutUpdated	The application fires this event whenever the Silverlight visual tree layout changes.
Loaded	The application fires this event after it constructs the `FrameworkElement` object and adds it to the object tree.
LostFocus	The application fires this event whenever a `UIElement` object loses focus.
MediaEnded	The application fires this event after the `MediaElement` object has stopped audio or video play.
MediaFailed	The application fires this event any time the media source has an error.
MediaOpened	The application fires this event after it validates and opens the media stream, and the code has read the file headers.
SizeChanged	This event fires whenever either the `ActualHeight` or the `ActualWidth` property changes value on a `FrameworkElement`.
Unloaded	The application fires this event when the `MediaElement` object has lost its connection with the main object tree.

If the `AutoPlay` attribute shown in Listing 9-1 is set to `True`, the application plays the audio when the page loads. You can specify the audio source statically in the XAML file. If the audio source is a file (such as an .mp3 file or a .wav file) packaged within your application, you should add it to your project first. You can right-click the project entry in Solution Explorer and choose Add ⇨ Existing Item from the context menu. You'll see the Add Existing Item dialog box where you can locate the audio file you want to add. After you select the item you want to add, click Add and the IDE will add it to Solution Explorer for you. After that, you should set the `BuildAction` property to `Content`. For example, the following code adds `JackieSound.WAV` to the `MediaElement`.

```
<MediaElement x:Name="Audio1" AutoPlay="False" Source="JackieSound.WAV"
MediaOpened="Audio1_MediaOpened"/>
```

You don't have to add the media source statically to the XAML file. It's also possible to add the media source dynamically using code. Compared with statically specifying the media source file, the dynamic load method enables the application to choose the file on the fly if the application prompts the user with a dialog box or other UIs. It becomes more flexible but the side effect is you may lose some performance.

Available for download on Wrox.com

```
private void AudioStart_Click(object sender, RoutedEventArgs e)
{
    Audio1.Source = new Uri("JackieSound.WAV", UriKind.Relative);
    Audio1.Position = new TimeSpan(0);
}
```

Code snippet WP7AudioPlayerDemo\WP7AudioPlayerDemo\MainPage.xaml.cs

Because setting the source will reset the audio's position (the current progress through the audio's playback time) to 00:00:00, if you want to locate a specific start point you can set the `Position` property after you set the source.

In addition to an audio file, the source can also be an audio streaming specified by an absolute or relative source URI. An absolute URI is a complete reference to the resource, and a relative URI depends on a previously defined base URI. Here's an example of defining an absolute URI source, where the media position is defined precisely:

```
private void AudioStart_Click(object sender, RoutedEventArgs e)
{
   Audio1.Source = new
      Uri("http://dc232.4shared.com/img/462949607/ed97b821/dlink__2Fdownload_
            2FiqCUthcU_3Ftsid_3D20101227-45361-f962eca/preview.mp3",
            UriKind.Absolute);
   Audio1.Position = new TimeSpan(0);
}
```

You've been shown several `MediaElement` properties, and Table 9-2 provides a summary of those properties.

TABLE 9-2: MediaElement Main Properties

PROPERTY	DEFINITION
ActualHeight	Retrieves `FrameworkElement` object's height. This property is not applicable to audio playback.
ActualWidth	Retrieves `FrameworkElement` object's width. This property is not applicable to audio playback.
AutoPlay	Plays the media described by the `MediaElement Source` property automatically when set to `True`, otherwise not.
Balance	Retrieves and assigns a ratio of volume between stereo speakers. Some OEM Phones have stereo speakers.

PROPERTY	DEFINITION
BufferingProgress	When loading multimedia content into your media player application, you need to create a buffer to store the content. You can define the application to play back the real time content by streaming or using the buffered content in the buffer. When using content in buffer, the application can monitor how much content is in the buffer. This property indicates the current buffering development process state in the buffer.
BufferingTime	Retrieves and assigns the amount of time to buffer.
CacheMode	Caches the rendered content when set to `True`. Caching improves rendering performance and reduces network traffic as well.
CanPause	Shows whether the application can pause the media when it calls the `Pause()` method. After the application raises the `MediaOpened` event, and if the `CanPause` property is `False`, the application can change the display characteristics of the Pause button.
CanSeek	Gives a value that indicates whether the application can reposition the media. This property can be set as `False` or `True`. For live streaming media, it is set to `False`. This API can be used to determine whether user capabilities are available at a point in time so the application can be shown the proper UI.
CurrentState	Indicates the `MediaElement` status. This property affects application behavior for applications that use `MediaElement` methods such as `Play()` and `Stop()`. For example, when the player is in the `Buffering` state, the application can't make a call to `Play()` until the state changes to another state such as `Playing` or `Paused`.
DownloadProgress	Indicates the remote content's downloaded percentage.
IsMuted	Sets the audio state to mute or indicates whether the audio is muted.
Name	Sets or gets the `MediaElement` object name.
NaturalDuration	Obtains the length of the opened media file. For live media, this property returns a value of `Automatic`.
Position	Sets or gets the `MediaElement` content's play position.
Source	Obtains the `MediaElement` source data, usually by URI value.
Volume	Sets or gets the `MediaElement` volume. The application's volume slider can use this property to control volume.

When the application plays music, the user can increase or decrease the volume as needed. The `VolumeUp_Click()` and `VolumeDown_Click()` methods contain the code required to change the volume. The player uses the `AudioStart_Click()` and `AudioStop_Click()` methods to start or stop audio playback. Additionally, the player calls the `AudioPause_Click()` and `AudioResume_Click()` methods to pause or resume play. The code for supporting these functions is shown in Listing 9-2.

> *This example also shows a potential failing of media support for WP7. The pause function works fine for .mp3 files, but not for .wav files.*

Available for download on Wrox.com

LISTING 9-2: Main implementations for the sample audio player , WP7AudioPlayerDemo\WP7AudioPlayerDemo\MainPage.xaml.cs

```
namespace WindowsPhoneAudio1
{
    public partial class MainPage : PhoneApplicationPage
    {
        double volumeChange = 0.1;
        double volumeMax = 1.0;
        double volumeMin = 0;
        TimeSpan timeSpan;

        private void Audio1_MediaOpened(object sender, RoutedEventArgs e)
        {
            Audio1.Play();
        }

        private void AudioStart_Click(object sender, RoutedEventArgs e)
        {
            Audio1.Source = new Uri("JackieSound.WAV", UriKind.Relative);
            Audio1.Position = new TimeSpan(0);
        }

        private void AudioStop_Click(object sender, RoutedEventArgs e)
        {
            Audio1.Stop();
        }

        private void AudioPause_Click(object sender, RoutedEventArgs e)
        {
            Audio1.Pause();
            timeSpan = Audio1.Position;
        }

        private void AudioResume_Click(object sender, RoutedEventArgs e)
```

```
            {
                Audio1.Position = timeSpan;
                Audio1.Play();
            }

            private void VolumeUp_Click(object sender, RoutedEventArgs e)
            {
                if (Audio1.Volume < volumeMax)
                    Audio1.Volume += volumeChange;
            }

            private void VolumeDown_Click(object sender, RoutedEventArgs e)
            {
                if (Audio1.Volume > volumeMin)
                    Audio1.Volume -= volumeChange;
            }
        }
    }
```

The `AudioStart_Click()` defines where to load the media file, and where the start position is. When the Pause button is clicked, the method `AudioPause_Click()` is invoked. Similarly, when the resume, stop, volume up, and volume down buttons are clicked, the corresponding `AudioResume_Click()`, `AudioStop_Click()`, `AudioVolumeUp_Click()`, and `AudioVolumeDown_Click()` are called.

Table 9-3 contains a summary of the methods you'll use most often. You can find the complete listing of all the events, properties and methods at `http://msdn.microsoft.com/library/system.windows.controls.mediaelement.aspx`.

TABLE 9-3: MediaElement Main Methods

METHOD	DEFINITION
Pause	Pauses the play action when the media is playing. Use the `Play()` method to resume play.
Play	Starts playing the media content at the given state. If the media player is stopped, calling `Play()` will start playing the media from the beginning. When the media player is paused, calling `Play()` will resume playing the media from the paused position.
SetSource(MediaStreamSource)	Sets the source used to load media for the media player from a `MediaStreamSource` object. The `MediaStreamSource` object will have a `MediaElement` subclass when the user sets it. If you define the `MediaElement Source` in the .xaml file and set the `MediaElement.Source` property in your C# code, the source set in the C# code wins.

continues

TABLE 9-3 *(continued)*

METHOD	DEFINITION
SetSource(Stream)	Sets the source used to load media for the media player from a `Stream` object. Unlike the `Source` property, which relies on a URI, a `Stream` object can accept an existing media stream provided by the user. You may load the media stream using the `WebClient` APIs and `SetSource()` can leverage the existing content stream.
Stop	Stops playing the media content. When media is playing, user can use this method to stop the media play and reset the media position to point at the beginning of the media stream.

Playing Sounds Using SoundEffect

The `SoundEffect` class in the `Microsoft.Xna.Framework.Audio` namespace provides another way to play audio on WP7. The usage difference between `SoundEffect` and `MediaElement` is that `SoundEffect` is not a UI element which can be defined in the XAML. Thus you can create the `SoundEffect` object directly in code. Although it is provided by the XNA assembly, you can use this class when building a Silverlight application.

`SoundEffect` supports only .wav files, but you can also convert .mp3 and .mp4 files into the .wav file format. There are a lot of conversion tools available, for example Audacity (`http://audacity.sourceforge.net/`). It's important to realize that the .wav format file is larger than that of the .mp3 or .mp4 because the .mp3 and .mp4 files are compressed. Therefore if you aren't using the media purely for game purposes, you can use the `MediaElement` object to play the media as described in the "Playing Sounds Using MediaElement" section of the chapter to gain a performance benefit.

Just as you added the audio source file into the application in the last sample project, you can add the .wav files to your application project for the `SoundEffect` object. Then you can use the `Load()` method of the `ContentManager` to load the sound effect. The `Play()` method of the `SoundEffect` object can be used to play the sound:

```
// Create SoundEffect object
SoundEffect sound;
// Load sound
sound = Content.Load<SoundEffect>("busy");
// Play sound
sound.Play();
```

You can also open a stream to a .wav file by calling `TitleContainer.OpenStream()` with the name of the .wav file. And then you can call the `SoundEffect.FromStream()` method to read the stream object you obtain from the `TitleContainer.OpenStream()`.

Table 9-4 contains a summary of the main `SoundEffect` properties and methods. You can see the entire list at `http://msdn.microsoft.com/library/dd282429.aspx`.

TABLE 9-4: SoundEffect Main Properties and Methods

PROPERTY	DEFINITION
Duration	Gets the media length. This is a read-only property.
MasterVolume	Gets and sets the volume.
SpeedOfSound	Gets and sets the media playing speed to simulate different environments. Using higher speeds decreases Doppler effects.
METHOD	**DEFINITION**
FromStream	Sets the data source used to feed the `SoundEffect` object.
Play	Launches the audio play action.

Sound, Picture, and Graphics Integration

In many cases, especially for games, you need to integrate audio elements with other media elements such as pictures and graphics. It is useful to understand how you can bundle these media elements together in your applications.

In this section, you'll use a sample application, `MediaPicker`, to show you the tight integration between sound, picture, and graphics elements by playing two pictures on the XNA game canvas with a sound when the pictures collide. The main purpose of this sample is to demonstrate how to manipulate these media elements instead of really introducing how to develop a game. Thus you will focus on handling the media elements. Figure 9-5 is a screenshot of this sample application. Because this is an XNA application, you can create the MediaPicker project by following the steps introduced in Chapter 8. The game loop is handled by the `Media` class as defined in Listing 9-3.

> *As mentioned in Chapter 8, to run XNA graphics on the WP7 emulator, the graphics card of your computer needs to support Direct X version 10 or above and the Driver mode needs to be WDDM 1.1 or above. Otherwise, you will still be able to compile the code but cannot test it on the emulator.*
>
> *Note that all the sample code in this chapter will NOT run on an emulator that does not meet the driver requirements. However, you will be able to run the sample applications on a physical Windows Phone 7 device.*

The classes in XNA to handle pictures, graphics, and audio can be found in the `Microsoft.Xna.Framework.Media`, `Microsoft.Xna.Framework.Graphics`, and `Microsoft.Xna.Framework.Audio` namespaces. To access the media resources available on the WP7, such as pictures, songs, and videos, you can use `MediaLibrary` class in the `Microsoft.Xna.Framework.Media` namespace. For example, the `Pictures` property of the `MediaLibrary` class provides the access to all the pictures in the Picture Hub of your WP7 device. In addition, the `Microsoft.Xna.Framework.Audio` namespace provides access to the `SoundEffect` class described in the "Playing Sounds Using SoundEffect" section to which you were just introduced. Listing 9-3 shows the properties you're going to use in this sample application.

FIGURE 9-5: WP7 sound and picture play

LISTING 9-3: Property definitions used in the game, MediaPicker\MediaPicker\Media.cs

Available for download on Wrox.com

```
namespace MediaPicker
{
    /// <summary>
    /// SoundEffect and Picture Demo Game
    /// </summary>
    public class Media : Microsoft.Xna.Framework.Game
    {
        GraphicsDeviceManager mediaGraph;
        SpriteBatch mediaCanvas;
        MediaLibrary mediaLibrary = new MediaLibrary();

        Texture2D mediabox1;
        Texture2D mediabox2;
        Vector2 mediaPosition1;
        Vector2 mediaPosition2;
        Vector2 mediaSpeed1 = new Vector2(50.0f, 50.0f);
        Vector2 mediaSpeed2 = new Vector2(100.0f, 100.0f);
        int media1Height;
        int media1Width;
        int media2Height;
        int media2Width;

        bool isPicture;
        SoundEffect sound;
        Random random;

        public Media()
```

```
            {
                mediaGraph = new GraphicsDeviceManager(this);
                Content.RootDirectory = "Content";
                random = new Random();

                // Frame rate is 30 fps by default for Windows Phone.
                TargetElapsedTime = TimeSpan.FromTicks(333333);
            }
        }
    }
```

The `mediaGraph` object is used to handle the configuration and management of the graphics device. The `mediaCanvas` object is used to draw sprites on the screen. The `mediaLibrary` object is used to get access to the pictures stored on the device. In this sample application, two pictures at time are displayed in two `Texture2D` objects (`mediabox1` and `mediabox2`). These pictures will move randomly on the screen. The `sound` object is used to play a sound when the two pictures collide. The application uses a combination of the `Initialize()` and `LoadContent()` methods to initialize the game, as shown in Listing 9-4.

LISTING 9-4: Initialize the game, MediaPicker\MediaPicker\Media.cs

Available for download on Wrox.com

```
protected override void Initialize()
{
   base.Initialize();
}

protected override void LoadContent()
{
   // Create a new mediaCanvas, which can be used to draw textures.
   mediaCanvas = new SpriteBatch(GraphicsDevice);

   ShowMedia();

   // Load Sound
   sound = Content.Load<SoundEffect>("busy");

   // Set starting positions of the media.
   mediaPosition1.X = 0;
   mediaPosition1.Y = 0;

   mediaPosition2.X =
      mediaGraph.GraphicsDevice.Viewport.Width - mediabox1.Width;
   mediaPosition2.Y =
      mediaGraph.GraphicsDevice.Viewport.Height - mediabox1.Height;

   // Calculate the height and width of the media.
```

continues

LISTING 9-4 *(continued)*

```
    media1Height = mediabox1.Bounds.Height;
    media1Width = mediabox1.Bounds.Width;

    media2Height = mediabox2.Bounds.Height;
    media2Width = mediabox2.Bounds.Width;
}
```

These methods provide a place to query required services and to load non-graphic content. The `Initialize()` method calls `base.Initialize()` to enumerate any components and initialize them. The application calls `LoadContent()` once the game starts and is the place to query required services to load media-related content.

To prepare the pictures that will be displayed on the screen, the method `ShowMedia()` is defined as shown in Listing 9-5.

Available for download on Wrox.com

LISTING 9-5: Prepare the pictures to display on the screen, MediaPicker\MediaPicker\Media.cs

```
void ShowMedia()
{
    PictureCollection picGroup;

    picGroup = mediaLibrary.Pictures;

    // Set the current photo to display
    int index1 = random.Next(picGroup.Count);
    int index2 = random.Next(picGroup.Count);

    while (index1 == index2)
    {
        index2 = random.Next(picGroup.Count);
    }

    Stream stream1 = picGroup[index1].GetThumbnail();
    mediabox1 = Texture2D.FromStream(GraphicsDevice, stream1);
    Stream stream2 = picGroup[index2].GetThumbnail();
    mediabox2 = Texture2D.FromStream(GraphicsDevice, stream2);
}
```

The `picGroup` object holds the access to the picture collections by using the `Pictures` property of the `MediaLibrary` class. The application randomly chooses two pictures from the picture collection, and loads their thumbnails into the sprites which will be drawn on the screen. As mentioned in Chapter 8, the game logic is implemented in the `Update()` method as shown in Listing 9-6.

Available for download on Wrox.com

LISTING 9-6: Update the media movement position in the game, MediaPicker\MediaPicker\Media.cs

```
/// <summary>
/// Allows the MediaPicker game to run logic such as updating the world,
/// checking for collisions, gathering input, and playing audio.
/// </summary>
/// <param name="mediaRun">Provides a snapshot of timing values.</param>
protected override void Update(GameTime mediaRun)
{
    // Allow the game to exit.
    if (GamePad.GetState(PlayerIndex.One).Buttons.Back ==
          ButtonState.Pressed)
        this.Exit();
    // Move the media around.
    Updatemedia(mediaRun, ref mediaPosition1, ref mediaSpeed1);
    Updatemedia(mediaRun, ref mediaPosition2, ref mediaSpeed2);
    IsIntersect();

    base.Update(mediaRun);
}

/// <summary>
/// Update the medias position.
/// </summary>
void Updatemedia(GameTime mediaRun,
                 ref Vector2 mediaPosition,
                 ref Vector2 mediaSpeed)
{
    // Move the media by speed, scaled by elapsed time.
    mediaPosition +=
        mediaSpeed * (float)mediaRun.ElapsedGameTime.TotalSeconds;
    int MaxX =
        mediaGraph.GraphicsDevice.Viewport.Width - mediabox1.Width;
    int MinX = 0;
    int MaxY =
        mediaGraph.GraphicsDevice.Viewport.Height - mediabox1.Height;
    int MinY = 0;
    // Check for bounce.
    if (mediaPosition.X > MaxX)
    {
        mediaSpeed.X *= -1;
        mediaPosition.X = MaxX;
    }
```

continues

LISTING 9-6 *(continued)*

```
        else if (mediaPosition.X < MinX)
        {
            mediaSpeed.X *= -1;
            mediaPosition.X = MinX;
        }
        if (mediaPosition.Y > MaxY)
        {
            mediaSpeed.Y *= -1;
            mediaPosition.Y = MaxY;
        }
        else if (mediaPosition.Y < MinY)
        {
            mediaSpeed.Y *= -1;
            mediaPosition.Y = MinY;
        }
    }
    /// <summary>
    /// See if the mediabox are intersecting each other, and play sound
    /// and swap pictures if true
    /// </summary>
    void IsIntersect()
    {
        BoundingBox mb1 =
           new BoundingBox(
              new Vector3(
                 mediaPosition1.X - (media1Width / 2),
                 mediaPosition1.Y - (media1Height / 2), 0),
              new Vector3(
                 mediaPosition1.X + (media1Width / 2),
                 mediaPosition1.Y + (media1Height / 2), 0));
        BoundingBox mb2 =
           new BoundingBox(
              new Vector3(
                 mediaPosition2.X - (media2Width / 2),
                 mediaPosition2.Y - (media2Height / 2), 0),
              new Vector3(mediaPosition2.X + (media2Width / 2),
                 mediaPosition2.Y + (media2Height / 2), 0));
        if (mb1.Intersects(mb2))
        {
            sound.Play();
            ShowMedia();
        }
    }
```

The main game loop keeps updating the pictures' state such as position and speed through the call to the `Updatemedia()` method. It also checks whether the pictures overlap each other or not through the call to the `IsIntersect()` method. If two pictures collide, a sound is played and the new pictures will be selected randomly again. After all these updates have been processed, if it is time to redraw itself, the game will call the `Draw()` method as shown in Listing 9-7 to draw a new frame.

Available for download on Wrox.com

LISTING 9-7: Redraw the screen, MediaPicker\MediaPicker\Media.cs

```
/// <summary>
/// This is called when the MediaPicker game's media should draw itself.
/// </summary>
/// <param name="mediaRun">Draw media canvas in
///    a snapshot of timing values.</param>
protected override void Draw(GameTime mediaRun)
{
    mediaGraph.GraphicsDevice.Clear(Color.Coral);
    // Draw the mediaCanvas.
    mediaCanvas.Begin(SpriteSortMode.BackToFront,
                      BlendState.AlphaBlend);
    mediaCanvas.Draw(mediabox1, mediaPosition1, Color.White);
    mediaCanvas.End();

    mediaCanvas.Begin(SpriteSortMode.BackToFront, BlendState.Opaque);
    mediaCanvas.Draw(mediabox2, mediaPosition2, Color.Gray);
    mediaCanvas.End();

    base.Draw(mediaRun);
}
```

This sample game has illustrated how to use the `SoundEffect`, `PictureCollection`, `MediaLibrary`, `GraphicsDevice`, and `Texture2D` objects together to build a simple game. By running this example you can see the audio and picture integration in WP7. In the next section you'll discover the video replay component of WP7.

PLAYING VIDEO ON WP7

There are two ways to programmatically play videos on WP7. The first approach is to user a launcher called `MediaPlayerLauncher` to let the system handle and control the playback. The use of this launcher is similar to other launchers that were introduced in Chapter 3. The other approach is to use the `MediaElement` class just introduced in the last section.

Playing Video Using MediaPlayerLauncher

The `MediaPlayerLauncher` class enables an application to launch the system media player to play the video with the given URI. Figure 9-6 shows the scenario when the system media player is launched to handle the video play. In this case, your application should follow the application life cycle principles and manage its own state as was discussed in Chapter 3. In addition, because the video playback is handled by the system media player, you cannot change the UI look-and-feel. Furthermore, the player state transitions totally depend on the user interactions and cannot be controlled programmatically.

FIGURE 9-6: Using the WP7 media player launcher

Listing 9-8 shows the code for how to use the MediaPlayerLauncher class, which is defined in the Microsoft.Phone.Tasks namespace.

Available for download on Wrox.com

LISTING 9-8: Use media player launcher to play video, WP7VideoPlayerDemo\WP7VideoPlayerDemo\MainPage.xaml.cs

```
using Microsoft.Phone.Tasks;

public partial class MainPage : PhoneApplicationPage
 {
     MediaPlayerLauncher Demolauncher;
     // Constructor
     public MainPage()
     {
         InitializeComponent();
         Demolauncher = new MediaPlayerLauncher();
     }

     private void MediaLaunchStart_Click(object sender, RoutedEventArgs e)
     {
         Demolauncher.Controls = MediaPlaybackControls.All;
         Demolauncher.Location = MediaLocationType.Install;
         Demolauncher.Media = new Uri("MVI_3086.wmv", UriKind.Relative);
         Demolauncher.Show();
     }
}
```

In this example, you declare a MediaPlayerLauncher object, Demolauncher, and initialize it in the constructor. When you need to use the launcher to play a video, you just need to set the Controls, Location, and Media properties, and then call Show() to launch the system media player. The Controls property sets the flags that determine which controls are displayed in the system media player by using the bitwise combination defined by the MediaPlaybackControls type. The Location property sets the location of the media file to be played, and this property can take values from MediaLocationType enumeration. MediaLocationType.Install means the media file is in the application's installation directory. MediaLocationType.Data should be used if the media file is in isolated storage or in the case of streaming for the network. The Media property sets the media played with the system media player and it takes a URI.

Playing Video Using MediaElement

Using the MediaElement class to play video is similar to playing audio. Compared to using the MediaPlayerLauncher, the benefit of using the MediaElement is that you can embed the video player inside your application and have full control of the player state.

Like the sample application used for audio play, you can build a sample application to demonstrate how to use MediaElement to play videos inside your application. Figure 9-7 shows the UI for this sample application, WP7VideoPlayerDemo. You can use the following steps to build this sample application:

1. Create a Windows Phone project and type **WP7VideoPlayerDemo** as the project name. For details of how to build a WP7 application, refer to Chapter 2.
2. Drag UI components for the application to the UI designer canvas. The main UI components used in this sample application are `Button` controls and a `MediaElement` control. As shown in Figure 9-7, the button label clearly presents the button function.
3. Create event handlers for action commands or events associated with each button. Listing 9-9 shows the event handler for each button.

FIGURE 9-7: Sample application using MediaElement to play video

From the code listed in Listing 9-9 you can see that the basic operations such as start, stop, pause, and resume the video playback, as well as increasing and decreasing the volume, are the same used in the audio play sample application.

Available for download on Wrox.com

LISTING 9-9: Using MediaElement to play video, WP7VideoPlayerDemo\WP7VideoPlayer Demo\MainPage.xaml.cs

```
private void VideoStart_Click(object sender, RoutedEventArgs e)
{
    Video1.Source = new Uri("MVI_3086.wmv", UriKind.Relative);
    Video1.Position = new TimeSpan(0);
    Video1.Play();
}

private void VideoStop_Click(object sender, RoutedEventArgs e)
{
    Video1.Stop();
}

private void VideoPause_Click(object sender, RoutedEventArgs e)
{
    Video1.Pause();
    timeSpan = Video1.Position;
}

private void VideoResume_Click(object sender, RoutedEventArgs e)
{

    Video1.Position = timeSpan;
    Video1.Play();
```

continues

LISTING 9-9 *(continued)*

```
}

private void VolumeUp_Click(object sender, RoutedEventArgs e)
{
    if (Video1.Volume < volumeMax)
        Video1.Volume += volumeChange;
}

private void VolumeDown_Click(object sender, RoutedEventArgs e)
{
    if (Video1.Volume > volumeMin)
        Video1.Volume -= volumeChange;
}
```

Unlike audio play, you can specify the height and width of the video display surface using the properties `Height` and `Width`. If you don't, once you specify a video source the media will display at its natural size and the layout will recalculate the size.

Reusable Media Player Controls

From the sample application using the `MediaElement` class to play audio and video, you can see that the basic media handling operations are similar. This suggests that it may be a good idea to build a customized user control like that introduced in Chapter 4. Doing so will enable you to reuse your code for media handling and increase your productivity. In addition, you can always match your own media player control's look-and-feel with different applications by making style changes — adjusting the height, width, font, or other customizable features.

Figure 9-8 shows an example of a reusable media player user control. This media player control includes functions for start, stop, pause/resume, fast forward (using a slider), rewind (using the same slider), volume slider controls (using another slider), and total play length display. Because you have already seen how to create a user control in Chapter 3, and also showed the implementation of media handling functions, there's no need to dig into details of the control itself. The complete source code of this sample media player can be found in the source code package of this book at `WP7EnrichedMoviePlayerDemo`.

FIGURE 9-8: A reusable media player user control

SUMMARY

This chapter began with a discussion of the common characteristics of the multimedia architecture for WP7, iOS, and Android. All three platforms are built with a solid object-oriented design model, multimedia data flow, and state information preservation and transition concepts. The different models and examples presented in this section demonstrate that WP7 is easier to use, has richer APIs, and provides better management of audio, pictures, and video. WP7 also leverages its Silverlight framework and XNA framework to provide diverse programming models.

This chapter also has explained how to use the `System.Windows.Controls.MediaElement` class and XNA multimedia object model in the `Microsoft.Xna.Framework.Media` namespace to develop sound playback, picture display, and video playback code. The WP7 APIs are as easy to use as those provided by the iOS and Android development environment, if not easier.

The next chapter discusses the peripherals such as a microphone to record sound, a camera to take pictures, an accelerometer to play media files, and an FM radio sensor to listen to broadcasts.

10

Utilizing System Hardware

WHAT'S IN THIS CHAPTER?

- An overview of smartphone hardware
- Accessing the microphone
- Accessing the camera
- Using sensors
- Utilizing FM radio and Bluetooth

This chapter discusses a few system hardware components that are frequently used by popular mobile applications. These peripherals include microphone, camera, resistive and capacitive touch controller, and various sensors such as accelerometer, light sensor, magnetic sensor, and so on.

Before you delve into the details, it's important to obtain an architectural overview of the common hardware architecture across the three software systems: iOS, Android, and WP7. Then the chapter will discuss general methods and required APIs surrounding the following four topics on the three platforms: accessing microphone, accessing camera, accessing sensors, and utilizing FM radio and Bluetooth.

AN OVERVIEW OF SMARTPHONE HARDWARE

Mobile devices have revolutionized the way the world communicates. More and more people use mobile devices for their communication and entertainment needs on a daily basis — needs which are primarily fulfilled by a variety of mobile applications.

One of the main driving forces behind this revolution is the mass of new sensors and peripherals. People can use mobile devices to take pictures, record videos, and play games that leverage an accelerometer, a digital compass (magnetic sensor), and a Global Positioning System (GPS) module. Some devices have a frequency modulation (FM) radio module so one can listen to radio with it. In addition, the multi-finger touch ability enables users to play games that simulate a guitar, a piano, and other instruments, with the support of the capacitive multi-touch screen.

Moreover, with Bluetooth, Wireless Fidelity (Wi-Fi), and speech recognition capabilities, a mobile device can provide a rich set of connectivity options between devices, as well as between the device and a user.

Furthermore, an increasing number of mobile applications use common sensors as well as special-purpose sensors to provide functionality that's otherwise not possible with a PC. For example, there are mobile applications that work as a heartbeat monitor, a pedometer, and even an app for builders that measures horizontal balance.

Before diving into the software aspect of sensors and peripherals on mobile devices, it's necessary to present an overview of mobile hardware architecture. The following section will discuss a general ARM processor architecture that serves as the foundation for many of today's smartphones, including WP7 phones.

> *One of the biggest challenges with mobile device hardware design and integration is power consumption, since embedded processors and additional sensors and peripherals all draw power from the battery in the device, which can usually supply from 1000 mAh (milliamps per hour) to 1600mAh. Generally, GPS, Wi-Fi chips, mobile processors, and the display panel are among the top consumers of battery power. Historically, X86 mobile processors use more power than Advanced RISC Machine (ARM) processors. This is the main reason why today's smartphones predominantly have ARM processors.*

ARM Processor Architecture

ARM is a 32-bit reduced instruction set computer (RISC) with instruction set architecture (ISA) developed by ARM Holdings. Processor manufacturers purchase licenses from ARM Holdings and make ARM processors. Today, about 99 percent of the billions of mobile phones sold each year use at least one ARM processor. In addition, ARM processors are used in many other devices, from home appliances like microwave controllers, oven timers, home Wi-Fi routers, and video game controllers to industrial instruments like factory electronic scales, chemical content electronic analyzers, electronic telescopes, and other precision instruments.

The ARM architecture reduces the complicated translation layer of an x86 processor into a simple set of instructions. The integration of components and the use of dedicated controllers for different functions make the ARM processor different from an x86 processor, where most processing is performed by the main processor. The use of dedicated controllers in ARM processors is aimed at performing tasks directly in hardware, whereas an x86 processor relies on software to achieve the same functionality. That's why ARM processors perform the same task with few transistors and fewer CPU cycles, which translates into lower power consumption.

Furthermore, the relative simplicity of ARM processors makes the chips suitable for low-power applications. Even with the latest ARM processors manufactured by such silicon vendors as Qualcomm, ST-Ericsson, Samsung, and TI, optimizing device power consumption is still a critical issue, since not only sensors and peripherals need to be integrated properly with the chipset's power management modules, but the operating system and applications also need to be designed with power efficiency in mind.

Smartphone Hardware Components

This section discusses common hardware components that appear as part of a mobile device. Some Original Equipment Manufacturers (OEMs) add some special hardware components to their devices, such as a gyroscope to differentiate their products from the rest of the devices in the market. Since Apple controls the entire supply chain of its iPhone devices, it decides what hardware components an iPhone contains. In contrast, when building Android devices and WP7 devices, OEMs have the freedom to choose hardware components as needed. As a result, Android devices and WP7 devices have diverse hardware configurations.

Figure 10-1 shows the networking architecture that iOS, Android, and WP7 all share. All mobile phones come with a cellular radio frequency (RF) chip to connect the device to the cellular network for voice and messaging communication. Each mobile OS supports the use of a General Packet Radio Service (GPRS) or High Speed Packet Access (HSPA) chip to drive the data communication pipe between a device and cloud services. A Wi-Fi chip is a standard component on mobile devices that provides connections to a Wi-Fi hotspot at a much faster data rate than that of cellular data pipes. A Bluetooth chip allows the device to connect to other Bluetooth devices within short distances (usually less than 10 meters).

In addition, a GPS receiver chip is commonly used to provide GPS-based location data for such applications as maps and turn-by-turn navigation. There is often an FM radio chip to enable FM broadcast reception on the device.

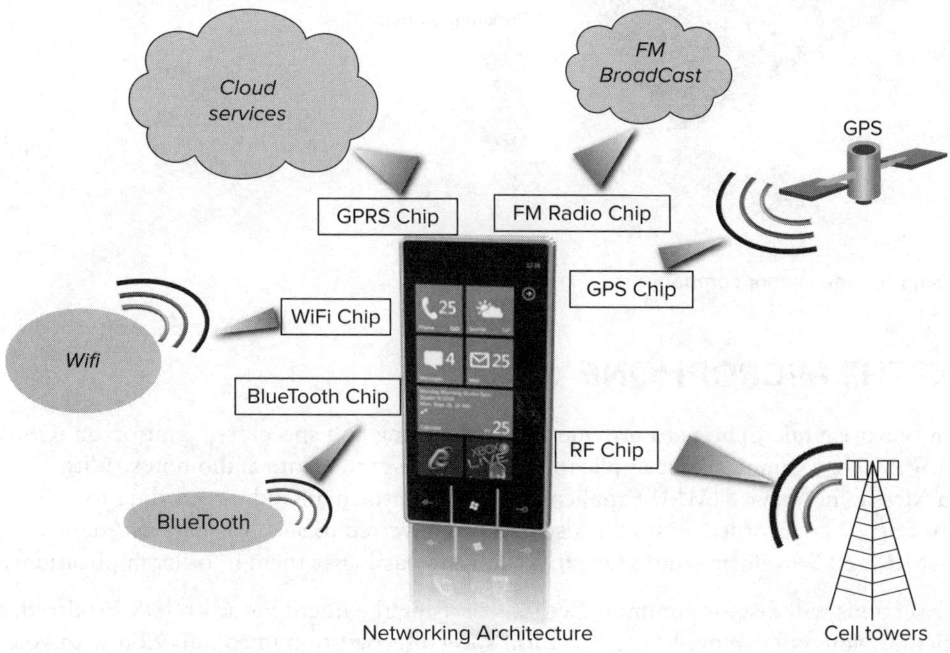

FIGURE 10-1: Smartphone networking components

A camera is a key component in mobile devices. Other than typical features such as supporting 5 or 8 megapixels and autofocus, some device OEMs have been able to provide a wealth of advanced features with their cameras, such as High Dynamic Range (HDR) and panorama camera photography enabled by photo-stitching technology.

Figure 10-2 shows common sensors used on smartphones that all three OSs support. A wide variety of applications can leverage these sensors to enable specific scenarios for a user. For example, the light sensor automatically adjusts the display's brightness so the user sees a brighter screen when the ambient light is strong and a dimmer screen in a dark setting. The proximity sensor enables the smartphone to detect the presence of nearby objects without any physical contact. The compass chip is a magnetic sensor that can provide device orientation. The accelerometer sensor and multitouch sensor can be used in a game or applications that detect a user's gesture.

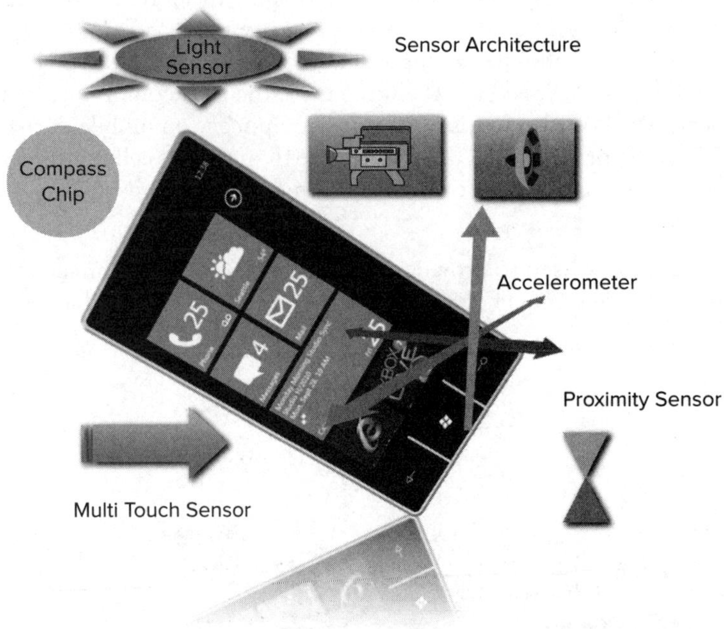

FIGURE 10-2: Smartphone sensor components

ACCESSING THE MICROPHONE

Many applications use a microphone to provide audio recording and speech recognition on iOS, Android, and WP7. For example, some applications allow a user to create audio notes. With a Multimedia Messaging Service (MMS) application one can attach an audio recording to a multimedia message. The recorded audio file also can be converted to such formats as Adaptive Multi-Rate (AMR) and Waveform Audio (.wav), so you can easily use them in other applications.

The following sections will discuss common methods of using the microphone on iOS, Android, and WP7. In particular, you will explore how to perform speech recognition in conjunction with voice recording across different platforms.

Accessing the Microphone on iOS

iOS relies on the `AVAudioRecorder` class to access iPhone's microphone and headset microphone to record sound. For complete details on accessing the microphone and on recording voice with various formats available on iOS, please refer to `http://developer.apple.com/library/ios/#documentation/AVFoundation/Reference/AVAudioRecorder_ClassReference/Reference/Reference.html`.

Starting with iOS 3.0 and newer versions, developers can leverage the UI accessibility programming interface VoiceOver to create applications that allow a user with visual impairments to navigate through the application's views and controls using speech and sound. For more information see `http://developer.apple.com/library/ios/#documentation/UserExperience/Conceptual/iPhoneAccessibility/Introduction/Introduction.html`.

Accessing the Microphone on Android

Android provides the `Android.media.MediaRecorder` class to perform audio and video recording tasks. This section will discuss how to use the `MediaRecorder` class to record audio. Further details can be found at `http://developer.android.com/guide/topics/media/index.html`.

The first step is to create a `MediaRecorder` object, set up its audio source, and define the output format (the container). The following code shows how to use the microphone as a source with the default format:

```
MediaRecorder myRecorder = new MediaRecorder();
myRecorder.SetAudioSource(MediaRecorder.AudioSource.MIC);
myRecorder.SetOutputFormat(MediaRecorder.OutputFormat.DEFAULT);
```

The next step is to define the encoder method, output filename, and file location:

```
myRecorder.SetAudioEncoder(MediaRecorder.AudioEncoder.DEFAULT);
FILE myFile = File.createTempFile("nexusVoice", ".wav");
myRecorder.SetOutputFile(myFile.getAbslutePath());
```

The final step prepares the system to use the microphone and starts recording:

```
myRecorder.prepare();
myRecorder.start();
```

To stop the recording, you call `myRecorder.stop()`. You can refer to the "Playing Audio" section of Chapter 9 for details of audio manipulations.

Android uses the `android.speech` package to perform speech-related tasks such as using the system's speech recognition service. Note that there is a similar speech assembly available for WP7. Conversely, you can use Android Text To Speech (TTS) package `android.speech.tts` to convert text to voice.

To use speech recognition, you need to first query the package manager to get a list of activities for the `RecognizerIntent.ACTION_RECOGNIZE_SPEECH` intent. Once a voice speech activity is found, the application can launch the activity with the `RecognizerIntent.ACTION_RECOGNIZE_SPEECH` intent, and a user can start to use speech recognition with the device. The following code snippet shows an example of such a procedure:

```
PackageManager pm = getPackageManager();
List activities = pm.queryIntentActivities(
   new Intent(RecognizerIntent.ACTION_RECOGNIZE_SPEECH), 0);
if (activities.size() != 0) {
   startVoiceRecognitionActivity();
}
private void startVoiceRecognitionActivity() {
   Intent intent = new Intent(RecognizerIntent.ACTION_RECOGNIZE_SPEECH);
   intent.putExtra(RecognizerIntent.EXTRA_LANGUAGE_MODEL,
      RecognizerIntent.LANGUAGE_MODEL_FREE_FORM);
   intent.putExtra(RecognizerIntent.EXTRA_PROMPT,
      "Speech recognition demo");
   startActivityForResult(intent, VOICE_RECOGNITION_REQUEST_CODE);
}
```

Note that the `startActivityForResult` method will launch the corresponding intent with a request code of `VOICE_RECOGNITION_REQUEST_CODE`. The result of the activity will be handled in the following method:

```
protected void onActivityResult(int requestCode, int resultCode, Intent data) {
   if (requestCode == VOICE_RECOGNITION_REQUEST_CODE
         && resultCode == RESULT_OK) {
// Use an ArrayList to hold resulted strings
      ArrayList<String> matches = data.getStringArrayListExtra(
           RecognizerIntent.EXTRA_RESULTS);
}
```

Accessing the Microphone on WP7

The simplest way to record audio on WP7 is by using the `Microsoft.Xna.Framework.Audio.Microphone` class. The following is excerpted from the sample Silverlight project WindowsPhoneMic, which is an example of using the `Microphone` class to record voice.

Besides common system assemblies, the following assemblies need to be added to the code:

```
using Microsoft.Xna.Framework;
using Microsoft.Xna.Framework.Audio;
```

Listing 10-1 shows a complete example of recording voice on WP7.

LISTING 10-1: Recording voice, WindowsPhoneMic\WindowsPhoneMic\MainPage.xaml.cs

Available for download on Wrox.com

```
namespace WP7Microphone
{
   public partial class MainPage : PhoneApplicationPage
   {
      private Microphone myMic = Microphone.Default;
      private byte[] cacheBuf;
```

```
        private MemoryStream memoryStorage = new MemoryStream();

        public MainPage()
        {
            InitializeComponent();

            DispatcherTimer dt = new DispatcherTimer();
            dt.Interval = TimeSpan.FromMilliseconds(33);
            dt.Tick += new EventHandler(dt_Tick);
            dt.Start();

            myMic.BufferReady += new
                EventHandler<EventArgs>(microphone_BufferReady);
        }

        void dt_Tick(object sender, EventArgs e)
        {
            try { FrameworkDispatcher.Update(); }
            catch { }
        }

        void microphone_BufferReady(object sender, EventArgs e)
        {
            myMic.GetData(cacheBuf);
            memoryStorage.Write(cacheBuf, 0, cacheBuf.Length);
        }

        private void recordButton_Click(object sender, EventArgs e)
        {
            myMic.BufferDuration = TimeSpan.FromMilliseconds(1000);
            cacheBuf = new byte[myMic.GetSampleSizeInBytes(myMic.BufferDuration)];
            memoryStorage.SetLength(0);
            myMic.Start();
        }

        private void stopButton_Click(object sender, EventArgs e)
        {
           myMic.Stop();
        }
    }
}
```

The `MainPage` class has private members such as `myMic` of type `Microphone`, a byte array `cacheBuf`, and a `MemoryStream` object named `memoryStorage`. The constructor of the `MainPage` class sets the timer to simulate the Xbox/DirectX New generation Architecture (XNA) framework game loop, since `Microphone` is part of the XNA framework. In addition, the constructor sets an event handler for the `BufferReady` event, which occurs when the audio capture buffer of the `Microphone` object is ready to be processed.

The `dt_Tick()` event handler is called at regular intervals by the application. It uses the XNA `FrameworkDispatcher` to see if a sound is playing. The `Update()` method updates the framework component status and also raises events as needed.

The next step is to create a microphone buffer-ready event handler, `microphone_BufferReady()`. In this example, the `microphone_BufferReady` method obtains the audio data from `myMic` by calling the `GetData()` method and stores the data in `cacheBuf`. The code then writes data in `cacheBuf` to the `MemoryStream` `memoryStorage` by calling the `Write()` method for a later playback.

The `recordButton_Click()` method is wired to the `MicOn` control to handle the click event for the Record button. This event handler configures `myMic` and writes audio data to `cacheBuf`, which is set to hold one second of audio data. It then starts `myMic`.

The `stopButton_Click()` method is wired to the `MicOff` control. This method handles the click event to stop the microphone.

Unlike Android, there is no speech processing API available in Silverlight or XNA for WP7. However, there are third-party solutions to work around this limitation, such as using software by Siri (`http://siri.com/`). The other solution is to use a web service that can convert uploaded voice data from a device to text and send back. For instance, to create a speech recognition application, a developer can build a web service that uses the .NET `System.Speech` API and perform speech recognition.

To implement text to speech in a WP7 application, you can utilize the speech service provided by the Microsoft Translator API (`http://www.microsofttranslator.com/dev/`). In essence, an application can send text to Bing web services and obtain corresponding speech data as a byte array. The application can save the byte array to its isolated storage and load it into a `MediaElement` object. For more information about using `MediaElement`, refer to the "Playing Audio on WP7" section of Chapter 9. This approach should be considered a work-around rather than a good solution, as its performance and stability are largely affected by those of the Translator web service.

ACCESSING THE CAMERA

Almost every smartphone is equipped with a camera that has 5 to 8 megapixel resolution, or even up to 10 megapixels. Users can take pictures or videos and upload them to their Facebook pages, online photo sites, or blogs. It is also natural to integrate camera capabilities into other applications such as photo and video editors and barcode scanners. On iOS, Android, and WP7, camera access is available to differing degrees. iOS and Android both provide API support for picture taking and video recording using the camera hardware. On the other hand, WP7 provides API only for picture taking at the time of writing. The following sections start by discussing different ways to access the camera on the three platforms and then provide some examples.

Accessing the Camera on iOS

On iOS, the `UIImagePickerController` class can be used to take pictures or record video using the built-in camera. To take a picture, first create a `UIImagePickerController` object and set its source type to Camera. The `cameraCaptureMode` property should be set to `UIImagePickerControllerCameraCaptureModePhoto`. If taking a video, it should be set to `UIImagePickerControllerCameraCaptureModeVideo`. The following code shows an example of performing video capture on iOS. Note that the `Camera` object's `sourceType` property and `cameraCaptureMode` property are set for video capture.

```
(void) camClicked: (UIButton *) button {
   UIImagePickerController *Camera =[[ UIImagePickerController alloc] init];
   Camera.sourceType = UIImagePickerControllerSourceTypeCamera;
   Camera.delegate = self;
   Camera.allowsImageEditing = YES;
   Camera.cameraCaptureMode = UIImagePickerControllerCameraCaptureModeVideo;
   [self presentModaViewController:Camera animated:YES]
}
```

For complete details on accessing the camera on iOS, refer to `http://developer.apple.com/library/ios/#documentation/uikit/reference/UIImagePickerController_Class/UIImagePickerController/UIImagePickerController.html`.

Accessing the Camera on Android

On Android, you access the camera using the `Android.hardware.Camera` class described at `http://developer.android.com/reference/android/hardware/Camera.html`.

This section shows briefly how to control the camera. The following code begins by defining the `myCamera` view class, which is extended from the `SurfaceView` class. The class has a `Camera` class instance named `AndroidCamera`. When the surface of the view is created, the `AndroidCamera` object is initialized and linked to the newly created surface for preview:

```
public class myCamera extends SurfaceView
   implements Surface SurfaceHolder.Callback
{
   private Camera AndroidCamera;

   private void surfaceCreated(SurfaceHolder AndroidHolder)
   {
      AndroidCamera = Camera.open();
      AndroidCamera.setPreviewDisplay(AndroidHolder);
   }
}
```

In the `surfaceChanged()` method, the following code sets the `Camera` object parameters and calls `startPreview()`, which is required before taking a picture or video:

```
private void surfaceChanged(SurfaceHolder AndroidHolder, int style,
                             int width, int height)
{
   Camera.Parameters Androidparams = AndroidCamera.getParameters();
   Androidparams.SetPictureFormat(PixelFormat.JPEG);
   Androidparams.setPreviewSize(width, height);
   AndroidCamera.setParameters(Androidparams);
   AndroidCamera.startPreview();
}
```

The following `shootPicture()` method shows how to actually take a picture. The `mycameraCallback()` method will be used to store the captured image data.

```
private void shootPicture()
{
   if(AndroidCamera != null)
      AndroidCamera.takePicture(null, null, mycameraCallback);
}
```

If you want to write code to create a video recording, the application must create a `Camera` object, initialize parameters, and start the preview, using the same techniques as shown for pictures. Additionally, you have to call the `unlock()` method to allow the media process to access the camera, pass the `Camera` object to the `setCamera(Camera)` method, and use the `MediaRecorder` class to record the video.

FIGURE 10-3: WP7 camera sample

Accessing the Camera on WP7

On WP7, you access the built-in camera using the `Microsoft.Phone.Tasks.CameraCaptureTask` class. WP7 does not support video capture at the time of writing.

Figure 10-3 shows a screenshot of the sample project WPCamera, which will be discussed in this section.

Listing 10-2 shows the main page of the sample WPCamera project.

LISTING 10-2: WP7 camera example, WPCamera\WPCamera\MainPage.xaml.cs

Available for download on Wrox.com

```
namespace WPCamera
{
   public partial class MainPage : PhoneApplicationPage
   {
      Stream capturedImage;
      // Constructor
      public MainPage()
      {
         InitializeComponent();
      }

      private void WPCamBtn_Click(object sender, RoutedEventArgs e)
      {
         CameraCaptureTask cct = new CameraCaptureTask();
         cct.Completed += new EventHandler<PhotoResult>(OnCameraTaskReturn);
         cct.Show();
      }

      public void OnCameraTaskReturn(object sender, PhotoResult e)
      {
         try
```

```
                {
                    e.ChosenPhoto.Position = 0;
                    BitmapImage bmi = new BitmapImage();
                    capturedImage = e.ChosenPhoto;
                    MediaLibrary ml = new MediaLibrary();

                    Picture p = ml.SavePicture(
                        Guid.NewGuid().ToString(), capturedImage);
                    }
                    catch (Exception err)
                    {
                        PageTitle.Text = err.ToString();
                    }
               }
          }
      }
```

As shown in Listing 10-2, the use of `CameraCaptureTask` is quite straightforward. When the Capture button is clicked, its event handler creates a `CameraCaptureTask` object. When the camera task returns, the picture data is carried in the `PhotoResult` object, which is passed back in the `CameraCaptureTask`'s Completed event. The sample code obtains the picture and saves it to the media library of the device.

USING SENSORS

Many smart phones are equipped with such sensors as an ambient light sensor, an accelerometer, a proximity sensor, and a magnetic sensor. These sensors are generally supported by iOS, Android, and WP7.

Android supports accelerometer, light sensor, magnetic sensor, orientation sensor, temperature sensor, and proximity sensor. On iOS, applications can access accelerometer, proximity sensor, and orientation sensor using the `UIDevice` object. WP7 provides accelerometer and orientation support through XNA. Support for other sensors, such as magnetic sensor and proximity sensor, is not available at the time of writing but could be available in future releases.

This section focuses more on accelerometer development. As the device moves, the accelerometer sensor reports linear acceleration changes along the primary axes in three-dimensional space as shown in Figure 10-4. The application can use the data to detect both the current orientation of the device (relative to the ground) and any instantaneous changes to that orientation. Such information can be leveraged by games and applications to enable rich user interactions with the device.

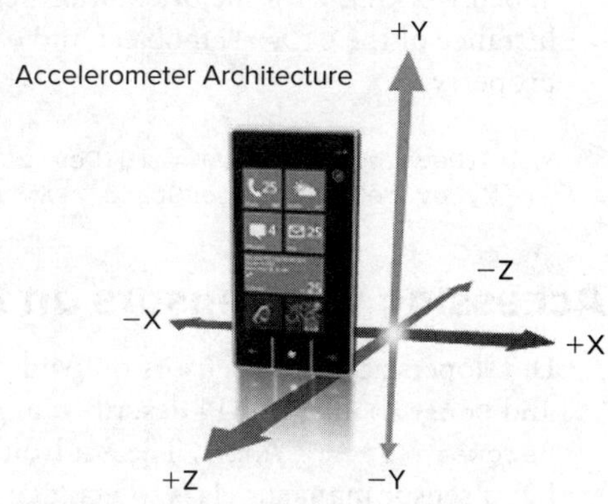

FIGURE 10-4: Accelerometer

Accessing the Sensors on iOS

Accelerometer support on iOS is provided with the `UIAccelerometer` class. The sensor measures the linear accelerational changes along three primary axes, and the data is relative to free fall. A value of 1 indicates that the device is experiencing 1 g of force on it. Details about the iOS accelerometer API and `UIAccelerometer` object are found at `http://developer.apple.com/library/ios/#documentation/UIKit/Reference/UIAccelerometer_Class/Reference/UIAccelerometer.html`.

To access the proximity sensor on iOS, you should first set the `proximityMonitoringEnabled` property of the `UIDevice` object to `YES`.

```
UIDevice *mydevice = [UIDevice currentDevice];
mydevice.proximityMonitoringEnabled = YES;
```

The application registers a callback method, `proximityChanged`, when a proximity change occurs as shown here:

```
if (mydevice.proximityMonitoringEnabled = =YES)
[[NSNotificationCenter defaultCenter] addObserver:self
   selector:@selector(proximityChanged:)
   name:@"UIDeviceProximityStateDidChangeNotification" object:mydevice];
```

When you hold the device to your ear, the proximity monitoring process triggers notifications. The observer callback of such proximity notifications can check the `proximityState` property of the `UIDevice` object to obtain the proximity state:

```
(void) proximityChanged:(NSNotification *)notification {
   UIDevice *device = [notification object];
   NSLog(@"iOS my application proximity state: %i", mydevice.proximityState);
}
```

When working with the orientation sensor, the application can read orientation by obtaining an instance of the `UIDevice` object and can obtain the orientation by reading the object's `orientation` property:

```
UIDevice *mydevice = [UIDevice currentDevice];
MyDeviceOrientationState = mydevice.orientation;
```

Accessing the Sensors on Android

Developers accessing sensors on Android use the `Android.hardware.Sensor`, `SensorEvent`, and `SensorManager` API described at `http://developer.android.com/reference/android/hardware/Sensor.html`. The Android sensor object model is different from iOS and WP7 — it has a sensor manager class to manage all supported sensors. It is relatively easy to access sensors because of the uniform access methods, and it also provides flexibility for future sensor additions.

The following `FreeFallSample` class shows how to use a `SensorManager` object named `AndroidSensorManager` to access the accelerometer on the device. The class implements the `onSensorChanged()` method, with which accelerometer data can be obtained:

```
public class FreeFallSample extends Activity, implements SensorEventListener {
   private final SensorManager AndroidSensorManager;
   private final Sensor AndroidAccelerometer;
   private float AndroidAccel ;
   private float AndroidPresentAccel;
   private float AndroidPreviousAccel;

   public FreeFallSample() {
      AndroidAccelerometer =
         AndroidSensorManager.getDefaultSensor(Sensor.TYPE_ACCELEROMETER);
      String Android_sensor_service = Context.SENSOR_SERVICE;
      AndroidSensorManager = (SensorManager)
         getSystemService(Android_sensor_service);
      AndroidSensorManager.registerListener(this,
         AndroidSensorManager.getDefaultSensor(Sensor.TYPE_ACCELEROMETER),
         SensorManager.SENSOR_DELAY_NORMAL);

      AndroidPresentAccel = SensorManager.GRAVITY_EARTH;
      AndroidPreviousAccel = SensorManager.GRAVITY_EARTH;
      AndroidAccel = 0.00f;
   }

   protected void onResume() {
      super.onResume();
      AndroidSensorManager.registerListener(this, AndroidAccelerometer,
         SensorManager.SENSOR_DELAY_NORMAL);
   }

   protected void onPause() {
      super.onPause();
      AndroidSensorManager.unregisterListener(this);
   }

   public void onAccuracyChanged(Sensor sensor, int accuracy) {
   }

   public void onSensorChanged(SensorEvent event) {
      float x = se.values[0];
      float y = se.values[1];
      float z = se.values[2];
      AndroidPreviousAccel = AndroidPresentAccel;
      AndroidPresentAccel = (float) Math.sqrt((double) (x*x + y*y + z*z));
      float delta = AndroidPresentAccel - AndroidPreviousAccel;
      AndroidAccel =  AndroidAccel * 0.6f + delta;
    }
}
```

`SensorManager` enables developers to change the data refresh rate of the accelerometer when the `registerListener` method of the `SensorManager` object is called. Depending on the required sensitivity of accelerometer data, an application can choose from `SENSOR_DELAY_FASTEST`, `SENSOR_DELAY_GAME`, or `SENSOR_DELAY_UI` from `SENSOR_DELAY_NORMAL` for this setting.

The Android sensor object model makes it easy to work with other sensors by simply replacing `Sensor.TYPE_ACCELEROMETER` with other sensor types including `Sensor.TYPE_LIGHT`, `Sensor.TYPE_MAGNETIC_FIELD`, and `Sensor.TYPE_ORIENTATION`.

Accessing Sensors on WP7

On WP7, applications can access supported sensors using the `Microsoft.Devices.Sensors` namespace. The `Accelerometer` class in this namespace can be used to retrieve value changes when the device is in acceleration. To control data acquisition from the sensor, the `Accelerometer` class provides `Start()` and `Stop()` methods. The `ReadingChanged()` event of the class can be used to handle the data. In addition, the `State` property indicates the state of the sensor, which can be `NotSupported`, `Disabled`, `Initializing`, `Ready`, `NoData`, and so on.

You can refer to the online documentation at `http://msdn.microsoft.com/library/microsoft.devices.sensors.aspx` to read more details.

The following shows some examples from the MediaPicker project that uses an `Accelerometer` object. This sample allows users to move pictures on the screen. The movement will speed up when the user shakes the device with more than 1.5 g force in the x or y axis. If the user applies 1 g of force in z direction, the application resets the display of pictures. A user can press and hold each picture icon to move it around on the screen. Figure 10-5 shows a screenshot of the application.

FIGURE 10-5: Using an accelerometer in the MediaPicker sample application

This sample application is an XNA application. Therefore it requires the following namespaces:

```
using System;
using System.IO;
using System.Collections.Generic;
using Microsoft.Devices.Sensors;
using Microsoft.Xna.Framework;
using Microsoft.Xna.Framework.Audio;
using Microsoft.Xna.Framework.Graphics;
using Microsoft.Xna.Framework.Input;
using Microsoft.Xna.Framework.Media;
using Microsoft.Xna.Framework.Input.Touch;
```

The `Media` class shown in Listing 10-3 is derived from the `Microsoft.Xna.Framework.Game` and has the following data members, including an `Accelerometer` object and `MediaElement` objects for picture manipulation. The `Vector3` object `accelReading` will be used to save accelerometer data along three axes.

LISTING 10-3: The Media class, MediaPicker\MediaPicker\Media.cs

Available for download on Wrox.com

```
namespace MediaPicker
{
    /// <summary>
    /// This is the main type for your game
    /// </summary>
    public class Media : Microsoft.Xna.Framework.Game
    {
        // Game related
        GraphicsDeviceManager mediaGraph;
        SpriteBatch spriteBatch;
        Random randomGen;

        //Accelerometer
        Accelerometer accelSensor;
        bool accelActive;
        Vector3 accelReading = new Vector3();

        // Game data
        MediaElement pinchedItem;
        MediaElement draggedItem;

        List<MediaElement> mediaElements = new List<MediaElement>();
        const int MediaElementWidth = 200;
        const int MediaElementHeight = 150;

        Vector2 delta1 = new Vector2(0, 0);
        Vector2 delta2 = new Vector2(0, 0);
        Vector2 position1 = new Vector2(0, 0);
        Vector2 position2 = new Vector2(0, 0);
        public static SpriteFont spriteFont;
        ...
    }
}
```

Listing 10-4 shows the `Initialize()` method of the `Media` class, which calls the `StartAccelerometer()` method defined in the same `Media` class as well. This `Initialize()` method also enables touch gesture support.

LISTING 10-4: The Initialize method of the Media class, MediaPicker\MediaPicker\Media.cs

Available for download on Wrox.com

```
protected override void Initialize()
{
    this.StartAccelerometer();

    // Setup touch
    TouchPanel.EnabledGestures = GestureType.Tap | GestureType.DoubleTap |
       GestureType.Pinch | GestureType.PinchComplete | GestureType.FreeDrag |
       GestureType.DragComplete;

    base.Initialize();
}
```

Listing 10-5 shows the `StartAccelerometer()` method of the `Media` class. This method creates an `Accelerometer` object and sets the `AccelerometerReadingChanged()` method to be the event handler of the `ReadingChanged` event for the `Accelerometer` object.

Available for download on Wrox.com

LISTING 10-5: The StartAccelerometer method of Media class, MediaPicker\MediaPicker\Media.cs

```
private void StartAccelerometer()
{
   accelSensor = new Accelerometer();

   // Add the accelerometer event handler to the accelerometer sensor
   accelSensor.ReadingChanged += new
      EventHandler<AccelerometerReadingEventArgs>(AccelerometerReadingChanged);

   // Start the accelerometer
   try
   {
      accelSensor.Start();
      accelActive = true;
   }
   catch (AccelerometerFailedException e)
   {
      // the accelerometer couldn't be started.  No fun!
      accelActive = false;
   }
   catch (UnauthorizedAccessException e)
   {
      // This exception is thrown in the emulator-which doesn't support an
      // accelerometer.
      accelActive = false;
   }
}
```

Listing 10-6 shows the `Update()` method of the `Media` class. This method reacts upon user gestures to perform a UI update. Gestures such as single tap, double tap, pinch, and free drag are supported. In addition, this method uses the `Vector3` object `accelReading` of the class to access accelerometer data and change the moving speed of the pictures. Note that the data in the `accelReading` object is obtained from the `AccelerometerReadingChanged` method shown earlier in Listing 10-5.

Available for download on Wrox.com

LISTING 10-6: The Update method of the Media Class, MediaPicker\MediaPicker\Media.cs

```
protected override void Update(GameTime gameTime)
{
   // Allow the game to exit.
   if (GamePad.GetState(PlayerIndex.One).Buttons.Back == ButtonState.Pressed)
```

```
    this.Exit();

// Check for touch

while (TouchPanel.IsGestureAvailable)
{
    GestureSample gs = TouchPanel.ReadGesture();

    switch (gs.GestureType)
    {
    case GestureType.Tap:
        foreach (MediaElement ele in this.mediaElements)
        {
            if (ele.IsPositionWithin(gs.Position))
            {
                ele.Tapped();
                break;
            }
        }
        break;
    case GestureType.DoubleTap:
        foreach (MediaElement ele in this.mediaElements)
        {
            if (ele.IsPositionWithin(gs.Position))
            {
                ele.DoubleTapped();
                break;
            }
        }
        break;
    case GestureType.Pinch:
        if (this.pinchedItem == null)
        {
            foreach (MediaElement ele in this.mediaElements)
            {
                if (ele.IsPositionWithin(gs.Position))
                {
                    this.pinchedItem = ele;
                    break;
                }
            }
        }
        else
        {
            this.pinchedItem.Pinched(GraphicsDevice, gs);
            this.position1 = gs.Position;
            this.position2 = gs.Position2;
            this.delta1 = gs.Delta;
            this.delta2 = gs.Delta2;
        }

        break;
    case GestureType.FreeDrag:
```

continues

LISTING 10-6 *(continued)*

```
            if (this.draggedItem == null)
            {
                foreach (MediaElement ele in this.mediaElements)
                {
                    if (ele.IsPositionWithin(gs.Position))
                    {
                        this.draggedItem = ele;
                        position1 = gs.Position;
                        break;
                    }
                }
            }
            else
            {
                this.draggedItem.Dragged(gs.Delta);
            }

            break;

        case GestureType.PinchComplete:
            if (this.pinchedItem != null)
            {
                this.pinchedItem.PinchComplete();
                this.pinchedItem = null;
            }
            break;
        case GestureType.DragComplete:
            if (this.draggedItem != null)
            {
                this.draggedItem = null;
            }
            break;
    }
}

if (accelReading.X > 1.5 || accelReading.Y > 1.5)
{
    foreach (MediaElement ele in this.mediaElements)
    {
        ele.SetInMotion();
        ele.ChangeSpeed(50);
    }
}
else if (accelReading.Z > 1)
{
    foreach (MediaElement ele in this.mediaElements)
    {
        ele.Reset();
    }
```

```
        }

        // Move the sprite around.
        foreach (MediaElement ele in this.mediaElements)
        {
        if (ele.State == MediaElementState.InMotion)
            {
                this.UpdateSprite(gameTime, ele);
            }
        }

        // The time since Update was called last.
        float elapsed = (float)gameTime.ElapsedGameTime.TotalSeconds;

        RotationAngle += elapsed;
        float circle = MathHelper.Pi * 2;
        RotationAngle = RotationAngle % circle;

        base.Update(gameTime);
    }
```

Listing 10-7 shows the `AccelerometerReadingChanged()` method, which obtains accelerometer data and saves it to the `accelReading` object.

Available for download on Wrox.com

LISTING 10-7: An example of obtaining accelerometer data in the Media class, MediaPicker\MediaPicker\Media.cs

```
public void AccelerometerReadingChanged(object sender,
      AccelerometerReadingEventArgs e)
{
    accelReading.X = (float)e.X;
    accelReading.Y = (float)e.Y;
    accelReading.Z = (float)e.Z;
}
```

To obtain device display orientation on WP7, the `GraphicsDeviceManager` class in the `Microsoft.Xna.Framework` namespace can be used. The `SupportedOrientations` property of the `GraphicsDeviceManager` class can be used to get or set display orientation if automatic rotation and scaling is enabled. To get orientation change notifications, create an event handler for the `OrientationChanged` event of the `GameWindow` class, which is also in the `Microsoft.Xna.Framework` namespace. For more details, refer to `http://msdn.microsoft.com/library/microsoft.xna.framework.graphicsdevicemanager.supportedorientations.aspx`.

API support for other sensors on WP7 was not available at the time of writing. But it's likely Microsoft will make that available with a later release, given that WP7's sensor object model and the `Devices.Sensors` namespace are already in place and can be extended to support other sensors that are already used by the operating systems and Microsoft applications.

UTILIZING FM RADIO AND BLUETOOTH

Neither iOS nor Android provides support for on-device FM radio. Some applications on iOS and Android use web services to enable an online radio experience. Some Android devices are equipped with an FM radio chip, and the device OEMs can build a radio application that accesses the FM radio using unpublished APIs. On the other hand, WP7 provides full FM radio support as part of the standard API, since every WP7 device has a built-in FM radio chip.

Utilizing the FM Radio On WP7

Developers can use WP7's FM radio API in the `Microsoft.Devices.Radio` namespace to access FM radio functionality. At the center of this namespace is the `FRRadio` class that developers can use to create an instance of the radio, turn the radio on and off, and tune it. The complete WP7 `FMRadio` object model is found at `http://msdn.microsoft.com/library/ff769541.aspx`.

Listing 10-8 shows an example of how to use the `FMRadio` class to access FM radio. This is excerpted from the sample project WP7FMRadio. The `WP_SetFrequency()` of the `MainPage` class method creates an `FMRadio` instance, turns its power on, and sets the region of the radio chip. It also sets the `Frequency` and does some exception handling as well. Figure 10-6 shows a screenshot of WP7FMRadio application.

LISTING 10-8: Using FM radio on WP7, WP7FMRadio\WP7FMRadio\MainPage.xaml.cs

Available for download on Wrox.com

```
namespace WP7FMRadio
{
   public partial class MainPage : PhoneApplicationPage
   {
      // Constructor
      public MainPage()
      {
         InitializeComponent();
      }

      private void WP_SetFrequency(object sender, RoutedEventArgs e)
      {
         try
         {
            FMRadio WPFMRadio = FMRadio.Instance;

            WPFMRadio.PowerMode = RadioPowerMode.On;
            WPFMRadio.CurrentRegion = RadioRegion.UnitedStates;
            WPFMRadio.Frequency = 99.9;

            if (FMRadio.Instance.SignalStrength == 0.0)
            {
               MessageBox.Show("You may not use FM on emulator");
            }
         }
         catch (Exception ex)
         {
```

```
            MessageBox.Show(String.Format(
               "Error: {0}\n FM Radio has error", ex.Message));
         }

      }
   }
}
```

Utilizing Bluetooth

Both iOS and Android provide API support for Bluetooth. iOS supports Bluetooth through the Gamekit framework described at `http://developer.apple.com/library/ios/#documentation/NetworkingInternet/Conceptual/GameKit_Guide/Introduction/Introduction.html`. On Android an application can access Bluetooth using the `Android.bluetooth` package discussed at `http://developer.android.com/reference/android/bluetooth/package-summary.html`. At the time of writing, WP7 didn't provide any Bluetooth API.

FIGURE 10-6: WP7 FM radio tuner application

SUMMARY

This chapter began with a discussion of the common characteristics of Smartphone hardware architecture for iOS, Android, and WP7. Then it explained and compared the differences in microphone support among the three OSs. In this case, both Android and WP7 appear to be easier to use and have richer APIs that make it easy to record audio. Camera access was discussed next with examples for the iOS, Android, and WP7.

The chapter also discussed sensor APIs on the three platforms. iOS provides quite a few sensors, whereas Android provides the richest sensor support. WP7 provides strong support for the touch sensor and accelerometer APIs, and these APIs are as easy to use as the iPhone and Android development environment, if not easier.

WP7 provides FM radio API. Neither iOS nor Android provides such capability. On the other hand, WP7 doesn't support Bluetooth access to applications, while the other two platforms do.

The next chapter will explore an increasingly critical issue on mobile devices: security. You'll discover the security model of each platform, and understand access and permission controls for applications.

11

What You Need to Do about Security

WHAT'S IN THIS CHAPTER

- Defining the mobile application security models
- Working with the Windows Phone 7 security APIs
- What are 'best practices' for security?

To many smartphone users, security is probably the last feature on their minds. Indeed, when consumers decide which phone to buy, they probably care more about applications, the display, battery life, wireless signal strength, and multimedia experiences. The user is likely to wonder why security is even important. Many will think that security is something that security-obsessed people worry about, but that doesn't matter in the real world.

The truth is that security threats are real. When the first mobile phone malware appeared back in February 2004, it didn't do much harm other than display a message, "Caribe," every time the user turned on the phone. But it quickly spread itself to other Symbian-based phones when they were within range of infected Bluetooth communication networks. Since then, cyber-criminals have created and modified more mobile malware to attack smartphones. The damage can include erasing user data, locking down the phone OS, stealing bandwidth, or even pilfering users' confidential information. In November 2010, it was reported that more than a million cell phones in China had been attacked by a "zombie" virus, which can send users' *Subscriber Identity Module (SIM)* information to the hackers. The zombie virus can also automatically send out text messages containing virus links to other people on the address book of infected phones. It was estimated that the zombie virus cost cell phone users at least $300,000 per day. The trade press has also reported that various attacks can successfully infiltrate both iPhones and Android devices. By Nov. 25, 2011, just a month after WP7 was released, three developers had already found a security flaw in the system and were able to figure out a way to unlock WP7 devices. If the unlocker program were made available, end users could install any applications to their phones without going through the Marketplace. This would have placed the security structure of WP7 in jeopardy. Luckily, Microsoft was able to reach an agreement with the developers. The security flaw was fixed, and the unlocker program was removed from the Internet in January 2011. With the rapid growth of the smartphone market, it is reasonable to believe that more sophisticated malware will target smartphones.

So after reading about all these attacks, you'll probably ask how security affects developers and wonder whether there is anything you should do about it. The bottom line is that you don't want to make headlines when your application becomes the victim of malicious attacks. It hurts your image and reputation and costs you both financially and legally. A virus can even put you out of business.

This chapter begins with an introduction to application security models and related components in the Windows Phone 7 (WP7), Android, and iOS environments. The concept of *sandbox* and *security chamber* will be reviewed, followed by a comparison of security features on the three platforms. After readers gain basic understanding of mobile application security, the .Net Security namespace APIs in the Windows Phone 7 development environment are reviewed. Examples will be given in detail to illustrate how to protect data confidentiality and integrity through encryption and hashing.

At the end of the chapter, a list of several security "best practices" is given to provide readers a guideline to developing secure WP7 applications.

UNDERSTANDING MOBILE APPLICATION SECURITY MODELS

It is imperative for developers to understand the nuances of mobile application security architecture and make informed decisions on writing secure code and deploying secure applications. The following sections discuss the security models of WP7, iOS, and Android.

Windows Phone 7 Security Overview

Like iPhone and Android, WP7 embraces the *sandbox* concept. The term sandbox refers to a set of controls that limits an application's access to system resources such as files, preferences, hardware, and so on. Security boundaries are also built around each application to prevent applications from negatively impacting each other or the system. Note that the sandbox can help reduce the attack surface, but it can't eliminate the risks of your application's being attacked. For example, if you don't handle user input carefully and your application is vulnerable to a buffer overflow or Cross-Site Scripting (XSS) attack, the sandbox can't help you safeguard your application.

In particular, WP7 defines different levels of application security, called *chambers*. Figure 11-1 explains the concept of the security chamber.

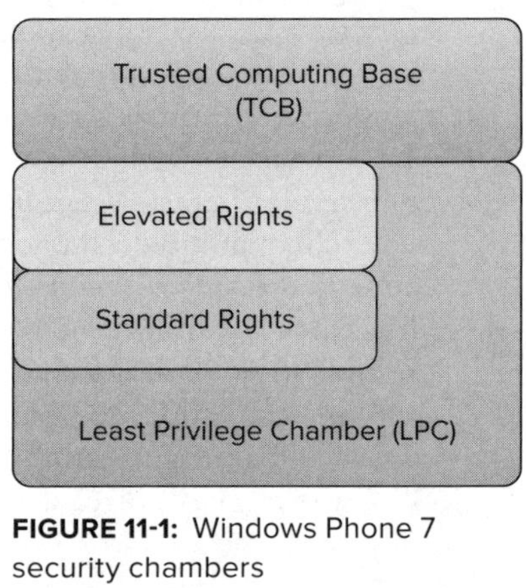

FIGURE 11-1: Windows Phone 7 security chambers

Trusted Computing Base (TCB) has the highest security privileges and can access almost all the system resources. TCB is the security level for OS kernel. Elevated Rights is the next chamber level, and applications at this level can access all resources except the security policy. Many built-in phone applications can run with Elevated Rights. Standard Rights is the default chamber for native-code applications. All applications that do not provide device-level services will run in the Standard Rights chamber. At the bottom of the diagram is the Least Privilege Chamber (LPC). The application you develop will run in this chamber by default.

Note that the scope of the LPC expands from minimal security to Elevated Rights. This approach is similar to Android in that an application starts with almost no permissions, but can request more permissions as required. In Microsoft terms, different permissions are called *capabilities*. Developers declare needed capabilities in the `WMAppMainifest.xml` file as needed. For example, if your application needs to access the location service and the network service, you can declare the needed capabilities as follows.

```
<App xmlns="" ... >
...
   <Capabilities>
     <Capability Name="ID_CAP_LOCATION" />
     <Capability Name="ID_CAP_NETWORKING" />
   </Capabilities>
...
```

If you don't request a capability in your manifest but try to use a feature restricted by it, you will get an Access Denied exception message. Table 11-1 describes the security capabilities provided by WP7 (`http://blogs.msdn.com/b/jaimer/archive/2010/04/30/windows-phone-capabilities-security-model.aspx`).

TABLE 11-1: List of WP7 Security Capabilities

CAPABILITY	DESCRIPTION	CLASS/NAMESPACE REQUIREMENT
ID_CAP_NETWORKING	Applications with access to network services. Disclosed because services could incur surcharge when roaming.	System.Net, Microsoft.Xna.Framework.GamerServices.GamerServicesComponent, WebBrowser class, Smooth Streaming Media Element (SSME) ID_CAP_PUSH_NOTIFICATION
ID_CAP_LOCATION	Applications with access to location services.	System.Device.Location
ID_CAP_MICROPHONE	Applications that access the microphone. The applications can record without displaying a visual indication that the recording is taking place.	Microsoft.Xna.Framework.Audio.Microphone

continues

TABLE 11-1 *(continued)*

CAPABILITY	DESCRIPTION	CLASS/NAMESPACE REQUIREMENT
ID_CAP_MEDIALIB	Apps that can access the media library.	MediaStreamSource, Microsoft.Devices.Radio, Microsoft.Xna.Framework.Media.MediaLibrary and MediaSource, Microsoft.Devices.MediaHistory
ID_CAP_GAMERSERVICES	Apps that interact with the Xbox live APIs. Disclosed due to privacy since data is shared with Xbox.	Xbox live API
ID_CAP_WEBBROWSERCOMPONENT	Applications that use the web browser component. This capability can create security risks when scripts are turned on.	WebBrowser class
ID_CAP_PHONEDIALER	Applications that place phone calls. In many cases, there is no visual indication to end-user.	Microsoft.Phone.Tasks
ID_CAP_PUSH_NOTIFICATION	Applications can receive push notifications from Internet service; disclosed because services could incur surcharge when roaming.	Microsoft.Phone.Notification
ID_CAP_SENSORS	Applications that need to access sensors, such as accelerometer, light sensor, etc.	Microsoft.Devices.Sensors

iOS Security Overview

Figure 11-2 illustrates the iPhone Operating System (iOS) security architecture. The security APIs are located in the Core Services layer, which uses the fundamental libraries in the Core OS layer. An iOS application can make a direct security API call without going through the Cocoa Touch layer and the Media layer. iOS also includes a Security Server daemon that implements common security protocols. But the Security Server daemon does not have a public interface for developers. Instead, it runs in the background to support applications that use keychain services, certificates, key, and trust services.

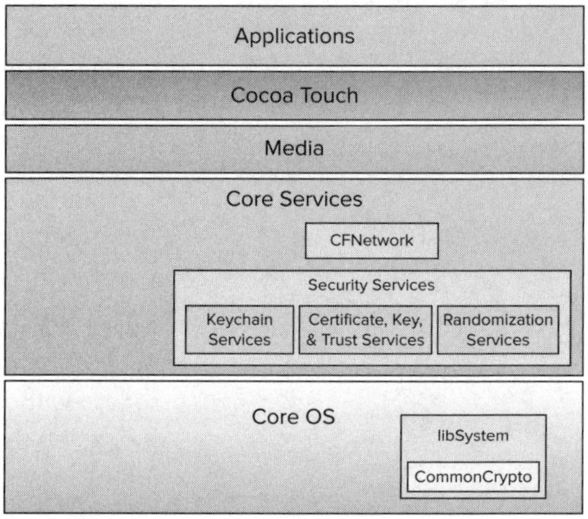

FIGURE 11-2: iOS security architecture

The *keychain service* is used to store passwords, keys, and certificates. The CFNetwork APIs are used to create and maintain secure data streams over the network. Randomization services can generate secure random numbers to further enhance the implementation of current encryption algorithms. The certificate, key, and trust services provide a number of important APIs to:

- Create, manage, and read certificates
- Add certificates to keychains
- Generate encryption keys and encrypt/decrypt data
- Generate and verify digital signatures
- Manage trust polices

iOS also provides several features to protect sensitive data and prevent applications from interfering with each other or the system. You'll notice that many of these features are also available on the Android and WP7 platforms.

Like WP7, iOS also supports sandboxing an application. In iOS, an application and its data are stored in a secure location that no other applications can tamper with. The keychain data is located outside the application sandbox so a vulnerable application can't steal the users' credentials.

Another important iOS feature is file protection. Files containing sensitive data can be marked as "protected" by iPhone owners. Protected files remain encrypted at all times. In addition, files can't be decrypted if the device is locked. This feature prevents someone from gaining access to protected files in an iPhone if the device is stolen.

Android Security Overview

Android also adopts the sandbox concept to better support mobile security. Similar to WP7, a basic Android application doesn't have any permissions associated with it and therefore can't access data from other applications or the system. If you want to share information between applications, the application code must explicitly declare corresponding permissions in the `AndroidManifest.xml` file. For example, if an application needs to monitor incoming SMS messages, it's necessary to declare `android.permission.RECEIVE_SMS` permission using the `<uses-permission>` tag as shown here:

```
<manifest xmlns:android="http://schemas.android.com/apk/res/android"
    package="com.android.app.myapp" >
    <uses-permission android:name="android.permission.RECEIVE_SMS" />
    ...
</manifest>
```

The permission an application requests is checked when you install it. If the system can't authenticate the application signature, or the user chooses not to grant the permission, the system will deny any request to access the requested resources, without prompting the user. The Android security model keeps the user informed about what an application can access and lets the user decide what information the application can use. But users may make poor decisions because they aren't well-informed or they're frustrated by applications not working the way they want.

Security Model Comparisons

In this section, we compare the WP7, iOS, and Android security models.

- **Application approval and distribution:** WP7 adopts a model that is similar to the iPhone. An application goes through a vigorous certification process before reaching the marketplace. Android on the other hand is more open to developers. You can install an application directly to an Android device for testing. You then sign your application with your private key and upload it to the Android Marketplace. In no time, it is published and made available to everyone. Since no lengthy certification process was involved, an information criminal has the potential to publish malicious applications and create some troubles for the users before those malwares can be stopped. However, it's important to understand that even though WP7 and iPhone apps are scrutinized before the developer releases them to the public, the certification process doesn't guarantee they are risk-free. The lengthy certification process, however, does reduce the chances of malware or phishing schemes reaching the end user.
- **Programming language:** WP7 supports only managed code and does not allow the developer to run native code using platform invoke (P/Invoke). Developers write iOS apps using Object C. It's possible to write Android applications using Java, native code, or hybrid code. From a security perspective, managed code certainly reduces vulnerabilities due to programming errors.
- **Permissions/capabilities:** Both WP7 and Android require an application to explicitly declare system resource and data needs. iOS doesn't have similar protection.
- **Sandboxing applications:** All three platforms implement sandbox techniques to lock applications down to their own isolated memory space.
- **Device security:** You can *jailbreak* an iPhone with ease so that you can install third-party applications that are not approved by Apple. Similarly, you can *root* an Android phone without much difficulty. Once your Android phone is rooted, you can overclock the processor, install additional features, and turn on Wifi and Bluetooth tethering. In both cases, the door to attack is wide open after users gain almost full control of the device and install whatever applications they like without thinking about security. It's too early to predict if users will be able to easily hijack WP7, but it would be naïve to believe that they won't find a way to do so.

USING WINDOWS PHONE 7 SECURITY APIS

In this section, you'll explore the security APIs that are available on the WP7 platform. A quick overview of the .NET security namespace is followed by encryption and hashing examples. If the main security concern of your application is to be able to securely pass user's login credentials from a WP7 device to the authentication server, you may also consider using a third-party API, such as the Hammock Open Authentication (OAuth) library discussed in Chapter 6.

.NET Security Namespace

The .NET Framework `System.Security` namespace supplies the underlying structure of the Common Language Runtime (CLR) security system. It provides a collection of security classes to developers to protect their applications. Due to hardware limitations, not all the child namespaces are supported on the WP7 platform. Table 11-2 lists commonly used security classes/APIs for WP7.

TABLE 11-2: List of .NET Security Namespaces for WP7

NAMESPACE	DESCRIPTIONS
System.Security	Supplies the underlying structure of the Silverlight security system
System.Security.Cryptography	Contains a collection of cryptographic services, such as symmetric encryption, hashing algorithms, and random number generators
System.Security.Cryptography.X509Certificates	Provides the X.509 v.3 certificate implementation
System.Security.Permissions	Defines policy-based operation and resources access control
System.Security.Principal	Defines the basic functionalities of identity and principal objects

The cryptography algorithm classes that WP7 supports are as follows:

- `AES` (Advanced Encryption Standard)
- `HMACSHA1` (Hash-based Message Authentication Code-Secure Hash Algorithm)
- `HMACSHA256`
- `Rfc2898DeriveBytes`
- `SHA1` (Secure Hash Algorithm)
- `SHA256`

Protecting Data Confidentiality with Encryption

The `System.Security.Cryptography` namespace offers three major class types: symmetric encryption algorithms, hash algorithms, and helper functions such as random number generator.

Symmetric encryption is a good choice to protect data confidentiality in the WP7 environment because smartphones have limited computing power compared to their desktop counterparts. Symmetric encryption can run about 1,000 times faster than asymmetric encryption when using a similar key strength (`http://www.cryptopp.com/benchmarks.html`). A well-designed and implemented symmetric encryption algorithm can provide reasonably good security without sacrificing performance.

In the current `Cryptography` namespace, the only symmetric encryption algorithm supported on WP7 is the `AesManaged` class. Advanced Encryption Standard (AES) uses the Rijndael cipher. It's the encryption standard the U.S. government adopted to replace the Data Encryption Standard (DES), and Triple-DES. DES was considered computationally impossible to crack when it was selected by the National Bureau of Standards in the 1970s. However, with the processing speed of modern computers becoming faster and faster, it was reported that a 56-bit DES key was broken within 22 hours in 1999 (`http://www.cryptool.de/download/PCP297.cryptography.pdf`). Triple-DES later replaced DES as the standard because the effective key strength of triple-DES can reach 112 bits. But it is a weaker encryption compared to its successor, AES encryption. The AES key size can be 128 bits, 192 bits, and 256 bits. The `AesManaged` class limits the block size to 128 bits and supports only the Cipher-Block Chaining (CBC) cipher mode.

The next example explains how to use this class to perform basic encryption and decryption. Use the following steps to create the solution:

1. Start Microsoft Visual Studio Express 2010 for Windows Phones. Windows starts the Visual Studio Express 2010 IDE.
2. Choose File ➪ New Project. You'll see the New Project dialog box.
3. Choose the Visual C# Silverlight for Windows Phone templates from the left panel. Highlight the Windows Phone Application template. You'll see a description of this template in the right pane of the New Product dialog box.
4. Type **aes** in the Solution Name field and the Name field. Click OK. Visual Studio creates a blank application for you.
5. Add a `TextBox`, `inputbox`; two `Button` controls, `encryptbtn` and `decryptbtn`; and a `TextBlock`, `resultbox`, to the content panel.
6. Add the click event handler to respond to user requests for both buttons.
7. Type the code shown in Listing 11-1 in `MainPage.xaml`.

LISTING 11-1: Layout of AES demo application, aes\aes\MainPage.xaml

Available for download on Wrox.com

```
<!--LayoutRoot is the root grid where all page content is placed-->
<Grid x:Name="LayoutRoot" Background="Transparent">
    <Grid.RowDefinitions>
        <RowDefinition Height="Auto"/>
        <RowDefinition Height="*"/>
    </Grid.RowDefinitions>

    <!--TitlePanel contains the name of the application and page title-->
    <StackPanel x:Name="TitlePanel" Grid.Row="0" Margin="12,17,0,28">
```

```
            <TextBlock x:Name="ApplicationTitle" Text="MY APPLICATION"
                       Style="{StaticResource PhoneTextNormalStyle}"/>
            <TextBlock x:Name="AES" Text="AES Demo" Margin="9,-7,0,0"
                       Style="{StaticResource PhoneTextTitle1Style}"/>
        </StackPanel>

        <!--ContentPanel - place additional content here-->
        <Grid x:Name="ContentPanel" Grid.Row="1" Margin="12,0,12,0">
            <TextBox Height="72" HorizontalAlignment="Left" Margin="0"
                    Name="inputbox" Text="" VerticalAlignment="Top" Width="460" />
            <Button Content="Encrypt" Height="72" HorizontalAlignment="Left"
                    Margin="0,72,0,0" Name="encryptbtn" VerticalAlignment="Top"
                    Width="220" Click="encryptbtn_Click" />
            <Button Content="Decrypt" Height="72" HorizontalAlignment="Left"
                    Margin="230,72,0,0" Name="decryptbtn" VerticalAlignment="Top"
                    Width="220" IsTabStop="False" Click="decryptbtn_Click" />
            <TextBlock Height="420" HorizontalAlignment="Left" Margin="0,150,0,0"
                    Name="resultbox" Text="" VerticalAlignment="Top" Width="456"
                    TextWrapping="Wrap" />
        </Grid>
    </Grid>
```

Figure 11-3 illustrates the layout of the controls.

FIGURE 11-3: AES demo UI layout

Now add the button click event handlers to both buttons, as described in Listing 11-2 in `MainPage.xaml.cs`.

LISTING 11-2: WP7 AES encryption demo, aes\aes\MainPage\xaml.cs

Available for download on Wrox.com

```
using System;
using System.Collections.Generic;
using System.Linq;
using System.Net;
using System.Windows;
using System.Windows.Controls;
using System.Windows.Documents;
using System.Windows.Input;
using System.Windows.Media;
using System.Windows.Media.Animation;
using System.Windows.Shapes;
using Microsoft.Phone.Controls;
//This is needed for AES encryption:
using System.Security.Cryptography;

namespace aes
{
   public partial class MainPage : PhoneApplicationPage
   {
      //WP7 supports 128-bit AES. key and iv should be 128-bit (16-Byte) long
      //The key and the iv in this example are defined as follows:

      private byte[] key=
         new byte[16] {1,2,3,4,5,6,7,8,9,10,11,12,13,14,15,16};
      private byte[] iv=
         new byte[16] {16,15,14,13,12,11,10,9,8,7,6,5,4,3,2,1};

      byte[] encodedbytes;
      byte[] cleartext;

      //use utf8 encoding to convert between bytes and string
      System.Text.UTF8Encoding utf8 = new System.Text.UTF8Encoding();

      // Constructor
      public MainPage()
      {
         InitializeComponent();
      }

      private void encryptbtn_Click(object sender, RoutedEventArgs e)
      {
         if ( inputbox.Text.Length == 0)
         {
            resultbox.Text =
               "The message to be encrypted is empty. Please input your message\n";
            return;
         }

         resultbox.Text = "";
         cleartext = utf8.GetBytes(inputbox.Text);
         AesManaged aesmanaged = new AesManaged();
```

```
        ICryptoTransform encryptor = aesmanaged.CreateEncryptor(key, iv);

        System.IO.MemoryStream ms = new System.IO.MemoryStream();

        CryptoStream cs =
           new CryptoStream(ms, encryptor, CryptoStreamMode.Write);

        cs.Write(cleartext, 0, cleartext.Length);
        cs.FlushFinalBlock();

        ms.Position = 0;
        encodedbytes = new byte[ms.Length];
        ms.Read(encodedbytes, 0, encodedbytes.Length);

        ms.Close();

        cs.Clear();
        cs.Close();

        resultbox.Text = "The encrypted message in byte format:\n";

        for (int i = 0; i < encodedbytes.Length;i++ )
        {
           resultbox.Text += encodedbytes[i];
           resultbox.Text += " ";
        }

     }

     private void decryptbtn_Click(object sender, RoutedEventArgs e)
     {
        if (encodedbytes == null)
        {
           resultbox.Text =
               "There is no encrypted message. Please encrypt a message first";
           return;
        }

        AesManaged aesmanaged = new AesManaged();
        ICryptoTransform decryptor = aesmanaged.CreateDecryptor(key,iv);

        System.IO.MemoryStream ms = new System.IO.MemoryStream();

        CryptoStream cs =
           new CryptoStream(ms, decryptor, CryptoStreamMode.Write);

        cs.Write(encodedbytes, 0, encodedbytes.Length);
        cs.FlushFinalBlock();

        ms.Position = 0;
        cleartext = new byte[ms.Length];
```

continues

LISTING 11-2 *(continued)*

```
            ms.Read(cleartext, 0, cleartext.Length);

            ms.Close();

            cs.Clear();
            cs.Close();

            resultbox.Text = "The decrypted message is:\n";
            resultbox.Text += utf8.GetString(cleartext, 0, cleartext.Length);

            //empty encoded bytes
            encodedbytes = null;

        }
    }
}
```

When the user clicks the Encrypt button, this application encrypts the string the user inputs and displays the result in the TextBlock. It can also decrypt a ciphered message stored in memory when the Decrypt button is clicked.

To use the AES encryption, both parties must know a secret key prior to the communication. In addition, the code also requires an initialization vector (IV) for the cipher operation. Recipients must also know the IV to be able to decipher the message. For the sake of simplicity, this example defines a pair of 16-byte (128-bit) arrays as the key and the iv, respectively, as shown in the following code:

```
//WP7 supports 128 bit AES, the length of key and iv should be 128bit, or
16Bytes long
//The key and the iv in this example are defined as follows (16Bytes)

private byte[] key= new byte[16]{1,2,3,4,5,6,7,8,9,10,11,12,13,14,15,16};
private byte[] iv= new byte[16]{16,15,14,13,12,11,10,9,8,7,6,5,4,3,2,1};
```

In real-life applications, the application can generate a key using password-based key derivation functionality, such as the Rfc2898DeriveBytes class in the System.Security namespace. And the application can generate the IV using a random number generator, which it transmits along with the cipher text.

The clear text and ciphered text are stored in byte arrays. The code creates an encoding object to convert the data between a string and a byte array:

```
        byte[] encodedbytes;
        byte[] cleartext;
        //use utf8 encoding to convert between bytes and string
        System.Text.UTF8Encoding utf8 = new System.Text.UTF8Encoding();
```

In the click event handler, encryptbtn_Click(), the Text property of inputbox is first converted to a byte array:

```
cleartext = utf8.GetBytes(inputbox.Text);
```

To encrypt `cleartext`, a new instance of `AesManaged` is initialized, and the code creates a symmetric `encryptor` object using the `key` and `iv` defined earlier.

```
AesManaged aesmanaged = new AesManaged();
ICryptoTransform encryptor = aesmanaged.CreateEncryptor(key, iv);
```

To transform a clear text message into a ciphered message, the code needs a `CryptoStream` object to link the `encryptor` with a targeted data stream. In this example, we use a `MemoryStream` object to store transformed information. You must declare the `CryptoStreamMode` as `Write` because the code writes information to the targeted data stream.

```
System.IO.MemoryStream ms = new System.IO.MemoryStream();
CryptoStream cs =
   new CryptoStream(ms, encryptor, CryptoStreamMode.Write);
```

The `CryptoStream.Write()` method transforms the bytes from a buffer and writes information to the current `cryptostream`. It takes three parameters: a buffer it reads from, the starting position within the buffer, and the number of bytes to read from the buffer. The `CryptoStream.FlushFinalBlock()` method updates the target data stream, `MemoryStream ms`, and syncs it with the transformed results.

```
cs.Write(cleartext, 0, cleartext.Length);
cs.FlushFinalBlock();
```

The ciphered message is now stored in `MemoryStream` and the application can read it into the pre-defined byte array `encodedbytes`.

```
ms.Position = 0;
encodedbytes = new byte[ms.Length];
ms.Read(encodedbytes, 0, encodedbytes.Length);
```

At this point, the application has finished the encryption and can close `MemoryStream` and `CryptoStream`. To display the encoded message in a readable format, the value of each byte is shown in the `TextBlock` object.

To illustrate the decryption process, this example assumes the encrypted message is already stored in the byte array, `encodedbytes`, and the values of `key` and `iv` are also known. In a real-world application, you can use a password string as the secret that is known only to the sender and receiver. You can then use encryption helper functions, such as `Rfc2898DeriveBytes`, to derive the byte value of the AES encryption key.

Since AES is a symmetric encryption algorithm, the decryption method is very similar to the encryption method. You still need to create a new instance of the `AesManaged` class, and create a new `decryptor` object based on the `key` and `iv`. Then, link the `decryptor` to `MemoryStream ms` with `CryptoStream cs`. Next, decrypt `encodedbytes` to clear text, store it in `ms`, and copy the results to the `cleartext` byte array. After the decryption is completed, the code can close both streams and convert the `cleartext` from a byte array to a string. Finally, the application can display the decoded message in the `TextBlock`.

From this example, it is clear that implementing AES encryption is a fairly simple and straightforward process on the WP7 platform. For simplicity, this demo does not implement any exception handling. A real-world application should implement full exception handling, though, because cryptography tends to be very error-prone.

Ensuring Data Integrity with Hashing

Hash functions perform a one-way transformation and map an arbitrary length of binary string to a small fixed length of binary string. A well-crafted hash algorithm is designed in such a way that two different inputs will not result in the same hash value, and a small change to the input will result in a significant and unpredictable change to the output. In addition, it would be computationally impossible to figure out the original message from its hash value. This feature makes hash algorithms a perfect technology for protecting data integrity.

Three abstract classes are currently supported in the `System.Security.Cryptography.HashAlgorithm` namespace. They are `KeyedHashAlgorithm`, `SHA1`, and `SHA256`. The first requires using a key to perform the hash calculation while the other two do not. The benefit of including a key in the hash function is that it can provide authenticity, because only those who have the key can calculate the hash value.

In current Microsoft implementation, `HMACSHA1` and `HMACSHA256` are the two classes that developers can use to generate message authentication code. One visible difference between the SHA-1 and SHA-256 hashing algorithms is the length of the hash value. The output of SHA-1 is 160 bits long, while the output of SHA-256 is 256 bits long. Another thing to notice is that SHA-1 was claimed to have been broken by some researchers back in February 2005 (`http://www.openauthentication.org/pdfs/Attacks%20on%20SHA-1.pdf`). This means that you should choose SHA-256 whenever possible.

Using the hash function on WP7 is a fairly simple process. The main method you need to call is the `ComputeHash() method`. The next sample demonstrates how to use hashing to protect data integrity:

1. Start Microsoft Visual Studio Express 2010 for Windows Phones. Windows starts the Visual Studio Express 2010 IDE.
2. Choose File ➪ New Project. You'll see the New Project dialog box.
3. Choose the Visual C# Silverlight for Windows Phone Templates from the left panel. Highlight the Windows Phone Application template. You'll see a description of this template in the right pane of the New Product dialog box.
4. Type **hash** in the Solution Name field. Click OK. Visual Studio creates a blank application for you.
5. Add a `TextBox` control `cleartext`; two `Button` controls, `generatebtn` and `verifybtn`; and a `TextBlock` control `outputblock` to the content panel.
6. Add the click event handlers to respond to users' requests for both buttons.
7. Type the code shown in Listing 11-3 in `MainPage.xaml`.

LISTING 11-3: Layout of hash demo application, hash\hash\MainPage.xaml

Available for download on Wrox.com

```
<!--LayoutRoot is the root grid where all page content is placed-->
<Grid x:Name="LayoutRoot" Background="Transparent">
    <Grid.RowDefinitions>
        <RowDefinition Height="Auto"/>
        <RowDefinition Height="*"/>
    </Grid.RowDefinitions>

    <!--TitlePanel contains the name of the application and page title-->
    <StackPanel x:Name="TitlePanel" Grid.Row="0" Margin="12,17,0,28">
        <TextBlock x:Name="ApplicationTitle" Text="MY APPLICATION"
                   Style="{StaticResource PhoneTextNormalStyle}"/>
        <TextBlock x:Name="PageTitle" Text="Hash Demo" Margin="9,-7,0,0"
                   Style="{StaticResource PhoneTextTitle1Style}"/>
    </StackPanel>

    <!--ContentPanel - place additional content here-->
    <Grid x:Name="ContentPanel" Grid.Row="1" Margin="12,0,12,0">
        <TextBox Height="96" HorizontalAlignment="Left" Name="cleartext"
                 Text="" VerticalAlignment="Top" Width="456" />
        <Button Content="Generate Hash" Height="70" HorizontalAlignment="Left"
                Margin="0,88,0,0" Name="generatebtn" VerticalAlignment="Top"
                Width="220" Click="generatebtn_Click" />
        <Button Content="Verify Hash" Height="70" HorizontalAlignment="Left"
                Margin="230,88,0,0" Name="verifybtn" VerticalAlignment="Top"
                Width="220" Click="verifybtn_Click" />
        <TextBlock Height="400" HorizontalAlignment="Left" Margin="0,180,0,0"
                   Name="outputblock" Text="" VerticalAlignment="Top" Width="456"
                   TextWrapping="Wrap" />
    </Grid>
</Grid>
```

The control layout is shown in Figure 11-4.

FIGURE 11-4: Hash demo UI layout

The C# source code of this application is shown in Listing 11-4.

LISTING 11-4: WP7 hash demo, hash\hash\MainPage.xaml.cs

Available for download on Wrox.com

```
using System;
using System.Collections.Generic;
using System.Linq;
using System.Net;
using System.Windows;
using System.Windows.Controls;
using System.Windows.Documents;
using System.Windows.Input;
using System.Windows.Media;
using System.Windows.Media.Animation;
using System.Windows.Shapes;
using Microsoft.Phone.Controls;
//This is needed for Cryptography
using System.Security.Cryptography;


namespace hash
{
   public partial class MainPage : PhoneApplicationPage
   {

      byte[] hashvalue = new byte[32];
      byte[] key =
         new byte[16] {1,2,3,4,5,6,7,8,9,10,11,12,13,14,15,16};
      //Hash value of string "I love this book"
      byte[] storedhash= new byte[32] {83,128,177,83,242,174,7,22,144,81,33,3,253,
         104,214,233,232,179,171,200,65,201,253,128,166,62,91,31,27,240,38,78};

      System.Text.UnicodeEncoding encoding = new System.Text.UnicodeEncoding();


      // Constructor
      public MainPage()
      {
          InitializeComponent();

      }

      private void generatebtn_Click(object sender, RoutedEventArgs e)
      {
         //clear outputblock
         outputblock.Text = "";

         //Convert the string to a byte array
         byte[] textarray = encoding.GetBytes(cleartext.Text);

         HMACSHA256 ha = new HMACSHA256(key);

         //Calculate the SHA256 hash value
```

```
            hashvalue = ha.ComputeHash(textarray);
            ha.Clear();

            //Output the hash value to the textblock
            for (int i = 0; i < hashvalue.Length; i++ )
            {
               outputblock.Text += (hashvalue[i].ToString() + " ");
            }

         }

         private void verifybtn_Click(object sender, RoutedEventArgs e)
         {

            //Convert the string to a byte array
            byte[] textarray = encoding.GetBytes(cleartext.Text);

            HMACSHA256 ha = new HMACSHA256(key);

            //Calculate the SHA256 hash value
            hashvalue = ha.ComputeHash(textarray);

            ha.Clear();
            //Compare hashvalue to the stored hashvalue
            bool match = true;
            for (int i = 0; i < hashvalue.Length; i++ )
            {
               if ( hashvalue[i] != storedhash[i])
               {
                  match = false;
                  break;
               }
            }

            if (match) outputblock.Text= "Yes, the hash value of the text you
             + "input matches the pre-stored hash value!";
            else outputblock.Text = "Sorry, the hash value of the text you input"
                +"does not match the pre-stored hash value";
           }
       }
   }
```

The key value and hash value of a known string are hard-coded in the code using a technique similar to the encryption example earlier in the chapter.

```
byte[] hashvalue = new byte[32];
byte[] key =
   new byte[16] {1,2,3,4,5,6,7,8,9,10,11,12,13,14,15,16};
//Hash value of string "I love this book"
byte[] storedhash= new byte[32] {83,128,177,83,242,174,7,22,144,81,33,3,253,
   104,214,233,232,179,171,200,65,201,253,128,166,62,91,31,27,240,38,78};

System.Text.UnicodeEncoding encoding = new System.Text.UnicodeEncoding();
```

The code defines a 32-byte array, `hashvalue`, to store the result; a 16-byte array, `key`, is used as a 128-bit key for encryption; and a 32-byte array, `storedhash`, contains the hash value of the string "I love this book." Again, the application uses an encoding object, `encoding`, to convert the data between a byte array and a string.

When a user clicks Generate Hash, the code clears the `Text` field of `outputblock`. It then converts the string `cleartext.Text`, which is the user input, to a byte array by calling the `GetBytes()` method of `encoding`. A new instance of `HMASHA256` class `ha` is initialized with the `key`. To calculate the hash value, the code calls `ComputeHash()` and the results are stored in `hashvalue`. Finally, the value of each byte in `hashvalue` is displayed in `outputblock`.

When the user clicks Verify Hash, this program repeats the process to calculate the hash value of a current message. It the hash value matches the stored hash value, a confirmation message appears in `outputblock`. Otherwise, the application displays a mismatched message.

From this example, we can see that WP7 provides full support for the hashing function. In a practical sense, the hashing function key should be a secret between the communication parties. You can pick a secret string, such as a password, to use as a seed, and use methods such as `Rfc2898DeriveBytes()` to generate a random key. It's suggested that the length of the key for a hash algorithm also plays an important role in the overall security of a hash implementation. Don't use a key size less than 128 bits in a real-world application.

SECURITY BEST PRACTICES

Information security is really about three major goals: confidentiality, integrity, and availability.

- Confidentiality protects sensitive data and user privacy.
- Integrity means that the information is authentic and no one has modified it.
- Availability ensures that information is always available to the authorized users at any time.

One mistake many junior programmers make is to think that security is merely added onto an application; they always push it back as late as possible in the development process. Sooner or later, they realize that in order to implement security features when most of the coding is done they must completely break a large portion of the application. It's devastating to make any architectural change to accommodate security at the end of the development cycle. The following security guide tips will help you deliver robust and safe applications:

- **Follow the Security Development Lifecycle (SDL) guideline:** The SDL is a security assurance process for software development. It is a collection of security activities software developers have learned over the years. You need to follow these guidelines throughout the entire software development cycle.
- **Minimize permissions:** The least-privilege principle is a common practice in the security field. When writing WP7 applications, don't declare capabilities your application doesn't really need.
- **Use encryption to protect sensitive data:** On WP7, you can achieve confidentiality through data encryption. The challenge, however, is how to manage the secret key in your

application. In addition, you should clear the memory space as soon as the encryption/decryption is finished by explicitly calling the `clear()` method. Don't wait on the garbage collector to free up the memory.

- **Use hashing to protect integrity:** Use hashing or hash-based message authentication code algorithms to support integrity.
- **Write solid code to enhance availability:** To increase the chance of your applications being more available to the end users, you need to publish robust applications and keep them updated as frequently as possible. And remember, only .NET managed code is allowed on WP7.

SUMMARY

This chapter explains and compares the security architectures of WP7, iPhone, and Android. All three platforms are built with a strong sense of security. However, WP7 appears to have higher standards when it comes to security. It follows iOS's lead by vigorously examining each application before it reaches the user. And it doesn't allow the original equipment manufacturer (OEM) to customize the OS to reduce any potential mistakes made by the phone manufacturer. WP7 uses a concept similar to Android permissions to grant an application just enough capabilities to perform a task. In addition, WP7 requires developers to write applications using managed code. The WP7 is probably the most secure mobile OS among the three.

You also discovered how to use the `System.Security.Crytography` APIs to encrypt and decrypt messages, and generate message authentication code. These APIs are as easy to use in the WP7 environment as they are in the iPhone and Android development environments, if not easier.

With mobile threats on the rise, developers have to keep updated about the new threats and new security technology, and must keep improving their skills to write secure code.

A

An Introduction to Smartphone Chipset

Smartphones are usually built on top of a silicon chipset (also known as System-On-Chip, or SOC) that serves as the foundation of the device hardware. Smartphone Original Equipment Manufacturers (OEMs) as system integrators can customize and configure the chipset, as well as add additional components to connect to it. For a specific chipset product, the chipset vendor usually provides a reference software platform, which can be Android, WP7, or some other operating systems. OEMs use the reference software platform as a foundation to build specific software for a device that uses the chipset and other components. Major chipset vendors are Qualcomm, Texas Instruments, Samsung, ST-Ericsson, and so on. As discussed in the "An Overview of Smartphone Hardware" section of Chapter 10, smartphone chipsets are predominantly ARM-based.

A smartphone chipset provides a core set of functions ranging from cellular communication, to Wi-Fi and Bluetooth communication, to general computing, to power management, to memory and storage interface, and to peripheral interfaces. This appendix aims at providing you a general overview of smartphone chipsets.

Figure A-1 shows an example of a high-level block diagram of a GSM (Global System for Mobile) smartphone chipset.

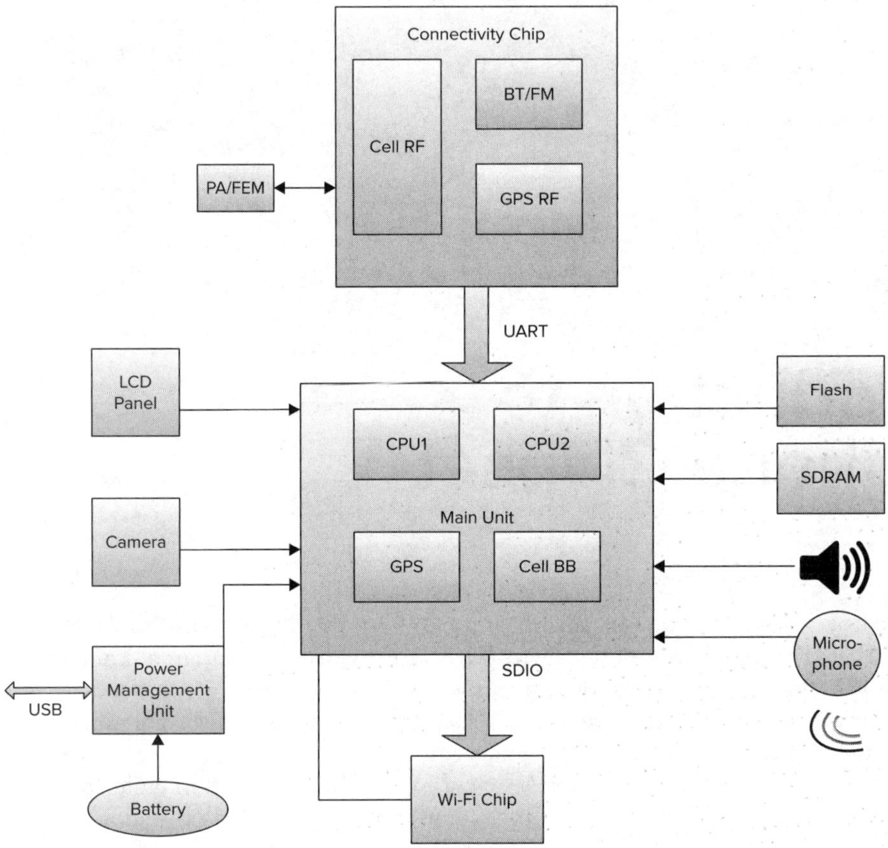

FIGURE A-1: A smartphone chipset block diagram

Despite the fact that smartphone chipsets vary significantly, most of them have a main unit that consists of CPUs and digital voice and data processors known as cellular baseband processors (the Cell BB block in Figure A-1). Other components connect to the main unit via various interfaces. The following is a list of the components as illustrated above:

- One CPU used for the modem and another used as an application processor to run the operating system. The latter can have dual CPU cores
- A cellular base band processor that provides cellular voice and data processing capabilities
- A GPS (Global Positioning System) processor that handles GPS data
- Flash memory used to store the operating system and application code
- Connectivity providers, such as a cellular Radio Frequency (RF) chip, a Bluetooth chip, and a GPS RF chip, which connect to the main unit via a Universal Asynchronous Receiver/ Transmitter (UART) bus
- A Synchronous Dynamic Random Access Memory (SDRAM) chip that provides RAM memory for the system to run

- A Liquid Crystal Display (LCD) panel that connects to the main unit via a special interface
- A digital camera
- A Power Management Unit (PMU) that connects to a battery, and a Universal Serial Bus (USB) interface
- A Wi-Fi component that connects to the main unit via a Secure Digital Input Output (SDIO) bus
- A speaker
- A microphone

In addition to the components listed above, there are other dedicated chips not shown in the figure. For example, as you'll see in the functional diagram of a chipset in Figure A-2, a Graphics Processing Unit (GPU) and some Digital Signal Processors (DSPs) are often used for 3D graphics and audio processing, respectively.

It's also helpful to understand a smartphone chipset from a functional breakdown standpoint. A smartphone generally provides powerful multimedia capabilities and complex data processing capabilities as well as strong extensibility to support peripherals such as sensors and microphones. Let's use a Qualcomm chipset as an example to discuss a general functional diagram of typical smartphone chipsets, as shown in Figure A-2.

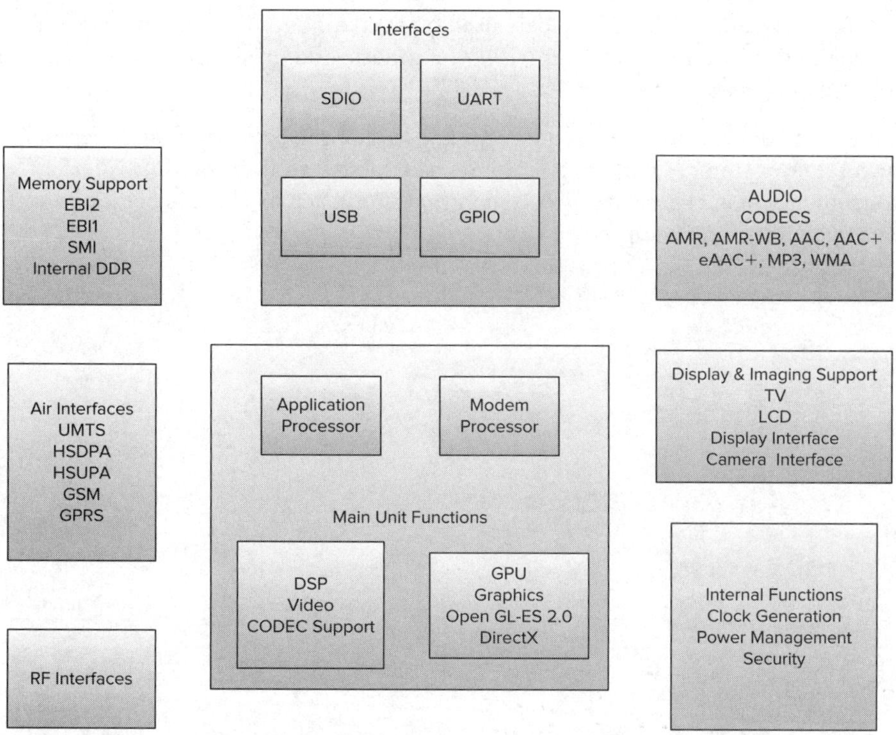

FIGURE A-2: A smartphone chipset function diagram

On a Qualcomm QSD8x50 chipset, the two CPUs are the application processor and the modem processor. The application processor can run at 1 GHz or even high frequency.

The DSP runs at 595 MHz and the High Definition (HD) video processor has maximum display resolution at a 24-bit color depth of 1280×720.

The GPU on the QSD8x50 chipset is the graphics engine that speeds the rendering of 3D graphics content. The GPU is driven by major hardware-accelerated graphics solutions such as Open Graphics Library for Embedded Systems (OpenGL ES) and DirectX. The same GPU on the QSD8x50 also supports a simultaneous encode/decode video telephony solution for high data rate connections with 3G networks such as Wideband Code Division Multiple Access (WCDMA), Universal Mobile Telecommunications System (UMTS), and High Speed Packet Access Plus (HSPA+).

For audio support, the QSD8x50 provides an array of multimedia codecs as listed in the diagram. These codecs are designed to bring high quality and high fidelity to voice conversations, music, game audio, streaming audio, and ringtones.

To drive the on-board display panel or one or two external LCDs, a chipset defines a specific digital display interface. For example, on Qualcomm 8x50, the display interface is Mobile Display Digital Interface (MDDI). There is also an interface to connect to a digital camera.

Air interfaces and RF interfaces refer to the typical cellular network access standard. There are variants of the same chipset that support both GSM/UMTS and CDMA technologies.

The on-board component interface varies depending on the technology of the component. Common interfaces such as USB, SDIO, and UART are widely used on smartphone chipsets. In addition, a chipset often uses General Purpose Input Output (GPIO), a generic and programmable interface to control some peripherals such as sensors.

Additionally, a chipset provides memory and a bus interface so that the processors and memory chips can interconnect. For example, Qualcomm chipsets usually use an External Bus Interface (EBI) to connect to memory chips. There are also system functions such as clock generations and power management on a smartphone chipset.

B

An Introduction to Microsoft Expression Blend for Windows Phone

One of the key advantages Microsoft has over its competitors is the feature-rich development tools it provides. The sample applications in this book are mainly developed using Microsoft Visual Studio 2010 Express for Windows Phone, which is an IDE environment that many software developers are familiar with. In this appendix, you're exposed to a different development environment: Microsoft Expression Blend 4 for Windows Phone. The purpose of this appendix is to walk you through the basics of Microsoft Expression Blend 4 and explain how you can unleash the power of this tool to deliver solid WP7 applications with stunning user experiences.

Microsoft Expression Blend 4 is a visual tool for designing and prototyping desktop, web, and mobile applications. The full version of this software isn't free. However, Microsoft Expression Blend for Windows Phone is offered to developers at no cost; it's included in the free full Windows Phone developer package. If for any reason you cannot find this tool on your computer, you can always visit the Microsoft App Hub to download it and install it (`http://create.msdn.com/en-us/home/getting_started`).

In the lineup of Microsoft development tools, Microsoft has positioned Expression Blend in an interesting spot. It's like Expression Design in the sense that you can easily create graphic designs. It's also like Visual Studio because you can edit Extensible Application Markup Language (XAML) code and run your applications directly from Expression Blend. As a matter of fact, you can use Expression Blend to open and edit Silverlight-based WP7 applications that are developed using Visual Studio and vice versa. But Expression Blend doesn't offer as many graphic tools as Expression Design does, and you don't have to export your vector artwork to Visual Studio. And unlike Visual Studio, Expression Blend doesn't offer powerful debugging capabilities. It doesn't support XNA at all. In many ways, Expression Blend isn't competing against Expression Design or Visual Studio. It complements both tools and works as a bridge that seamlessly connects an attractive user interface (UI) with back-end programming logic. And it's certainly a perfect tool for interactive designers and XAML architects.

MICROSOFT EXPRESSION BLEND IDE

In this section, you'll learn about the Microsoft Expression Blend for Windows Phone working environment. The program name appears as Microsoft Expression Blend 4 on your computer. To illustrate the basic tools, you can create an empty project as follows.

1. Start Microsoft Expression Blend 4. Windows starts the Expression Blend 4 IDE.
2. Choose File ➪ New Project. You'll see the New Project dialog box, as shown in Figure B-1. You'll also notice from this window that Microsoft Expression Blend 4 supports the same five Silverlight templates that Visual Studio 2010 Express for Windows Phone does.

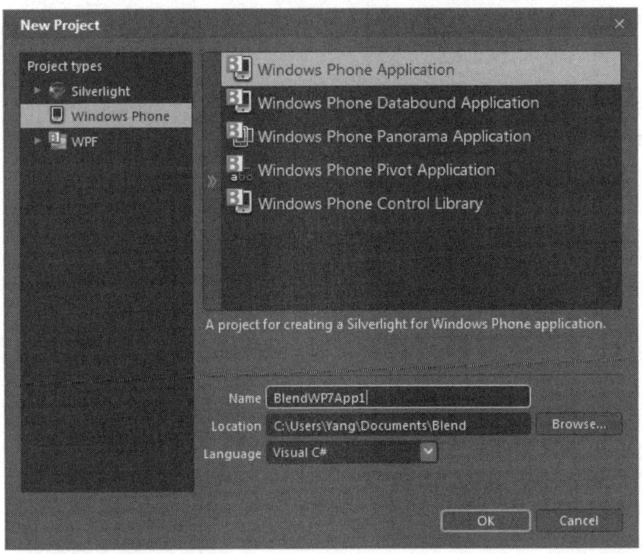

FIGURE B-1: New Project dialog box of Expression Blend

3. Choose Silverlight for Windows Phone templates from the left panel. Highlight the Windows Phone Application template. You'll see a description of this template shown below the template section box.
4. Type **BlendWP7App1** in the Name field and select Visual C# from the Language drop-box. Click OK. Expression Blend creates a blank application for you, as shown in Figure B-2.

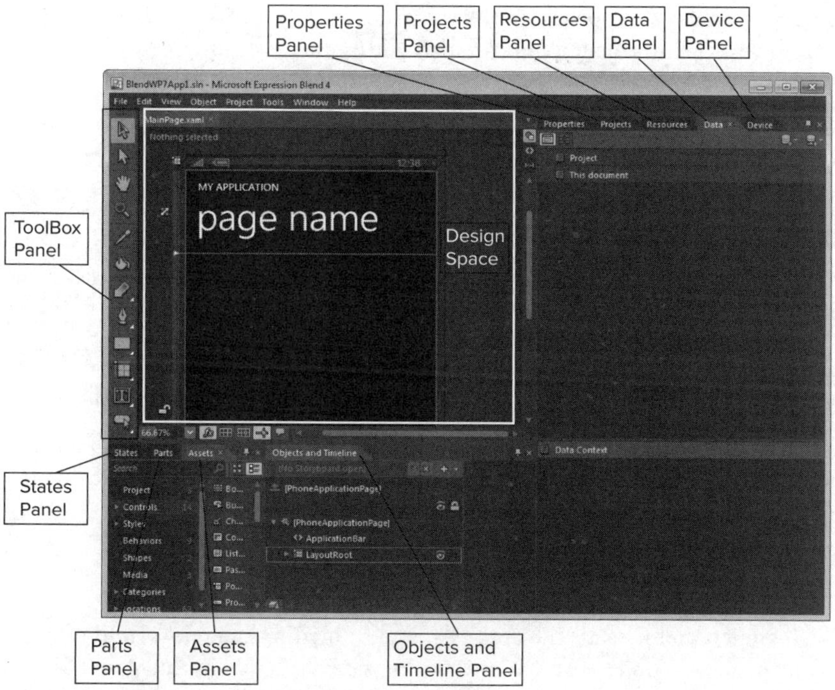

FIGURE B-2: Expression Blend IDE environment

On your computer, the position of each panel might not be identical to Figure B-2. But that isn't something you need to be concerned about because you can easily dock/undock, hide/unhide, or reposition those panels with ease. The following list describes each of these panels.

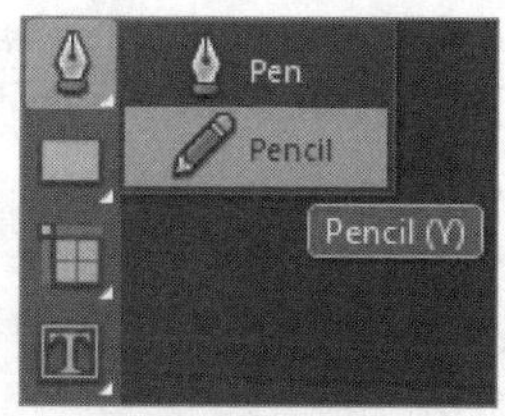

FIGURE B-3: Select a control from a control group

- **ToolBox panel:** This is the top left panel in Figure B-2. It contains tools such as Selection, Pen, and Rectangle. Note that the tiny triangle icon at the bottom right of a tool entry means it contains more than one tool. For example, if you click and hold (or right-click) on the Pen icon, you can select either the Pen control or the Pencil control, as shown in Figure B-3.
- **Design space:** This is the center of the IDE environment. You can directly design your artwork or edit the XAML/C# code from the design space.
- **Properties panel:** This is one of the panels on the right side of the IDE. You can edit the properties of a control by using this panel.
- **Projects panel:** Figure B-4 shows the Projects panel. It looks a lot like the Solution Explorer in Visual Studio.

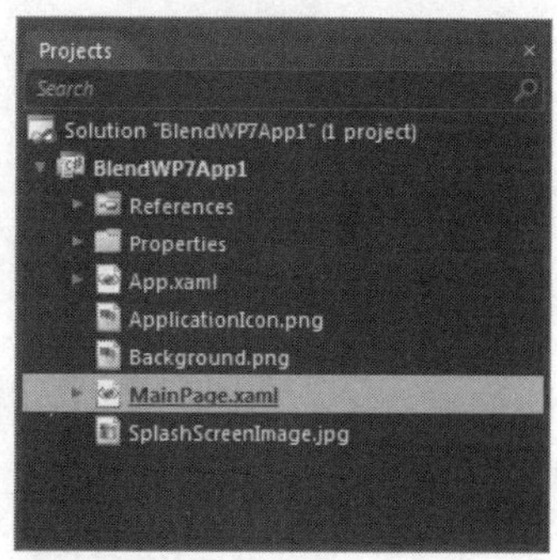

FIGURE B-4: Projects panel

- **Resources panel:** This panel also appears on the right side of Figure B-2. It lists all the resources in your current open project. You can also add/remove resources by using this panel.
- **Data panel:** Next to the Resources panel in Figure B-2, you'll find the Data panel. It lists the data sources in current open project. You can also add/remove data sources on the current project.
- **Device panel:** This panel also appears on the right side by default. You can change the default orientation and WP7 Theme colors from this panel. You can also choose whether your project connects to the emulator or to the real device from this panel. Figure B-5 illustrates the options in this panel.

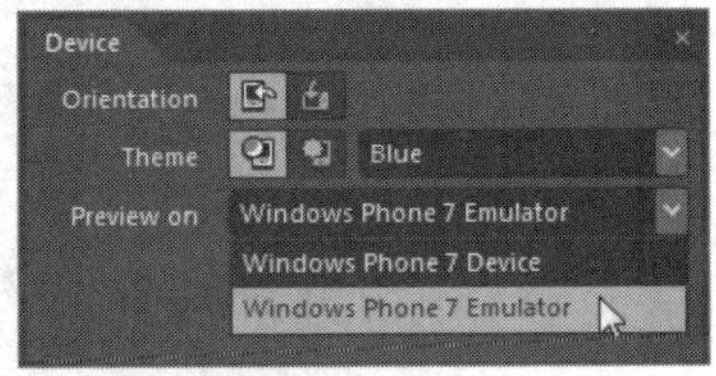

FIGURE B-5: Device panel

- **States panel:** This panel appears on the bottom left side of Figure B-2. It enables you to define state groups and state transitions. As a result, you can easily create animation effects triggered by events such as `mouse over`.
- **Parts panel:** Next to the States panel, you'll find the Parts panel. It provides a list of parts you can use to design control templates.
- **Assets panel:** The Assets panel lists all the controls, styles, media, behaviors, and effects that you use in your project. You can find Windows Phone 7 controls in this panel.
- **Objects and Timeline panel:** On the bottom of Figure B-2, you'll find this panel. It enables you to create storyboards with animation features.
- **Output panel:** By default, this panel isn't displayed in the IDE. If you start running a project in Expression Blend, build results and error information automatically show up in this panel.

If you accidentally close a panel, you can always re-enable it by selecting its entry in the Window menu. In the Expression Blend IDE, you can re-arrange all these panels and save the layout as a *workspace*. You can also minimize/restore those panels by pressing the F4 key. To rotate between different workforces, press the F6 key.

CREATING ANIMATION WITH EXPRESSION BLEND

Recall in Chapter 8 that you learned how to cycle through a portion of a sprite to generate an animation effect. In this section, you'll learn how to use the `Objects` and `Timeline` controls to create a storyboard with Expression Blend. If you have already closed the empty project `BlendWP7App1`, please re-open it. When you follow the instructions below, you'll make a textbox that moves, rotates, and changes its color.

1. Select the `PageTitle` object from the Properties panel, change its font size to 36pt, and change the `Text` property to `Blend Animation`.
2. From the Assets Panel, choose Controls ➪ TextBox. Drag and drop this `TextBox` control to the `ContentPanel` of `PhoneApplicationpage`.

3. From the Properties panel, change the color to `Red`, set `HorizontalAlignment` to `Center`, and change `Text` to `I love this book`, as shown in Figure B-6.

FIGURE B-6: Change the properties of the Textbox control.

4. Next, create a storyboard by clicking the plus sign on the top of the Objects and Timeline panel. You'll see a Create Storyboard Resource window pop up. Enter **animateTB** in the `Name (Key)` field, as shown in Figure B-7. Click OK to begin recording the animation. You'll see a red round button appear on the top left of the design space with the words "animateTB timeline recording is on" next to it.

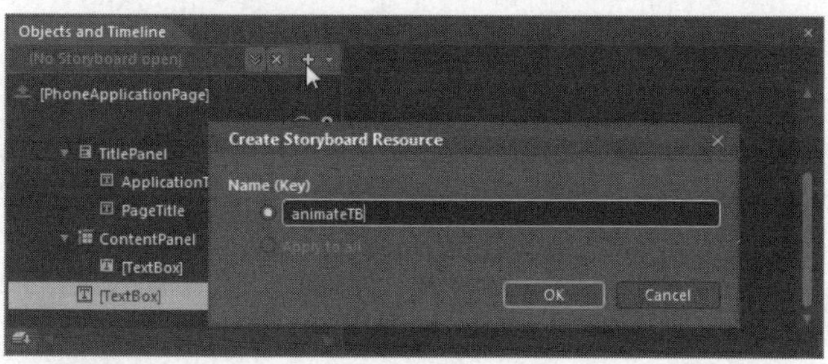

FIGURE B-7: Add a story board.

5. Set the timeline to 1s (second), and select `textBox`. Drag the `textBox` from the top to the bottom of the `ContentPanel`. From the Properties window, change the `Text` font to 36pt and its color to `Blue`, as shown in Figure B-8.

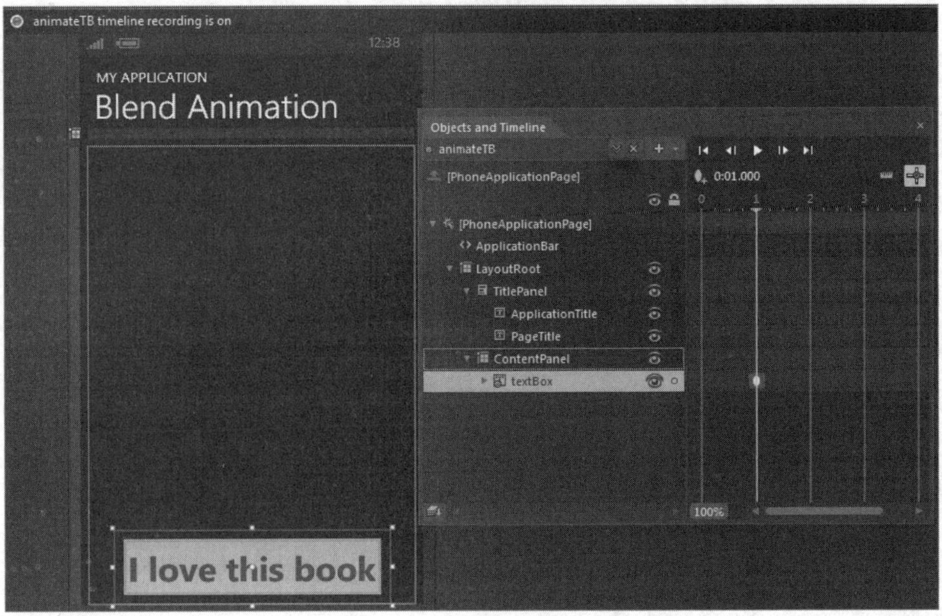

FIGURE B-8: Adding a key frame at the time value 1s

6. Set the timeline to 2 seconds, and select `textBox`. Rotate `textBox` 180 degrees to make it upside down and move the `textBox` to the center of the `ContentPanel`.
7. Set the timeline to 3 seconds, and select `textBox`. Reset `textBox` so that it has the same look at time 0s by dragging it to the top of the `ContentPanel`, rotating the control 180 degrees, changing the font size to 19pt, and changing the foreground color to red.
8. Click the round red button on the top left of the design space to stop recording the storyboard. The color of the button turns to gray, and the statement changes to "animateTB timeline recording is off."
9. Now you can enjoy the animation by clicking Play in the Objects and Timeline panel.

You've accomplished a lot so far without writing a single line of code. But if you press F5 to start running this application, you won't see any animations because the storyboard isn't hooked to any events. There are two ways you can start playing this animation. One is to call its `Begin()` method in an event handler, the other is to use `Behaviors`. In this section, you'll learn how to use `Behaviors` to trigger an event to play the animation.

From the Assets panel, click `Behaviors`. Select `ControlStoryBoardAction` and drop it onto the `LayoutRoot` in the Objects and Timeline panel, as shown in Figure B-9.

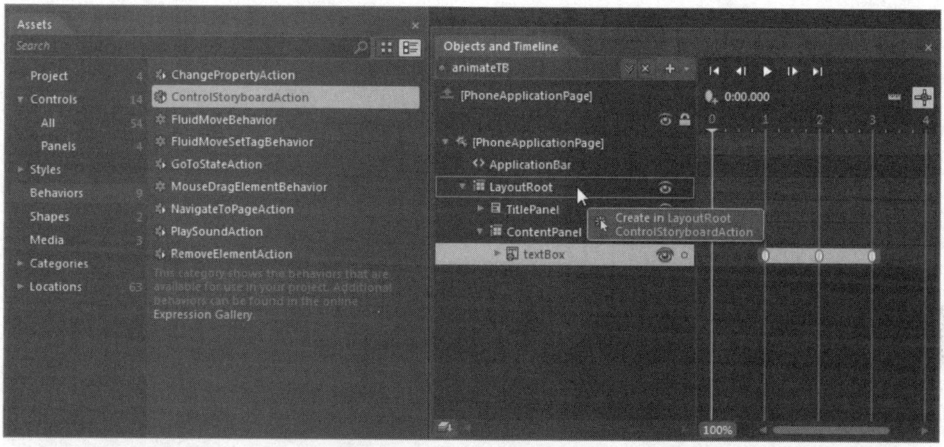

FIGURE B-9: Adding ControlStoryboardAction behavior to LayoutRoot

Now with `ControlStoryboardAction` selected in the Properties panel, select `Loaded` from the `EventName`, `Play` from `ControlStoryboardOption`, and `animateTB` from `Storyboard`, shown in Figure B-10.

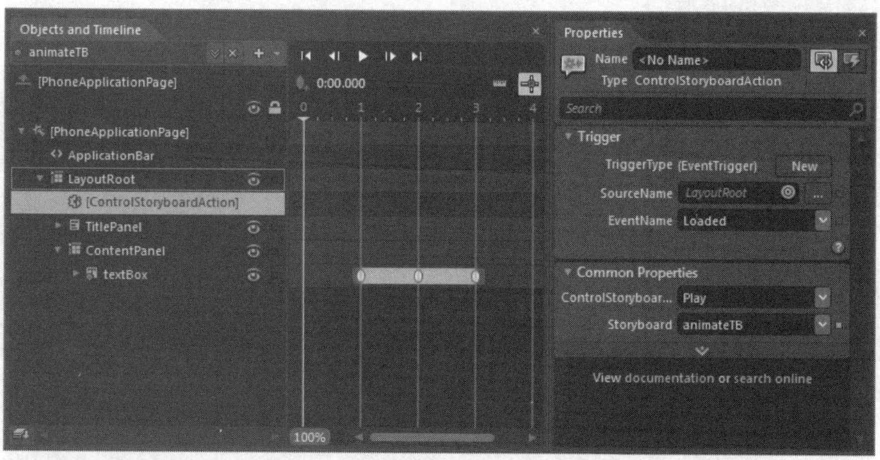

FIGURE B-10: Set ControlStoryboardAction properties to play animationTB

Press F5 to start this application. You'll see animation start once the page is loaded. And now you have created an animation storyboard without writing a single line of code: isn't that amazing!

In the next section, you'll learn how to customize a UI control and you'll also learn how to hook the animation with a control event handler.

CUSTOMIZING UI WITH EXPRESSION BLEND

Expression Blend provides many great tools that enable you to create eye-catching UI components. For ease of discussion, this section shows how to add a 3D perspective to a Button control. The following steps assume you created the example in the "Creating Animation with Expression Blend" section of this appendix.

1. Drag a `Button` from the Assets panel to the `ContentPanel`. Align it to the bottom left of `ContentPanel`.
2. Right-click the `Button` control, choose Edit Template ➪ Edit a Copy from the Context menu as shown in Figure B-11. You'll see a Create Style Resource window pop up.
3. Type **Btn3D** in the `Name (Key)` property and set the `Define In` property to `This document` as shown in Figure B-12. Click OK. You'll see `Btn3D (Button Template)` appear in the Objects and Timeline panel.

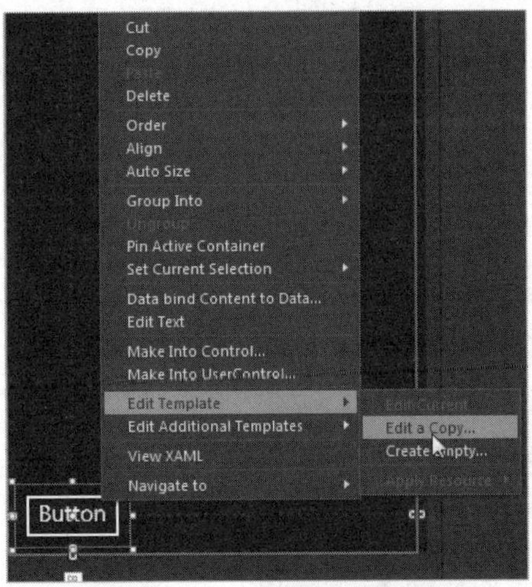

FIGURE B-11: Editing the template of a Button control

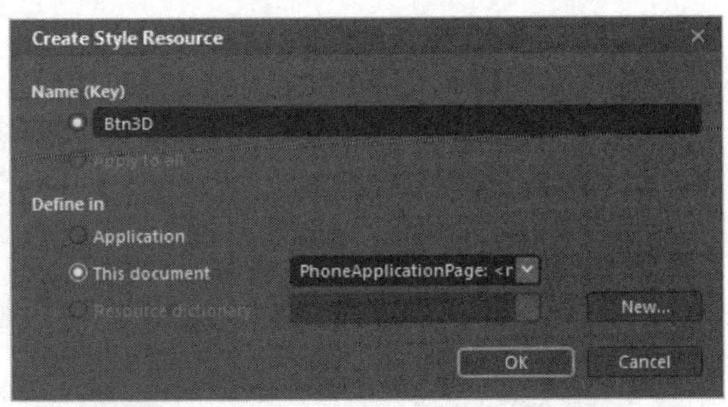

FIGURE B-12: Creating a new style resource

4. Select `[Grid]` from the Objects and Timeline panel. Select the `Gradient` brush and set the `Background` color to `Red`. Next find the `Projection` option in the `Transform` group on the Properties panel, and change the X value to -50 degree, as shown in Figure B-13.

FIGURE B-13: Changing the grid of a Button control

5. Click the up arrow key to the left of Btn3D in the Objects and Timeline panel to quit the template editing mode. This action returns the scope to the `PhoneApplicationPage`. If the storyboard recording mode is on, you can turn it off by clicking the round red recording button.

6. Now drag and drop another `Button` to the `ContentPanel`, and move it to the bottom right of `ContentPanel`. To apply the template you defined in Steps 1 through 5, right-click the new `Button` and choose Edit Template ➪ Apply Resource ➪ Btn3D from the Context menu.

7. Change the `Text` property of the left button to `Play` and the right button to `Pause`. You can then add an event handler to each button. You can perform this task within Expression Blend but it's probably a better idea to develop the application code in Visual Studio. Figure B-14 illustrates this process: right-click the `MainPage.xaml` file entry in the Projects panel and choose Edit in Visual Studio from the Context menu.

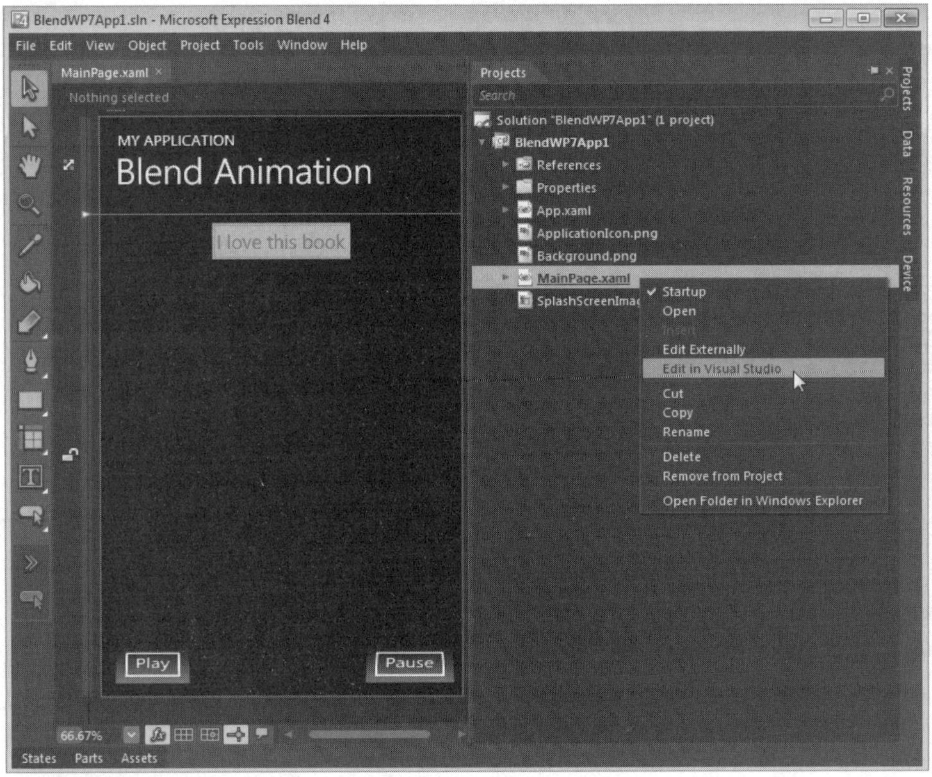

FIGURE B-14: Open project Item from Visual Studio

Within Visual Studio, add the button click event handlers, as described below.

Available for download on Wrox.com

```
//Play button
private void Button_Click(object sender, RoutedEventArgs e)
{
    if (this.animateTB.GetCurrentState() == ClockState.Filling)
        this.animateTB.Begin();
    else this.animateTB.Resume();
}

//Pause button
private void Button_Click_1(object sender, RoutedEventArgs e)
{
    this.animateTB.Pause();
}
```

Code snippet BlendWP7App1\BlendWP7App1\MainPage.xaml.cs

When the user clicks the Play button, the code checks the storyboard state. If the state is `ClockState.Filling`, it means the storyboard has finished playing, and you can call the `Begin()`

method. Otherwise, the code calls the `Resume()` method. When the user clicks the Pause button, the code calls the `animateTB.Pause()` method.

After you finish editing code in Visual Studio, you can save the changes and close Visual Studio. When you return to Expression Blend, the IDE warns you that external programs have modified project items. Click OK to reload the project. Press F5 again to start the application; you can now enjoy playing this animation over and over again.

SUMMARY

Expression Blend is a powerful tool that delivers a graphical environment in which to create UI components. The fact that it works seamlessly with Visual Studio makes developing WP7 applications a joyful experience. This appendix shows only the tip of the iceberg. You're encouraged to learn more about Expression Blend so you can deliver more attractive applications to the Windows Phone Marketplace.

C

Sample Applications Guide

This book includes a variety of sample applications that illustrate the core programming technologies on Windows Phone 7. Table C-1 lists the instructions for all the sample applications.

To open the sample projects in your IDE:

With 32-bit system using Visual Studio 2010 Express for Windows Phone or the full version, just double-click the solution file.

With 64-bit system using some variants of Visual Studio 2010, select File ➪ Open, then choose the solution file.

TABLE C-1: List of Sample Applications

CHAPTER	APPLICATION	FUNCTION
Chapter 2	HelloWP7	Demonstrate building basic application on WP7.
Chapter 3	WP7Lifecycles	Demonstrate how you can manage WP7 application life cycles and states.
Chapter 4	WP7Controls	Demonstrate the use of Pivot control and Panorama control.
	WP7Navigation	Demonstrate the page navigation mechanism on WP7.
Chapter 5	AppDataSample	Demonstrate application isolated storage usage and data serialization on WP7. This is also a client of the MyNoteService web service provided by the SampleCloudService sample.
	SampleCloudService	Demonstrate a WCF cloud service MyNoteService.

continues

TABLE C-1 *(continued)*

CHAPTER	APPLICATION	FUNCTION
Chapter 6	SocialDemo	Show a Twitter client implementation on WP7 with OAuth.
	PushNotification	Show a push notification client on WP7, and also a push server emulator on Windows desktop.
	AdControlDemo	Demonstrate using AdControl to display ads on WP7.
Chapter 7	CurrentLocationTxt	Display current location information in text blocks.
	SimpleMap	Demonstrate using Bing Maps Control on WP7.
	Street2Geo	Demonstrate using Bing Maps geocoding web service on WP7.
	TrackMe	Demonstrate using location service to show current location on a map on WP7.
Chapter 8	Cube3D	Demonstrate XNA 3D graphics on WP7.
	Rolling2D	Demonstrate XNA 2D graphics on WP7.
	ToyAnimation	Use Sprites to do animations on WP7.
Chapter 9	MediaPicker	Demonstrate sound, picture, and its icon movement and swap on WP7.
	WP7AudioPlayerDemo	Demonstrate audio player on WP7.
	WP7VideoPlayerDemo	Demonstrate WP7's own native MediaPlayerLauncher, and another video player based on WP7's `MediaElement`.
	WP7EnrichedMoviePlayerDemo	Demonstrate a multimedia player on WP7 with the user control component for reusability purpose.
Chapter 10	MediaPicker	Demonstrate pictures' movement speed which is managed by accelerometer on WP7.
	WindowsPhoneMic	Demonstrate Windows Phone microphone on WP7.
	WP7FMRadio	Demonstrate FM radio tuner on WP7.
	WPCamera	Demonstrate taking picture and video by camera on WP7.
Chapter 11	AES	Demonstrate AES encryption/decryption on WP7.
	Hash	Demonstrate hash algorithms on WP7.
Appendix B	BlendWP7App1	Demonstrate WP7 application development with Microsoft Expression Blend 4

CHAPTER 2

HelloWP7

Demonstrate building basic application on WP7.

Note

None.

Instructions

Double click the `HelloWP7.sln` file and open the solution file in Visual Studio 2010. Press F6 to build the solution and then press F5 to start debugging.

Click the WP7 Back button to quit the program.

CHAPTER 3

WP7Lifecycles

Demonstrate how you can manage WP7 application life cycles and states.

Note

None.

Instructions

Double click the `WP7Lifecycles.sln` file and open the solution file in Visual Studio 2010. Press F6 to build the solution and then press F5 to start debugging.

Click the Click button to increase the click count for both new clicks and total clicks. Click the Settings button for the option to reset the total clicks count after the program quits.

Click the WP7 Back button to quit the program. If the checkbox in settings is checked, it will reset the total clicks count. When you re-launch the application, the total clicks count will be 0. Otherwise, the total clicks count will be the same as the count number when the program quit last time.

CHAPTER 4

WP7Controls

Demonstrate the use of Pivot control and Panorama control.

Note

The Silverlight for Windows Phone Toolkit (`http://silverlight.codeplex.com/`) has to be installed to compile and run this application.

Instructions

Double-click the `WP7Controls.sln` file and open the solution file in Visual Studio 2010. Press F6 to build the solution and then press F5 to start debugging.

Click the labels under the "collections" Panorama item to go into the Pivot details page.

Click the WP7 Back button to quit the program if the back stack is empty.

WP7Navigation

Demonstrate the page navigation mechanism on WP7.

Note

None.

Instructions

Double-click the `WP7Navigation.sln` file and open the solution file in Visual Studio 2010. Press F6 to build the solution and then press F5 to start debugging.

Click the page number to navigate to different pages. Click the Back button to return to the previous page.

Click the WP7 Back button to quit the program if the back stack is empty.

CHAPTER 5

AppDataSample

Demonstrate application isolated storage usage and data serialization on WP7. This is also a client of the MyNoteService web service provided by the SampleCloudService sample.

Note

This sample should be run after the referenced web service is started. The web service is provided by the SampleCloudService project. Because the port number of the WCF web service may change when it is debugged in Visual Web Developer 2010 Express, you must make sure you have the correct web reference URL (such as `http://127.0.0.1:81`) configured in this project.

Instructions

The application will pull all the previously saved notes from the cloud service when it starts, and will push all locally saved notes to cloud service when it exits.

Once the UI appears, you can type in the note in the input box and then click the Save button (the one with a disc icon). Then you will see the note saved locally on the device. To enter more notes, press the "BACK" hardware key and do it again. When you exit the application, the notes will be pushed to the cloud service.

Alternatively, you can click the Options button (icon on the right) and click Settings, then enable "Sync to cloud right after saved locally." This will enable the feature that pushes a note to the cloud right after it is saved locally on the device.

On the main UI, you can also show a list of locally saved notes by clicking the folder icon.

SampleCloudService

Demonstrate a WCF cloud service MyNoteService.

Note

This sample must be run in Microsoft Visual Web Developer 2010 Express or full version with administrator privilege.

Instructions

Run the application. Then run the AppSampleProject.

CHAPTER 6

SocialDemo

Show a Twitter client implementation on WP7 with OAuth.

Note

The solution requires `Hammock.WindowsPhone.dll` to be added as a reference. Visit `http://hammock.codeplex.com/` to download the binary package and choose:

`Hammock-Binaries\.NET 4.0\Windows Phone 7\Hammock.WindowsPhone.dll.`

Instructions

When the application runs and the Twitter authorization page appears, enter a valid Twitter account to proceed. Then you will see a list of your tweets.

Push Notification

Show a push notification client on WP7, and also a push server emulator on Windows desktop.

Note

Make sure that when you run the solution, both projects are configured to run.

Instructions

To try the sample PushNotification solution, start to run the solution (which by default will run both projects) and make sure the desktop application runs on the PC, and the device application runs on the device emulator or a real device.

Copy the registered channel URI (for example, `http://sn1.notify.live.net/...`) from the console output of the PushNotificationClient project in the IDE (View ➪ Output) and paste it into one of the three channel textboxes in the PushServerEmulator UI.

From the PushServerEmulator application you can send three types of notifications to the application on the device. You can use the controls to customize the notification content.

To see tile notifications for the application, you must create a tile for it on the Home screen. You can do this by pressing and holding the application name in the program list and selecting "Pin to Start."

AdControlDemo

Demonstrate using AdControl to display ads on WP7.

Note

You must use your application ID and ad unit ID in order to try this sample. These are set in `MainPage.xaml.cs`:

```
AdControl adControl = new AdControl(
    "*******************", //ApplicationID
    "30234", //ad unit ID registered at Pubcenter
    AdModel.Contextual, //Contextual is the only supported model
    false); //Enable auto refreshing of ads every 60s by default
```

Instructions

When you run the application with your application ID and ad unit ID, you may see exceptions (if you enabled "catch unhandled exceptions" for the Common Language Runtime) such as "Invalid Operation IsolatedStorage." This is a bug in AdControl. Simply hit Continue and you will see the ads units.

CHAPTER 7

CurrentLocationTxt

Display current location information in text blocks.

Note

You will need to run the `GpsEmulator.exe` program in the `GPSEmulator` folder before debugging this program in the emulated environment.

Instructions

Double-click the `currentLocationTxt.sln` file and open the solution file in Visual Studio 2010. Press F6 to build the solution and then press F5 to start debugging.

Click the High button to start location service in high accuracy. Click the Low button to start location service in low accuracy. Click the Stop button to stop the location service.

Click the WP7 Back button to quit the program.

SimpleMap

Demonstrate using Bing Maps Control on WP7.

Note

To run this program, you will need to acquire a Bing Maps key first. Then enter your key in the line `ApplicationId=` in the `MainPage.xml` file.

Instructions

Double-click the `simpleMap.sln` file and open the solution file in Visual Studio 2010. Press F6 to build the solution and then press F5 to start debugging.

Click the Satellite/Road button to switch between Aerial mode and Road mode. Use the slider control to zoom in and out.

Click the WP7 Back button to quit the program.

Street2Geo

Demonstrate using Bing Maps geocoding web service on WP7.

Note

To run this program, you will need to acquire a Bing Maps key first. Then enter your key in the line `req.Credentials.ApplicationID=` in the `MainPage.xml.cs` file.

Instructions

Double-click the `street2Geo.sln` file and open the solution file in Visual Studio 2010. Press F6 to build the solution and then press F5 to start debugging.

Enter an address or a place in the textbox. Click the Find Geo Data button to get the current latitude and longitude information.

Click the WP7 Back button to quit the program.

TrackMe

Demonstrate using the location service to show current location on a Map on WP7.

Note

To run this program, you will need to acquire a Bing Maps key first. Then enter your key in the line `ApplicationId=` in the `MainPage.xml` file.

You will need to run the `GpsEmulator.exe` program in the `GPSEmulator` folder before debugging this program in the emulated environment.

Instructions

Double-click the `TrackMe.sln` file and open the solution file in Visual Studio 2010. Press F6 to build the solution and then press F5 to start debugging.

Click the Satellite/Road button to switch between Aerial mode and Road mode. Click the Start/Stop button to start or stop tracking your location on the map.

Click the WP7 Back button to quit the program.

CHAPTER 8

To run XNA graphics on the WP7 emulator, the graphic card of your computer needs to support Direct X version 10 or above and the Driver mode needs to be WDDM 1.1 or above. Otherwise, you will still be able to compile the code. But when you debug your program in the emulator, you will get the following error message: "The current display adapter does not meet the emulator requirements to run XNA Framework applications."

Note that all the sample code in this chapter will NOT run on an emulator that does not meet the driver requirements. However, you will be able to run the sample applications on a physical Windows Phone 7 device.

To find out the display adapter driver of a Windows computer, you can use the DirectX Caps Viewer tool (`http://msdn.microsoft.com/library/ee417852.aspx`). Or use the `DxDiag` command and then choose the Display tab. If the driver of your computer does not meet the requirements, a possible work-around is to update the display driver.

Cube3D

Demonstrate XNA 3D Graphics on WP7.

Note

Please see the chapter note.

Instructions

Double-click the `cube3D.sln` file and open the solution file in Visual Studio 2010. Press F6 to build the solution and then press F5 to start debugging.

Click the WP7 Back button to quit the program.

Rolling2D

Demonstrate XNA 2D Graphics on WP7.

Note

Please see the chapter note.

Instructions

Double-click the `rolling2D.sln` file and open the solution file in Visual Studio 2010. Press F6 to build the solution and then press F5 to start debugging.

Click the WP7 Back button to quit the program.

ToyAnimation

Demonstrate using Sprites to do animations on WP7.

Note

Please see the chapter note.

Instructions

Double-click the `ToyAnimation.sln` file and open the solution file in Visual Studio 2010. Press F6 to build the solution and then press F5 to start debugging. For the best viewing experience, change the orientation of the emulator/device to Landscape mode.

Click the WP7 Back button to quit the program.

CHAPTER 9

WP7AudioPlayerDemo

Demonstrate audio player on WP7.

Instructions

Double-click the `WP7AudioPlayerDemo.sln` file and open the solution file in Visual Studio 2010. Press F6 to build the solution and then press F5 to start debugging.

Click the WP7 Back button to quit the program.

MediaPicker

Demonstrate sound, picture, and its icon movement and swap on WP7.

Note

To run XNA graphics on the WP7 emulator, the graphic card of your computer needs to support Direct X version 10 or above and the Driver mode needs to be WDDM 1.1 or above. Otherwise, you will still be able to compile the code. But when you debug your program in the emulator, you will get the following error message: "The current display adapter does not meet the emulator requirements to run XNA Framework applications."

Note that this sample application will NOT run on an emulator that does not meet the driver requirements. However, you will be able to run the sample applications on a physical Windows Phone 7 device.

To find out the display adapter driver of a Windows computer, you can use the DirectX Caps Viewer tool (`http://msdn.microsoft.com/library/ee417852.aspx`). Or use the `DxDiag` command and then choose the Display tab. If the driver of your computer does not meet the requirements, a possible work-around is to update the display driver.

Instructions

Double-click the `MediaPicker.sln` file and open the solution file in Visual Studio 2010. Press F6 to build the solution and then press F5 to start debugging.

Click the WP7 Back button to quit the program.

WP7VideoPlayerDemo

Demonstrate WP7's own native `MediaPlayerLauncher`, and another video player based on WP7's `MediaElement`.

Note

None.

Instructions

Double-click the `WP7VideoPlayerDemo.sln` file and open the solution file in Visual Studio 2010. Press F6 to build the solution and then press F5 to start debugging. Click MediaPlayerLauncher to launch WP7's own media player application. For the best viewing experience, change the orientation of the emulator/device to Landscape mode. For the `MediaElement`-based video player, click Start, Stop, Resume, Pause and the Volumeup and Volumedown buttons to test the basic media playback control.

Click the WP7 Back button to quit the program.

WP7EnrichedMoviePlayerDemo

Demonstrate a multimedia player on WP7 with the user control component for reusability purposes.

Note

None.

Instructions

Double-click the `WP7EnrichedMoviePlayerDemo.sln` file and open the solution file in Visual Studio 2010. Press F6 to build the solution and then press F5 to start debugging.

Click the WP7 Back button to quit the program.

CHAPTER 10

WindowsPhoneMic

Demonstrate Windows Phone microphone on WP7.

Note

None.

Instructions

Double-click the `WindowsPhoneMic.sln` file and open the solution file in Visual Studio 2010. Press F6 to build the solution and then press F5 to start debugging.

Click the WP7 Back button to quit the program.

WPCamera

Demonstrate taking picture and video by camera on WP7.

Note

None.

Instructions

Double-click the `WPCamera.sln` file and open the solution file in Visual Studio 2010. Press F6 to build the solution and then press F5 to start debugging.

Click the WP7 Back button to quit the program.

WP7FMRadio

Demonstrate FM radio tuner on WP7.

Note

None.

Instructions

Double-click the `WP7FMRadio.sln` file and open the solution file in Visual Studio 2010. Press F6 to build the solution and then press F5 to start debugging.

Click the WP7 Back button to quit the program.

MediaPicker

Demonstrate pictures' movement speed which is managed by accelerometer on WP7.

Note

To run XNA graphics on the WP7 emulator, the graphic card of your computer needs to support DirectX version 10 or above and the Driver mode needs to be WDDM 1.1 or above. Otherwise, you will still be able to compile the code. But when you debug your program in the emulator, you will get the following error message: "The current display adapter does not meet the emulator requirements to run XNA Framework applications."

Note that this sample application will NOT run on an emulator that does not meet the driver requirements. However, you will be able to run the sample applications on a physical Windows Phone 7 device.

To find out the display adapter driver of a Windows computer, you can use the DirectX Caps Viewer tool (`http://msdn.microsoft.com/library/ee417852.aspx`). Or use the `DxDiag` command and then choose the Display tab. If the driver of your computer does not meet the requirements, a possible work-around is to update the display driver.

Instructions

Double-click the `MediaPicker.sln` file and open the solution file in Visual Studio 2010. Press F6 to build the solution and then press F5 to start debugging.

Click the WP7 Back button to quit the program.

CHAPTER 11

AES

Demonstrate AES encryption/decryption on WP7.

Note

None.

Instructions

Double-click the `aes.sln` file and open the solution file in Visual Studio 2010. Press F6 to build the solution and then press F5 to start debugging.

Clicking the Encrypt button will encrypt the text entered in the textbox and clicking the Decrypt button will decrypt the encrypted message in byte format to clear text.

Click the WP7 Back button to quit the program.

Hash

Demonstrate hash algorithms on WP7.

Note

None.

Instructions

Double-click the `hash.sln` file and open the solution file in Visual Studio 2010. Press F6 to build the solution and then press F5 to start debugging.

Clicking the Generate Hash key will perform a one-way hashing on the text in the textbox. Hash code in Byte format will be shown in the `Textblock`. Clicking the Verify Hash button will compare the hash key entered in the textbox to the hash code of string "I love this book" (83 128 177 83 242 174 7 22 144 81 33 3 253 104 214 233 232 179 171 200 65 201 253 128 166 62 91 31 27 240 38 78)

Click the WP7 Back button to quit the program.

BlendWP7App1

Demonstrate WP7 Application development with Microsoft Expression Blend 4.

Note

You can open this solution with either Visual Studio 2010 or Expression Blend 4. But you are encouraged to open it with both IDEs.

Instructions

To open it with Microsoft Expression Blend 4, right-click the `BlendWP7App1.sln` file and select Open With Microsoft Expression Blend 4. You can press F5 to start debugging.

To open it with Visual Studio 2010, double-click the `BlendWP7App1.sln` file. Press F6 to build the solution and then press F5 to start debugging.

Clicking the Play button will start/resume textbox animation. Clicking the Pause button will pause the animation.

Click the WP7 Back button to quit the program.

INDEX

A

AAC. *See* Advanced Audio Coding
AAC-LC. *See* Low Complexity AAC
AAX. *See* Audible Enhanced Audio
AAX+. *See* Audible Enhanced Audiobook File
`AbstractThreadedSyncAdapter`, 97
accelerometer, 245
  Android, 9
`Accelerometer`, 248–253
`AccelerometerReadingChanged()`, 250
`accelReading`, 250
Access Denied, 259
`ACCESS_COARSE_LOCATION`, 157
`ACCESS_FINE_LOCATION`, 157
Acorn RISC Machine (ARM), 2
Action Bar, 91
Activated, 46
Active
  Android, 48
  iOS, 50
ActiveSync, 2
activities, 68
Activity Manager, 11
`ActualHeight`, 218
`ActualWidth`, 218
ad unit ID, 150
`Adapter`, 81
Adaptive Multi-Rate Audio (AMR), 214, 238
`AdaptView`, 81
`AdControl`, 151
`AdControlDemo`, 151, 298
Add References, 161–162
AdMob, 149
ADT. *See* Android Development Tools
Advanced Audio Coding (AAC), 213
Advanced Encryption Standard (AES), 263, 264–268
Advanced RISC Machine (ARM), 236–237
advertising, 146–152
AES. *See* Advanced Encryption Standard
`aes`, 264–268, 304–305
`AesManaged`, 264, 269
AIFF. *See* Audio Interchange File Format
`AllList.DataContext`, 90
AMR. *See* Adaptive Multi-Rate Audio
Android
  Action Bar, 91
  Active, 48
  `AdaptView`, 81
  applications
    execution model and life cycles, 47–49
    framework, 10–12
    project structure, 42–43
  Back button, 45
  Bluetooth, 255
  camera, 243–244
  copy and paste, 7
  data storage, 96
  debugging, 24–25
  development environment, 20
  display, 65
  Google, 3
  graphics, 188
  hardware, 237
  Home screen, 45, 65
  IDE, 20
  `ImageView`, 206

Android (*continued*)
- iPhone, 8–14
- JSON, 117
- local files and directories, 97
- location, 157, 158–159
- microphone, 239–240
- multimedia, 209–212
  - codecs, 213–214
- operating system, 8–10
- Paused, 48
- publishing, 35–36
- push notifications, 133–135, 137
- security, 261
- sensors, 246–248
- SQLite, 96, 98
- Stopped, 48
- `SyncAdapter`, 97
- system requirements, 16
- Tab Layout, 76
- UI, 20
- web services, 116–117
- Windows Phone Emulator, 26
- XML, 20, 66, 92

Android Development Tools (ADT), 42
- Eclipse, 20

Android Market, 12
`Android Version/`, 42
`android.graphics.drawable`, 188
`Android.hardware.Camera`, 243–244
`Android.hardware.Sensor`, 246
`AndroidManifest.xml`, 42, 261
`android.media.MediaPlayer`, 211
`android.permission.RECEIVE_SMS`, 261
`android.speech.tts`, 239
`android.view.animation`, 188
`android.view.SurfaceView`, 212
`android.widget.VideoView`, 211
`animateTB.Pause()`, 291
animation
- Expression Blend, 284–288
- UI, 64
- XNA, 198–200

`animationDuration`, 208
`Apache`, 116–117
APIs
- CFNetwork, 261
- choosers, 59–60
- data storage, 95–96
- isolated storage, 98–99
- launchers, 59–60
- .NET Framework, 6
- security, 262–274
- `System.Device.Location`, 156
- Translator, 242

`.apk`, 211
`<Aplication.Resources/>`, 93
APNs. *See* Apple Push Notification service
`App`, 72
App Hub, 28
- Expression Blend, 281

App Widgets, 65
`AppDataSample`, 101, 296–297
Apple. *See also* iPhone; iPhone Operating System
- App Store, iPhone, 12–13
- iAd, 10, 148–149

Apple Lossless, 213
Apple Push Notification service (APNs), 132–133
`Application`, 52
- `Application.Current`, 72

applications
- data storage, 95–112
- databound, 41
- execution model and life cycles, 43–59
  - Android, 47–49
  - iOS, 49–50
- framework, 6–7, 10–12
- native development, 8
- project structure, 40–43
  - Android, 42–43
  - iOS, 43
- sample, 293–305
- templates, 41
- third-party, 7

Application Bar, 65
- UI, 91–92

application ID, 150
Application state, 51

Application Title, 19
Application_Activated, 46
Application_Activated(), 51
ApplicationBarIconButton, 92
ApplicationBarMenuItem, 92
Application_Closing, 46, 51
Application_Closing()
  SaveDataToIsolatedStorage(), 58
  tombstones, 57
Application.Current, 52
  Application, 72
Application_Deactivated, 46, 51
Application_Deactivated()
  SaveDataToIsolatedStorage(), 58
  tombstones, 57
ApplicationID, Bing Maps, 173–174
Application_Launching, 45, 52
App.xaml, 42
  styles, 93
App.xaml.cs, 42, 46
ARM. *See* Acorn RISC Machine; Advanced RISC Machine
aspectratio, 202
assets/, 42
asynchronous
  UART, 278
  web services, 117
  Windows Azure service, 108–110
Audible Enhanced Audio (AAX), 213
Audible Enhanced Audiobook File (AAX+), 213
audio
  integration, 223–229
  multimedia, 215–229
Audio Interchange File Format (AIFF), 213
AutoPlay, 218
AVAudioPlayer, 208
Azure. *See* Windows Azure service

B

Back button
  Android, 45
  buttonBack, 74
  iPhone, 68
  navigation, 68–69
  OnBackKeyPress(), 74
Background, 50
background services
  iOS, 132
  multi-tasking, 61–62
BackStack, 72
Balance, 218
banner ads, 147
BasedOn, 93
BasicHttpBinding_IGeocodeService, 172
Begin(), 286
beginSeekingBackward(), 208
beginSeekingForward(), 208
Behaviors, 286
Berkeley Software Distribution (BSD), 8
best practices, security, 274–275
BindChannelToTileToast(), 138–139
binding. *See also* data binding
  Cocoa, iOS, 81–82
  MPNS, 136–140
Bing, 4
  Windows Phone Emulator, 22
Bing Maps, 156–157, 173–180
  ApplicationID, 173–174
  key, 167
  MainPage.xaml.cs, 170–172
  REST, 157
  SOAP, 156–157
  web services, 157
Blender, 200
BlendWP7Appl.sln, 282–284, 305
Bluetooth, 255
  Android, 8
Board Support Packages (BSPs), 4
Boolean
  newPageInstance, 53
  Urgent, 87
breadth of use, 29
BSD. *See* Berkeley Software Distribution
BSPs. *See* Board Support Packages
btnClick, UI, 54
btnClick_click(), 52
btnSettings, 51, 52

btnSettings_Click(), 52
BufferingProgress, 219
BufferingProgressChanged, 216
BufferingTime, 219
BufferReady, 241
Build Type, 207
BuildRawNotification(), 145
Button, 51, 270
buttonBack, 70
  Back button, 74
buttonFwd, 70

C

C#, 6
cacheBuf, 242
CacheMode, 219
camera, 186, 238, 242–245
  Android, 243–244
  iOS, 242–243
CameraCaptureTask, 61, 245
Cancel, 74
CancelEventArgs, 74
CanGoBack, 69, 71
  NavigationService, 73
CanGoForward, 69, 71
  NavigationService, 73
CanPause, 219
CanSeek, 219
Cascading Style Sheets (CSS), 92
cbReset, 51, 55
C2DM. *See* Cloud To Device Messaging
CE. *See* Windows Compact Edition
cell-site triangulation, 156
certification, Marketplace, 4, 29, 30–38
CFNetwork, 261
chambers, 258
changeUIStatus(), 165
channels, MPNS, 136
  binding, 137–140
  establishment, 137–140
CheckBox, 51, 86
  TwoWay, 87
  Urgent, 87
chipset, 277–280
choosers, 59–60
class library, templates, 41
cleartext, 269, 270
Click, 92
click-through rate (CTR), 148
Closing, 46
cloud data, 96–97, 105–110
Cloud To Device Messaging (C2DM), 133–135
Cocoa
  binding, iOS, 81–82
  Objective-C, 187
Cocoa Touch, 10
  security, 260
codecs, multimedia, 212–215
Collection, 140
color depth, 184
ColumnDefinitions, 67
COM. *See* Component Object Model
command line, Windows Phone Emulator, 24
Completed, 59
Component Object Model (COM), 10
ComputeHash(), 270, 274
confidentiality, 263–270, 274–275
connectionDidFinishLoading, 116
Content, 207
Content Provider, 11
content, UI, 64
Content.Load<>(), 199
ContentPanel, 284
  MainPage, 67–68
  textbox, 286
Control, 206
controls, 74–76
  page, 69
  Segmented Control, 76
ControlStoryBoardAction, 287
Convert(), 88
ConvertBack(), 88
Converter, 82
converters, MVVM, 87–89
coordinates, 184–185
copy and paste, 7
CoreLocation.framework, 157

cost per action (CPA), 148
cost per click (CPC), 148
cost per mille (CPM), 148
cost per thousand (CPT), 148
Cryptography, 264
CryptoStream, 269
CryptoStream.FlushFinalBlock(), 269
CryptoStream.Write(), 269
.cs, 18
CSS. *See* Cascading Style Sheets
CTR. *See* click-through rate
cube3D, 200–202, 300–301
curLoc, 179
currentLocationTxt, 160–170, 298–299
currentPage, 70
CurrentState, 219
CurrentStateChanged, 216
customization, UI, 92–94
    Expression Blend, 288–291

D

Dalvik Debug Monitor Server (DDMS), 26
Dalvik virtual machine, 9
data binding
    Model, 87
    MVVM, 81–90
    ViewModel, 86
    XAML, 82
Data Encryption Standard (DES), 264
data serialization, 101–104
data sharing, navigation, 71–72
Data Source, 206
data storage. *See also* isolated storage
    applications, 95–112
    cloud data, 96–97, 105–110
    design, 110–112
    HTML, 111
    SQLite, 110
    XML, 110
databound application, 41
DataContext
    View, 83
    ViewModel, 89
DataContractJsonSerializer, 122
DataContractSerializer, 101
DataSource, 207
DataTemplate, 87
DateTime, 101
DDMS. *See* Dalvik Debug Monitor Server
Deactivated, 46
debugging, 24–27
    Android, 24–25
    GpsEmulator.exe, 166
    iOS, 24–25
    SayButton_Click(), 25
    Zune, 27
Debug.WriteLine(), 25
default.properties, 42
delegate, 59
DES. *See* Data Encryption Standard
Design Templates for WP7, 64
Design View, 19
development environment, 15–37
    Android, 20
    iOS, 20
Development Tools, 15–27
device.exe, 5
Devices.Sensors, 253
Digital Signal Processors (DSPs), 279
Direct Push, 2
DirectX Caps Viewer, 184
Disabled, 165
Dispatcher.BeginInvoke(), 165, 179
display, 65
    LCD, 279
    screen size, 9
DisplayOrientation.LandscapeLeft, 196
DLLs, 10
DownloadProgress, 219
DownloadProgressChanged, 216
DownloadStringAsync(), 118, 119
DownloadStringCompleted(), 118
Draw(), 192, 193, 194, 201, 228
    3D space, 203
DSPs. *See* Digital Signal Processors
dt_Tick(), 241
Duration, 223

E

EBI. *See* External Bus Interface
Eclipse, 20
effective CPM (eCPM), 148
Elevated Rights, 259
`EmailAddressChooserTask`, 61
Embedded MultiMediaCard (EMMC), 211
EMMC. *See* Embedded MultiMediaCard
Emulator. *See* Windows Phone Emulator
`encodedbytes`, 269
`encoding`, 274
`Encrypt`, 268
`encryptbtn_Click()`, 268
encryption, 274–275
  security, 263–270
`encryptor`, 269
`endSeeking()`, 208
events
  manipulation, 90
  navigation, 72–74
  system event hooks, 61
event handlers
  UI, 90–91
  Visual Basic, 19–20
`EventReminder`, 145
Exchange Server, 132
Expression Blend, 7, 16, 281–291
  animation, 284–288
  App Hub, 281
  IDE, 282–284
  Silverlight, 281
  UI, 282
    customization, 288–291
  Visual Studio, 282
  XAML, 281
Extensible Application Markup Language (XAML), 6
  data binding, 82
  Expression Blend, 281
  graphics, 187
  `PhoneApplicationPage`, 66–67
  Silverlight, 18, 42, 82, 187
  Solution Explorer, 18
  UI, 18, 66–67
  Visual Studio, 18–19
  WYSIWYG, 18
Extensible Hypertext Markup Language (XHTML), 187
eXtensible Markup Language (XML)
  Android, 20, 66, 92
  data storage, 110
  files, 97–98
  IDE, 42
  LINQ, 128–129
  parsing, 127–130
  web services, 115–117, 119
External Bus Interface (EBI), 280

F

Facebook, 114
files
  isolated storage, 99–100
  local, 97–98
  XML, 97–98
`FileInputStream()`, 97
`FileMode.CreateNew`, 100
`FileOutputStream()`, 97
`filesys.exe`, 5
`findBtn_Click()`, 172
fixed time-step game loops, 192–193
flash memory, 278
Flickr, 114
FM. *See* frequency modulation
FM radio, 254–255
`FMRadio`, 254–255
fonts, 65–66
Forward button, 74
frames, 69
FreeBSD, 8
`Frequency`, 254
frequency modulation (FM), 235–236
`FromStream`, 223

G

games
  library, 41
  loop, 190–194
  templates, 41

Game Kit, 10
Game Studio, 16
GameInviteTask, 61
Gen/, 42
General Packet Radio Service (GPRS), 237
General Purpose Input Output (GPIO), 280
generatebtn, 270
GeocodeAsync(), 172
GeocodeRequest, 172
GeocodeServiceClient, 172
geocoding, 166–172
GeoCoordinateWatcher, 162, 165
  watcher, 179
GeoPositionStatus, 165
gestures, 90
Global Positioning System (GPS), 7, 156, 159–166
  Android, 8
  chipset, 278
Global System for Mobile (GSM), 277–278
GLSurfaceView, 188
GoBack(), 70, 71
GoForward(), 70, 71
Google. *See also* Android
  AdMob, 149
  Financial Service, 115
  Maps, iOS, 157
  Places, 115
GotFocus, 217
GPIO. *See* General Purpose Input Output
GPRS. *See* General Packet Radio Service
GPS. *See* Global Positioning System
GPS_EMULATOR, 162
GpsEmulatorClient.dll, 159, 176
GpsEmulator.exe, 159
  debugging, 166
GPU. *See* Graphics Processing Unit
graphics, 183–204
  Android, 188
  GUI, 187
  integration, 223–229
  iOS, 187–188
  Silverlight, 187, 189
  3D space, chipset, 279
  XAML, 187
  XNA, 183–184, 189–204
graphical user interface (GUI), 2–3
  graphics, 187
Graphics Processing Unit (GPU), 279, 280
GraphicsDeviceManager, 253
Grid, 67
GSM. *See* Global System for Mobile
GUI. *See* graphical user interface
GWES.exe, 5

H

hardware, 4, 235–255. *See also specific types of hardware*
  Android, 237
  development environment, 16
  iPhone, 237
  security, 262
  UI, 64
hash, 305
Hash-based Message Authentication Code-Secure Hash Algorithm (HMACSHA), 263
hashing, 270–274
hashvalue, 274
HD. *See* High Definition
HE-AAC. *See* High-Efficiency Advanced Audio Coding
Header, 76
HelloWP7, 66–67, 295
High Definition (HD), 280
High Speed Packet Access (HSPA), 237
High Speed Packet Access Plus (HSPA+), 280
High-Efficiency Advanced Audio Coding (HE-AAC), 213, 214
HMACSHA. *See* Hash-based Message Authentication Code-Secure Hash Algorithm
Home screen, 65
  Android, 45
HorizontalAlignment, 285
HSPA. *See* High Speed Packet Access
HSPA+. *See* High Speed Packet Access Plus
HTML
  data storage, 111
  WebKit, 8
HTTP
  raw notifications, 145
  web services, 117–122

HttpChannelNotification.
    BindToShellTile(), 140
HttpClient, 116
HttpNotification Channel, 137
HttpNotificationChannel.
    ShellToastNotificationReceived, 144

I

iAd, 10, 148–149
icons, 29–30
ID_CAP_GAMERSERVICES, 260
ID_CAP_LOCATION, 158, 259
ID_CAP_MEDIALIB, 260
ID_CAP_MICROPHONE, 259
ID_CAP_NETWORKING, 259
ID_CAP_PHONEDIALER, 260
ID_CAP_PUSH_NOTIFICATION, 260
ID_CAP_SENSORS, 260
ID_CAP_WEBBROWSERCOMPONENT, 260
IDE. *See* Integrated Development Environment
IE. *See* Internet Explorer
IEnumerable<XElement>, 130
IMA ADPCM. *See* Interactive Multimedia
    Association ADPCM
Image, 206, 207
images, PNG, 29
ImageView, Android, 206
impressions, 148
IMyNoteService, 107, 108
Inactive, iOS, 50
Info.plist, 43
initialization vector (IV), 268
Initialize(), 192, 193
    Media, 249
initialize(), 225
Initializing, 165
INotifyCollectionChanged, 85
INotifyPropertyChanged, 83–84
inputbox, 268
INSArray, 209
instruction set architecture (ISA), 236
Integrated Development Environment (IDE), 15
    Android, 20
    Expression Blend, 282–284
    iOS, 20
    MainPage.xaml, 66
    templates, 46
    Visual Studio, AdControl, 151
    XML, 42
Interactive Multimedia Association ADPCM
    (IMA ADPCM), 214
Internet Explorer (IE), 8
    Windows Phone Emulator, 22
Interprocess Communication (IPC), 96
interstitial ads, 147
iOS. *See* iPhone Operating System
IPC. *See* Interprocess Communication
iPhone
    Android, 8–14
    Apple App Store, 12–13
    application framework, 10–12
    Back button, 68
    copy and paste, 7
    hardware, 237
    operating system, 8–10
    publishing, 35–36
iPhone Operating System (iOS), 1, 8
    Active, 50
    application execution model and life cycles,
        49–50
    application project structure, 43
    Background, 50
    background services, 132
    Bluetooth, 255
    camera, 242–243
    Cocoa bindings, 81–82
    data storage, 96
    debugging, 24–25
    development environment, 20
    display, 65
    Google Maps, 157
    graphics, 187–188
    Home screen, 65
    IDE, 20
    Inactive, 50
    local files and directories, 97
    location, 157, 158–159
    microphone, 239
    multimedia, 208–209

multimedia codecs, 213
MVC, 20, 71
Not running, 50
push notifications, 132–133
security, 260–261
Segmented Control, 76
sensors, 246
SQLite, 96, 98
Suspended, 50
system requirements, 16
Toolbar, 91
`UIImageView`, 206, 208
web services, 116
XNA, 20
ISA. *See* instruction set architecture
`IsChecked`, 87
`IsIntersect()`, 228
`IsMuted`, 219
isolated storage, 98–105
APIs, 98–99
data serialization, 101–104
files, 99–100
persisted state data, 57–59
saving settings, 104–105
structure, 99
`IsolatedStorage`, 99
`IsolatedStorageException`, 100
`IsolatedStorageFile.OpenFile()`, 100
`IsolatedStorageSettings`, 104–105
`IsolatedStorageSettings.Application.Settings.Add()`, 99
`IsolatedStorageSettings.Application.Settings.Save()`, 99
`isPlay()`, 210
`IsRunningSlowly`, 194
`ItemSource`, 80
IV. *See* initialization vector
`IValueConverter`, 87

J

jailbreak, 262
Java, 9–10
Java Native Invocation (JNI), 9
JavaScript Object Notation (JSON), 97
Android, 117
parsing, 122–127
Twitter, 122–127
web services, 115–117, 119
`javax.xml.parsers`, 97
JNI. *See* Java Native Invocation
JSON. *See* JavaScript Object Notation

K

`Key`, 93
keychain service, 260–261
`KeyDown/UP`, 217
`KeyedHashAlgorithm`, 270
Key-Value Coding (KVC), 82
Key-Value Observing (KVO), 82
KVC. *See* Key-Value Coding
KVO. *See* Key-Value Observing

L

`Landscape`, 91
`LandscapeLeft`, 197
`LandscapeRight`, 197
Language-Integrated Query (LINQ), 117
XML, 128–129
languages, 66
`LastUpdateTime`, 101
launchers, 59–60
Launching, 45
`LayoutRoot`, 67
`LayoutUpdated`, 217
LCD. *See* Liquid Crystal Display
Least Privilege Chamber (LPC), 259
Lees, Andy, 3
library
class, 41
game, 41
OPENGL ES, 188
life in motion, 4
Linear Pulse Code Modulation (LPCM), 214
LINQ. *See* Language-Integrated Query
Linux kernel, 9

Liquid Crystal Display (LCD), 279
ListBox
  DataTemplate, 87
  ItemSource, 80
listStatus, 129
LoadContent(), 192, 193, 201, 225
Loaded, 217
local data, 96–97
local databases, 97–98
local files, 97–98
location, 155–181
  Android, 8, 157, 158–159
  cell-site triangulation, 156
  current, 159–172
  geocoding, 166–172
  GPS, 7, 159–166
    Android, 8
  iOS, 157, 158–159
  maps, 173–180
  Wi-Fi positioning, 156
LocationCollection, 180
LostFocus, 217
Low Complexity AAC (AAC-LC), 214
LPC. *See* Least Privilege Chamber
LPCM. *See* Linear Pulse Code Modulation

M

MainPage, 51, 53
  adding code, 52
  ContentPanel, 67–68
  WP_SetFrequency, 254
MainPage.xaml, 19, 42
  IDE, 66
  UI, 68
MainPage.xaml.cs, 42
  Bing Maps, 170–172
  UI, 68
MainWindow, 43
MainWindow.xaml.cs, 145
Manager, 206
manager, 211
manipulation events, 90
ManipulationCompleted, 90
ManipulationDelta, 90
ManipulationStarted, 90
Map Kit, 10
MapLayer, 180
MapPolygon, 180
MapPolyline, LocationCollection, 180
maps, 173–181. *See also* Bing Maps
Marketplace
  certification, 4, 29, 30–38
  publishing, 28–36
  registration, 7, 12, 28
  security, 262
  submission, 30–38
MasterVolume, 223
matrix, 185–186
MatrixCreateLookAt(), 204
Matrix.CreateRotation(), 203
Matrix.CreateScale(), 203
MDDI. *See* Mobile Display Digital Interface
Media
  Initialize(), 249
  Update(), 250
Media Player Framework, 208
mediacallback(), 209
mediaCanvas, 225
MediaElement, 215–222
  Silverlight, 207
  video, 230–232
MediaEnded, 217
MediaFailed, 217
mediaGraph, 225
MediaLibrary, 226
mediaLibrary, 225
MediaOpened, 217
MediaPicker, 223–229, 301–302, 304
  Accelerometer, 248–253
MediaPlayer, 207, 209–211
MediaPlayerLauncher, 61
  video, 229–230
MediaStreamSource, 221
MemoryStream memory Storage, 242
Metro. *See* user interface
microphone, 238–242
  Android, 239–240

chipset, 279
iOS, 239
Microphone, 240–242
microphone_BufferReady, 242
Microprocessor without Interlocked Pipeline Stages (MIPS), 2
Microsoft AdCenter AdControl, 150
Microsoft Custom Technology Differential Pulse Code Modulation (MS ADPCM), 214
Microsoft Push Notification Service (MPNS), 136
binding, 136–140
channels, 136
binding, 137–140
establishment, 137–140
handling push notifications, 144–146
raw notifications, 137, 145
sending push notifications, 140–144
tile, 137
toast, 136–137
URI, 140
Microsoft Sync Framework, 109–110
Microsoft.Phone.Tasks.CameraCapptureTask, 244–245
MIDI. *See* Musical Instrument Digital Interface
MIPS. *See* Microprocessor without Interlocked Pipeline Stages
M-JPEG. *See* Motion Joint Photographic Experts Group
MMA. *See* Mobile Marketing Association
MMS. *See* Multimedia Messaging Service
MO. *See* mobile operators
Mobile Display Digital Interface (MDDI), 280
Mobile Marketing Association (MMA), 147
mobile operators (MO), 205
Mode, 82
Model
data binding, 87
View, 83
Model, 84
ModelBone, 202–203
ModelMesh, 202–203
Model-View-Controller (MVC), 20, 71
Model-View-ViewModel (MVVM), 78
converters, 87–89
data binding, 81–90
pattern, 82–83
Motion Joint Photographic Experts Group (M-JPEG), 213
Motion Picture Experts Group (MPEG), 213
MouseLeftButtonDown, 90
MouseLeftButtonUp, 90
MovementThreshold, 165
PositionChanged, 179
MoveNext(), 207
MovePrevious(), 207
MP3, 213, 214
MPEG. *See* Motion Picture Experts Group
MPEG-4, 213, 214–215
MPMoviePlayerController, 209
MPMusicPlayerController, 208
MPNS. *See* Microsoft Push Notification Service
MS ADPCM. *See* Microsoft Custom Technology Differential Pulse Code Modulation
multimedia, 205–233
Android, 209–212
architecture, 206
audio, 215–229
codecs, 212–215
Android, 213–214
iOS, 213
integration, 223–229
iOS, 208–209
Multimedia Messaging Service (MMS), 147, 238
multi-tasking, 60–61
third-party applications, 7
multi-touch, 21
Musical Instrument Digital Interface (MIDI), 214
MVC. *See* Model-View-Controller
MVVM. *See* Model-View-ViewModel
myanimationView, 209
myimagecollection, 209
MyNoteServiceClient, 108
MySpace, 114

N

Name, 219
NAND, 211

native application development, 8
Native Development Kit, 8
NaturalDuration, 219
Navigate(), 70, 74
navigation
  Back button, 68–69
  data sharing, 71–72
  events, 72–74
  Panorama, 69
  Pivot, 69
  Silverlight, 68–69
  UI, 68–74
NavigationContext, 72
NavigationService, 69, 70–71
  CanGoBack, 73
  CanGoForward, 73
  .NET Framework, 52
  PhoneApplicationPage, 69
  PivotPage, 80
NavigationService.Navigate, 56
NavigationService.Navigate(), 71
.NET Framework, 2, 16
  APIs, 6
  NavigationService, 52
  security, 263
NewClicks, 52
newPageInstance, 53
NoData, 165
Not running, 50
NoteEntry, 101, 102, 104
noteList, 104
NoteListView
  PivotPage, 88
  View, 86
NoteListView.xaml, 88
notes.xml, 104
Notification Manager, 11
NSArray, 97
NSMutableData, 116
NSMutableDictionary, 97
NSMutableURLRequest, 116
NSNotificationCenter, 209
NSSearchPathForDirectoriesInDomain(), 97
NSURL, 116
NSXMLParser, 97

O

OAuth. *See* Open Authorization
oauth_verifier, 130–131
Objective-C, 10
  Cocoa, 187
Objects, 284
ObservableCollection, 85
ODat. *See* Open Data
OData. *See* Open Data
OEMs. *See* original equipment manufacturers
Office Live, 4
OHA. *See* Open Handset Alliance
OnBackKeyPress(), 73
  Back button, 74
onCreate(), 49
onDestroy(), 49
onDraw(), 188
OneTime, 82
OnManipulationStarted(), 71
OnNavigatedFrom(), 50–51, 53, 73
  adding code, 55–56
  UI, 56
OnNavigatedTo(), 50–51, 53, 73
  adding code, 55–56
  UI, 56
OnNavigatingFrom(), 72
OnOrientationChanged(), 91
onPause(), 49
onPerformSync(), 97
onRestoreInstanceState(), 49
onResume(), 49
onSaveInstanceState(), 49
onSensorChanged(), 247
open(), 140
Open Authorization (OAuth), 117, 130–131, 262
Open Data (OData), 110
Open Graphics Library for Embedded Systems (OpenGL ES), 188
Open Handset Alliance (OHA), 3
OpenGL ES. *See* Open Graphics Library for Embedded Systems
OpenGLContext, 188
OpenReadAsync(), 118
OpenReadComplete(), 118

operating system, 8–10. *See also* Android; iPhone Operating System
- chipset, 278
- development environment, 16

org.xml.sax, 97
Orientation, 91
orientation, UI, 91
OrientationChanged, 253
original equipment manufacturers (OEMs), 4, 8
- chipset, 277
- hardware, 237
- multimedia, 205

OTA. *See* over-the-air
outputblock, 274
over-the-air (OTA), 29

P

page controls, 69
Page, OnManipulationStarted(), 71
Page state, 50–51
Page1.xaml, 70
Panorama, 4, 76–81
- navigation, 69
- SelectionChanged(), 78
- templates, 41

PanoramaItem, 77
- TextBlock, 79
- WrapPanel, 80

PanoramaPage
- PivotPage, 81
- ViewModel, 90

Panorama.xaml.cs, 79
panoramic, 30
parsing
- JSON, 122–127
- XML, 127–130

Path, 82
Pause, 221
pause(), 210, 211
Paused, 48
pay per click (PPC), 148
PCM. *See* Pulse-Code Modulation
permissions, 262, 274
persisted state data, 57–59
Perst Embedded Database, 98
PhoneApplicationFrame, 69
PhoneApplicationPage, 50, 69, 284
- NavigationContext, 72
- Properties window, 67
- State, 53
- XAML, 66–67

PhoneApplicationPage(), 59
PhoneApplicationService, 51, 55
PhoneNumberChooserTask, 61
PhotoChooserTask, 61
PhotoResult, 245
Photoshop, 64
picGroup, 226
Pictures, 226
Pivot, 4, 76–81
- navigation, 69
- SelectionChanged(), 78
- templates, 41

PivotItem
- Header, 76
- PivotPage, 86
- PivotPage.xaml, 80
- SelectedIndex, 81

PivotPage
- NavigationService, 80
- NoteListView, 88
- PanoramaPage, 81
- PivotItem, 86

PivotPage.xaml, 79
- PivotItem, 80

pixels, 184
Play, 221, 223
play(), 208
Player, 206, 207
player, 211
PMU. *See* Power Management Unit
PNG, 29
Pocket Outlook, 2
Pocket PC 2000, 2
POI. *See* Points of Interest
Points of Interest (POI), 157
Portrait, 91
PortraitOrLandscape, 91
Position, 219

PositionChanged, 179
positionChanged, 165
positionChanged(), 179
Power Management Unit (PMU), 279
PPC. *See* pay per click
PreferredBackBufferHeight, 196
PreferredBackBufferWidth, 196
prepare(), 210, 211
prepareAsync(), 210
projection, 186
Properties page, 42
Properties window, 67
Properties\AppManifest.xml, 41
Properties\AssemblyInfo.cs, 41
Properties\WMAppManifest.xml, 41
PropertyChanged, 85
Protected AAC, 213
proximity sensor, Android, 9
proximityState, 246
publishing
    Android, 35–36
    iPhone, 35–36
    Marketplace, 28–36
Pulse-Code Modulation (PCM), 213
push notifications, 131–146. *See also* Microsoft Push Notification Service
    Android, 133–135, 137
    architecture, 131–135
    iOS, 132–133
    sample application, 297–298
PushNotesData(), 108–109
PushNotesDataAsync(), 108–109
Pushpin, 159, 175

Q

Qualcomm, 280
Quartz Core Graphics, 187
QueryString(), 81
    NavigationContext, 72

R

random access memory (RAM), 29, 278
raw notifications
    HTTP, 145
    MPNS, 137, 145
ReadDataCallBack(), 120–121
ReadFromFile(), 104
ReadingChanged, 250
ReadObject(), 101
Ready, 165
Really Simple Syndication (RSS), 119
recordButton_Click(), 242
Redirection Service, 35
Reduced Instruction Set Computer (RISC), 2
reduced instruction set computer (RISC), 236
References, 42
registerListener, 247
registration, Marketplace, 7, 12, 28
rendering, 185
    speed, 29
Representational StateTransfer (REST)
    Bing Maps, 157
    web services, 114–115
req.Credentials.ApplicationId, 172
res/, 42
Research in Motion (RIM), 2
reset(), 210
Resource, 207
responsiveness, 29
REST. *See* Representational StateTransfer
Rfc2898DeriveBytes, 263, 268, 274
RIM. *See* Research in Motion
RISC. *See* Reduced Instruction Set Computer; reduced instruction set computer
rolling2D, 190–192, 301
rollingPic, 197
rotation, 202
RowDefinitions, 67
RSS. *See* Really Simple Syndication

S

sample applications, 293–305
SampleCloudService, 101–102, 297
sandbox, 258
SaveDataToIsolatedStorage(), 58
SAX. *See* Simple API for XML
SayButton_Click(), 25

screen size, 9
screenshots, 30
ScrollViewer, 50
SD. *See* Secure Digital
SDIO. *See* Secure Digital Input Output
SDK. *See* software development kit
SDL. *See* Security Development Lifecycle
SDRAM. *See* Synchronous Dynamic Random Access Memory
Secure Digital (SD), 10, 95, 211
Secure Digital Input Output (SDIO), 279
Secure Hash Algorithm (SHA), 263, 270–274
Secure Sockets Layer (SSL), 133
security, 257–275
  Android, 261
  APIs, 262–274
  best practices, 274–275
  Cocoa Touch, 260
  encryption, 263–270
  hardware, 262
  hashing, 270–274
  iOS, 260–261
  Marketplace, 262
  .NET Framework, 263
  permissions, 262
Security Development Lifecycle (SDL), 274
Segmented Control, 76
Segoe WP, 65–66
SelectedIndex, 81
SelectionChanged(), 78
sensors, 245–253
  Android, 246–248
  iOS, 246
SENSOR_DELAY_FASTEST, 248
SENSOR_DELAY_GAME, 248
SENSOR_DELAY_NORMAL, 248
SENSOR_DELAY_UI, 248
SensorEvent, 246
SensorManager, 246–247
serialization, 101–104
Service Reference Settings, 168
setDateSource(), 210, 211
setFixedSize(), 212
setMediaController(), 212
SetSource(), 221, 222
SettingPage, 51
SetupChannelEventHandlers(), 138–139
setVideoURI(), 212
SHA. *See* Secure Hash Algorithm
ShellToastNotificationReceived, 144
shootPicture(), 244
Short Message Service (SMS), 35
  advertising, 147
Show(), 59
showResult(), 170
Silverlight, 4, 6, 16
  Expression Blend, 281
  graphics, 187, 189
  MediaElement, 207
  navigation, 68–69
  templates, 40–41
  Toolkit, 75–76
  UI, 42
  XAML, 18, 42, 82, 187
Silverlight Application Package (XAP), 7
  maximum size, 29
SIM. *See* Subscriber Identity Module
Simple API for XML (SAX), 97
Simple Object Access Protocol (SOAP)
  Bing Maps, 156–157
  web services, 114–115
SimpleMap, 173–175, 299
SizeChanged, 217
skipToBeginning(), 208
skipToNextItem(), 208
skipToPreviousItem(), 208
Smooth Streaming Media Element (SSME), 259
SMS. *See* Short Message Service
SOAP. *See* Simple Object Access Protocol
SOC. *See* System-On-Chip
SocialDemo, 117–131, 297
software, 4
  UI, 64
software development kit (SDK), 16
  Windows Azure service, 106
Solution Explorer
  Add References, 161–162
  View, 86
  View Code, 18
  Visual Studio, 194

Solution Explorer (*continued*)
  XAML, 18
Solution Folder, 84
SoundEffect, 222–223
Source, 82, 219
SpeedOfSound, 223
SplashScreenImage.jpg, 92
sprites, 185
  XNA, 194–198
SpriteBatch.Begin(), 197
SpriteBatch.Draw(), 197–198
SpriteBatch.End(), 197
SQLite, 96, 98
  data storage, 110
SQLiteDatabase, 98
SQLiteQueryBuild, 98
src/, 42
SSL. *See* Secure Sockets Layer
SSME. *See* Smooth Streaming Media Element
StackPanel, 19
  Grid, 67
Standard Rights, 259
start(), 210, 211
  setMediaController(), 212
Start screen, 65
StartAccelerometer(), 249–250
startPreview(), 243
State
  NavigationService.Navigate, 56
  PhoneApplicationPage, 53
  TextBlock, 54
status, 128
Status Bar, 65
StatusChanged, 165
statuses, 128
Stop, 222
stop(), 208, 211
  MediaPlayer, 210
Stop Debugging, 25
stopButton_Click(), 242
Stopped, 48
storage. *See* data storage; isolated storage
storedhash, 274
Stream, 222
Stream.BeginRead(), 120
Street2Geo, 167–172, 299
string, 87
Style, 93
styles, 92–93
Subscriber Identity Module (SIM), 257
SupportedOrientations, 91, 253
  DisplayOrientation.LandscapeLeft, 196
surfaceChanged(), 243
SurfaceView, 212, 243
Suspended, 50
SyncAdapter, 97
Synchronous Dynamic Random Access Memory (SDRAM), 278
SyncResult, 97
system architecture, 4–5
system event hooks, 61
system requirements
  Android, 16
  iOS, 16
System.Device.Location, 159
  APIs, 156
System.IO.IsolatedStorage, 99–100
System.Net.HttpWebRequest, 118, 119–122
System.Net.WebClient, 118
System-On-Chip (SOC), 277
System.Runtime.Serialization, 101
System.Security, 263, 268
System.Security.Cryptography, 263
System.Security.Cryptography.HashAlgorithm, 270
System.Security.Cryptography.X509Certificates, 263
System.Security.Permissions, 263
System.Security.Principal, 263
System.Windows.Controls, 207
System.Xml.Linq, 127

T

Tab Layout, 76
TableLayout, 67
TargetType, 93
tbAll, 79

tbNewClicks, 51, 53
  tombstones, 54
tbText, 79
tbTotalClicks, 51, 53
  tombstones, 54
tbVideo, 79
TCB. *See* Trusted Computing Base
Telephony Manager, 11
templates
  application, 41
  class library, 41
  databound application, 41
  Design Templates for WP7, 64
  game library, 41
  games, 41
  IDE, 46
  Panorama, 41
  Photoshop, 64
  Pivot, 41
  Silverlight, 40–41
  Visual Studio, 46
  XNA, 40–41
testing
  with actual device, 26–27
  Windows Phone Emulator, 21–26
  Zune, 27
Text To Speech (TTS), 239
textBack, 70
TextBlock, 19, 51, 67, 93
  Encrypt, 268
  PanoramaItem, 79
  State, 54
  string, 87
  textBack, 70
  textFrom, 70
  Title, 86, 87
  Type, 86
  UI, 54
TextBox, 50, 270
textbox, 286
textFrom, 70
texture, 194–198
themes, 94
third-party applications, 7
3D space
  Draw(), 203
  graphics, chipset, 279
  projection, 186
  views, 186
  world space, 186
  XNA, 200–204
tile, 137
Timeline, 284
TimeSpan, 90
Title, TextBlock, 86, 87
TitlePanel, 19
toast, 136–137
tombstones, 46
  Application_Closing(), 57
  Application_Deactivated(), 57
  tbNewClicks, 54
  tbTotalClicks, 54
Toolbar, iOS, 91
Toolbox, 19
Toolkit, Silverlight, 75–76
TotalClicks, 52
touch target, 66
TouchJSON, 116
ToyAnimation, 198–199, 301
TrackMe, 176–180, 299–300
Translator API, 242
Triple-DES, 264
Trusted Computing Base (TCB), 259
TTS. *See* Text To Speech
Twitter, 114, 117
  JSON, 122–127
TwoWay, 82
  CheckBox, 87
Type, TextBlock, 86

U

UART. *See* Universal Asynchronous Receiver/Transmitter
UI. *See* user interface
UI Design and Interaction Guide, 64
UIAccelerometer, 246
UIColor, 187

UIDevice, 245, 246
UIElementCollection, 175
UIFont, 187
UIImage, 187
UIImagePickerController, 242–243
UIImageView
    iOS, 206, 208
    myanimationView, 209
UIKit, 10
UIRequired Device, 158
UIViewContentModeScaleAspectFit, 208
UMTS. *See* Universal Mobile Telecommunications System
Universal Asynchronous Receiver/Transmitter (UART), 278
Universal Mobile Telecommunications System (UMTS), 280
Universal Serial Bus (USB), 279
Unix, 8
UnloadContent(), 192, 193
Unloaded, 217
Update(), 192, 193, 194, 201, 227
    Media, 250
    Microphone, 241
update(), 202
Updatemedia(), 228
Urgent, 87
URI
    MPNS, 140
    NavigationService.Navigate(), 71
USB. *See* Universal Serial Bus
user interface (UI), 10–11, 63–94
    Android, 20
    animation, 64
    Application Bar, 91–92
    btnClick, 54
    building, 66–92
    customization, 92–94
    design, 64–66
    event handlers, 90–91
    Expression Blend, 282
        customization, 288–291
    gestures, 90
    MainPage.xaml, 68
    MainPage.xaml.cs, 68
    navigation, 68–74
    OnNavigatedFrom(), 56
    OnNavigatedTo(), 56
    orientation, 91
    platform characteristics, 64–66
    programmatically defining, 67–68
    sample application layout, 51
    Silverlight, 42
    TextBlock, 54
    Visual Basic, 19–20
    XAML, 18, 66–67
<uses-permissions>, 261

V

Variable Bit Rate (VBR), 213, 214
variable time-step game loops, 192
VBR. *See* Variable Bit Rate
vectors, 185
verifybtn, 270
video, 147
    integration, 223–229
    MediaElement, 230–232
    MediaPlayerLauncher, 229–230
View
    creating, 86–87
    DataContext, 83
    Model, 83
    ViewModel, 83, 89–90
View, 86
View Code, 18
View Details at My Dashboard, 35
View System, 11
views, 3D space, 186
ViewModel
    creating, 85
    data binding, 86
    DataContext, 89
    PanoramaPage, 90
    View, 83, 89–90
Virtual Private Network (VPN), 2
Visual Basic, 6
    event handlers, 19–20

UI, 19–20
Visual Studio, 7, 15, 17–21
Expression Blend, 282
IDE, AdControl, 151
Solution Explorer, 194
Stop Debugging, 25
templates, 46
Windows Phone Emulator, 22–23
XAML, 18–19
Visual Web Developer Express, 107
Voice over Internet Protocol (VoIP), 7
multi-tasking, 61
VoIP. *See* Voice over Internet Protocol
Volume, 219
VPN. *See* Virtual Private Network

W

watcher, 179
WAV, 213, 238
WCDMA. *See* Wideband Code Division Multiple Access
WCF. *See* Windows Communication Foundation
WDDM. *See* Windows Display Driver Model
Weather underground, 115
Web Service Description Language (WSDL), 114
web services, 113–131
Android, 116–117
Bing Maps, 157
HTTP, 117–122
iOS, 116
JSON, 115–117, 119
REST, 114–115
RSS, 119
SOAP, 114–115
XML, 115–117, 119
WebClient, 118
WebKit, 8
websites, 147
Wide Video Graphic Array (WVGA), 64
Wideband Code Division Multiple Access (WCDMA), 280
Wi-Fi
Android, 8
chipset, 279
positioning, 156
Window Phone Design System-Codename Metro, 64
Windows Azure service, 105–107
asynchronous, 108–110
Microsoft Sync Framework, 109–110
SDK, 106
Windows Communication Foundation (WCF), 106, 122
Windows Compact Edition (CE), 1
Windows Display Driver Model (WDDM), 16
Windows Manager, 11
Windows Media Audio (WMA), 214
Windows Mobile 2003, 2
Windows Phone Developer Register Tool, 26
Windows Phone Emulator, 16
Android, 26
command line, 24
testing, 21–26
Visual Studio, 22–23
XNA, 184
WindowsPhoneMic, 240–241, 303
winphone7db, 98
WMA. *See* Windows Media Audio
WMAppManifest.xml, 70, 259
work-around solutions, 60–61
workspace, 284
world space, 186
WP7AudioPlayerDemo, 215–221, 301
WPCamera, 244–245, 303
WP7Controls, 82–93, 295–296
WP7EnrichedMoviePlayerDemo, 232, 302–303
WP7FMRadio, 254–255, 303–304
WP7Lifecycles, 47–59, 295
WP7Navigation, 70–74, 296
WP_SetFrequency, 254
WP7VideoPlayerDemo, 230–232, 302
WrapPanel, 80
Write(), 242
WriteObject(), 101
writeToFile(), 97
WSDL. *See* Web Service Description Language
WVGA. *See* Wide Video Graphic Array
WYSIWYG, 18

X

XAML. *See* Extensible Application Markup Language
`.xaml`, 18
XAP. *See* Silverlight Application Package
Xbox LIVE, 3, 4
Xbox/DirectX New Generation Architecture (XNA), 4, 6–7
  animation, 198–200
  game loop, 190–194
  Game Studio, 16
  graphics, 183–184, 189–204
  iOS, 20
  `microphone`, 241
  sprites, 194–198
  templates, 40–41
  texture, 194–198
  3D space, 200–204
  Windows Phone Emulator, 184
`XElement`, 130
XHTML. *See* Extensible Hypertext Markup Language
XML. *See* eXtensible Markup Language
`XMLReader`, 127
XNA. *See* Xbox/DirectX New Generation Architecture

Y

Yelp, 115
Youtube, 114

Z

zombie virus, 257
`ZoomBarVisibility`, 174
Zune, 3, 4, 7
  debugging, 27
  Redirection Service, 35
  testing, 27